Y0-AAY-405

PLASTICITY AND REGENERATION OF THE NERVOUS SYSTEM

ADVANCES IN EXPERIMENTAL MEDICINE AND BIOLOGY

Editorial Board:

NATHAN BACK, *State University of New York at Buffalo*

IRUN R. COHEN, *The Weizmann Institute of Science*

DAVID KRITCHEVSKY, *Wistar Institute*

ABEL LAJTHA, *N.S. Kline Institute for Psychiatric Research*

RODOLFO PAOLETTI, *University of Milan*

Recent Volumes in this Series

Volume 289
NUTRITIONAL AND TOXICOLOGICAL CONSEQUENCES OF FOOD PROCESSING
Edited by Mendel Friedman

Volume 290
THE IDENTIFICATION OF THE CF (CYSTIC FIBROSIS) GENE: Recent Progress and New Research Strategies
Edited by Lap-Chee Tsui, Giovanni Romeo, Rainer Greger, and Sergio Gorini

Volume 291
FUEL HOMEOSTASIS AND THE NERVOUS SYSTEM
Edited by Mladen Vranic, Suad Efendic, and Charles H. Hollenberg

Volume 292
MECHANISMS OF LYMPHOCYTE ACTIVATION AND IMMUNE REGULATION III: Developmental Biology of Lymphocytes
Edited by Sudhir Gupta, William E. Paul, Max D. Cooper, and Ellen V. Rothenberg

Volume 293
MOLECULAR BIOLOGY AND PHYSIOLOGY OF INSULIN AND INSULIN-LIKE GROWTH FACTORS
Edited by Mohan K. Raizada and Derek LeRoith

Volume 294
KYNURENINE AND SEROTONIN PATHWAYS: Progress in Tryptophan Research
Edited by Robert Schwarcz, Simon N. Young, and Raymond R. Brown

Volume 295
THE BASAL FOREBRAIN: Anatomy to Function
Edited by T. Celeste Napier, Peter W. Kalivas, and Israel Hanin

Volume 296
PLASTICITY AND REGENERATION OF THE NERVOUS SYSTEM
Edited by Paola S. Timiras, Alain Privat, Ezio Giacobini, Jean Lauder, and Antonia Vernadakis

Volume 297
NEW ASPECTS OF HUMAN POLYMORPHONUCLEAR LEUKOCYTES
Edited by W. H. Hörl and P. J. Schollmeyer

A Continuation Order Plan is available for this series. A continuation order will bring delivery of each new volume immediately upon publication. Volumes are billed only upon actual shipment. For further information please contact the publisher.

[Institute of Developmental Neuroscience and Aging. Conference (3rd : 1990 : Turin, Italy)]

PLASTICITY AND REGENERATION OF THE NERVOUS SYSTEM

Edited by

Paola S. Timiras
University of California, Berkeley
Berkeley, California

Alain Privat
Unité 336 INSERM
Université de Montpellier II
Montpellier, France

Ezio Giacobini
Southern Illinois University
Springfield, Illinois

Jean Lauder
University of North Carolina
Chapel Hill, North Carolina

and

Antonia Vernadakis
University of Colorado
Health Sciences Center
Denver, Colorado

QP 363.5
I55
1990

PLENUM PRESS • NEW YORK AND LONDON

Library of Congress Cataloging-in-Publication Data

Institute of Developmental Neuroscience & Aging. Conference (3rd :
1990 : Turin, Italy)
Plasticity and regeneration of the nervous system / edited by
Paola S. Timiras ... [et al.].
p. cm. -- (Advances in experimental medicine and biology ; v.
296)
"Third Conference of the Institute of Developmental Neuroscience
and Aging ... held April 5-7, 1990, in Torino, Italy"--T.p. verso.
Includes bibliographical references and index.
ISBN 0-306-43933-6
1. Nervous system--Regeneration--Congresses. 2. Nervous system-
-Growth--Congresses. 3. Neuroplasticity--Congresses. I. Timiras,
Paola S. II. Title. III. Series.
[DNLM: 1. Aging--physiology--congresses. 2. Nerve Regeneration-
-physiology--congresses. 3. Neuronal Plasticity--physiology-
-congresses. 4. Nervous System--growth & development--congresses.
W1 AD559 v. 296 / WL 102 I56p 1990]
QP363.5.I55 1990
599'.0188--dc20
DNLM/DLC
for Library of Congress 91-24014
CIP

Proceedings of the Third Conference of the Institute of Developmental
Neuroscience and Aging: Plasticity and Regeneration of the Nervous System,
held April 5-7, 1990, in Torino, Italy

ISBN 0-306-43933-6

© 1991 Plenum Press, New York
A Division of Plenum Publishing Corporation
233 Spring Street, New York, N.Y. 10013

All rights reserved

No part of this book may be reproduced, stored in a retrieval system, or transmitted
in any form or by any means, electronic, mechanical, photocopying, microfilming,
recording, or otherwise, without written permission from the Publisher

Printed in the United States of America

GUIDO FILOGAMO

A graduate (M.D., 1940) from the Medical School of the University of Torino (Italy), Professor Guido Filogamo spent almost his entire academic career (except for two years at the University of Sassari, Italy) at the University of Torino. In 1964, he was appointed Director of the newly-created Institute of Histology and Embryology of the University of Torino and in 1966, first, Director of the Institute of Anatomy and, then, Chairman of the Department of Human Anatomy and Physiology. For his many contributions to the medical curriculum, his inspired teaching of medical and postdoctoral students and his valuable scientific achievements, he was elected in 1985, President of the Medical School Faculty, a position he still currently holds. He is a member of numerous Italian and international scientific societies and academies and a recipient of several medals, honors and awards.

Interested at the beginning of his career in the histophysiology of the connective tissue (to which he contributed some valuable data), he soon moved to the study of the nervous system, an already classic area of competence of the Anatomy Institute at Torino. His first observations dealt with the symmetry and development of the avian brain utilizing histochemical neurotransmitter markers. These studies on development led him to further research of neuromuscular ontogenesis and regeneration, development of iris muscles and myocardial Purkinje cells. He was among the first investigators to apply the cholinesterase method to study the development of neuromuscular junction and plaques and to formulate the hypothesis of an early function of cholinesterase in the nervous system. His more recent and current research of the intestinal intramural neural plexuses has demonstrated that cholinergic and adrenergic activities, morphologic changes and developmental events are associated with (and influenced by) a number of physiological and pathological conditions.

Professor Filogamo was an enthusiastic and effective member of the Institute of Developmental Neuroscience and Aging since its beginning. For his pioneering contributions to the field of neuro-embryology and his introduction of the concept of plasticity, it seemed fit to hold the Third Conference of the Society on "Plasticity and Regeneration of the Nervous System" in Torino and to dedicate its works to Professor Guido Filogamo.

PREFACE

One of the most impressive advances in the field of neuroscience over the last decade has been the accumulation of data on plasticity and regeneration in the nervous system of mammals. The book represents the contribution of a group of neuroscientists to this rapidly expanding field, through a Conference organized by the Institute of Developmental Neuroscience and Aging (IDNA). The meeting was held in Torino, Italy during April 1990 in honor of a great pioneer in the field of Neuroembryology, Professor Guido Filogamo. His introduction of the concept of neuroplasticity has had a significant impact on the study of neurobiology.

This volume is divided into six sections, each focusing on one of the subject areas covered during the meeting : Molecular and Cellular Aspects of Central and Peripheral Nervous System Development; Hormones, Growth Factors, Neurotransmitters, Xenobiotics and Development; In Vivo and in Vitro models of Development; Development and Regulation of Glia; Regeneration; and Aging.

Development is extensively covered in four out of six sections of this book. Indeed, in many ways, the problem of remodeling and plasticity in the mature nervous system are analogous to those seen in the development of the nervous system circuitry during ontogeny. Basic questions such as the determination of cell differentiation and migration in the peripheral nervous system, the regulation of axonal calibre, variability of DNA content in developing and reactive neurons are all meaningful in the scope of neuronal reactivity. The roles of glia in development and reactivity of the nervous system appear increasingly diversified : cell proliferation, synthesis of trophic and inhibitory substances, uptake and release of neurotransmitters are only some of them. The influence of endogenous substances (hormones, growth factors), neurotransmitters (biogenic amines, GABA and others), and exogenous substances (xenobiotics) on neuronal and glial development, though known for a long time, is now gaining full recognition, and this topic is of major importance in our understanding of neuronal homeostasis.

Regeneration of the nervous system is under the influence of multiple factors : trophic substances such as NGF, and FGF, matrix components and specific glycoproteins such as gangliosides. Glial cells, either central or peripheral, may be permissive or obstructive to regeneration. The possible use or disuse of these factors in neurodegenerative diseases is a point of great interest.

Brain aging has been studied along two lines : "normal aging", considered as a physiological process, with its molecular and cellular characteristics, and "pathological aging", represented by Alzheimer disease. Though they may differ in several respects, these two lines converge to a common end : cell death, the mechanisms of which are crucial to the issue of regeneration of the nervous system.

Through the diversity of the contributions to this volume, it is clear that there is not, at least to date, a common thread which would logically connect all the issues of plasticity and regeneration in the nervous tissue. There is rather a multiplicity of factors, which reflect the complexity of the nervous system. The goal of the Contributors and Editors of the volume is to provide a composite image of this rapidly advancing field.

The Editors

CONTENTS

IN VIVO AND IN VITRO MODELS OF DEVELOPMENT

DEVELOPMENT AND REGULATION OF GLIA

REGENERATION

AGING

NEW MOLECULAR INSIGHTS ON THE DEVELOPMENT OF THE PERIPHERAL NERVOUS SYSTEM

C. Dulac, P. Cameron-Curry, O. Pourquié and N. M. Le Douarin

Institut d'Embryologie Cellulaire et Moléculaire
49 bis Avenue de la Belle Gabrielle
94130 Nogent-sur-Marne. FRANCE

INTRODUCTION

In the study of cell differentiation, a necessary step is to determine the onset of specific gene activities, in the various cell types emerging during embryogenesis. Production of monoclonal antibodies (Mab) able to distinguish specific antigenic determinants carried by differentiated cells, constitutes a mean to fulfill such a goal.

We have applied this approach to study the ontogeny of the central and peripheral nervous system (CNS, PNS) in the avian embryo.

We present here three new molecular markers. Two of them are instrumental in the study of central and peripheral glial cell differentiation : 4B3 and SMP (for Schwann Cell Myelin Protein), the third is a surface molecule expressed by neurons during certain phases of their differentiation (Dulac et al., 1988 ; Cameron-Curry et al., 1989 ; Pourquié et al., 1990).

A. SMP : A MARKER OF THE EARLY SCHWANN CELL DIFFERENTIATION

We have been interested in our laboratory for several years in trying to understand how the different cell lineages arising from the neural crest become segregated in the course of ontogeny. The development of the various types of peripheral glial cells has particularly attracted our attention together with the question of how and when, during the development of the PNS, the glial and neuronal lineages diverge.

A point in case concerns the phenotypic variability of the peripheral glia which falls so far into three main categories, the satellite cells of the sensory, and autonomic ganglia, the Schwann cells lining the peripheral nerves and the enteric glia. The progressive commitment of crest derived precursors to a glial fate and their differentiation into a particular glial cell type can be followed during development through the synthesis of cell- and stage-specific markers. The earliest molecular markers of neural crest cells such as NC1/HNK1 (Abo and Balch 1981; Vincent et al., 1983; Tucker et al., 1984) or GLN1 (Barbu et al., 1986) are not specific for glial cells. On the other hand, bona fide glial marker like the glial fibrillary acidic protein (GFAP) and several myelin constituents e.g; : Protein zero (Po), myelin basic protein (MBP), myelin associated glycoprotein (MAG), 2', 3'-Cyclic Nucleotide 3'-Phosphodiesterase (CNPase), start to be expressed at the terminal steps of glial differentiation. Our attempts to define more precocious and specific molecular markers for peripheral glial cells led us to select the SMP and 4B3

Plasticity and Regeneration of the Nervous System
Edited by P.S. Timiras *et al.*, Plenum Press, New York, 1991

Mabs obtained after immunizing mice with purified myelin glycoproteins from adult quail sciatic nerves and brachial plexuses.

The expression of SMP is restricted in vivo to the Schwann cell subset of peripheral glial cells and to oligodendrocytes

The first original feature of the SMP antigen is its strict specificity for myelinating and non myelinating Schwann cells in the PNS and for oligodendrocytes in the CNS. Peripheral myelinated fibers (recognized by their birefringence under polarised light and the presence of nodes of Ranvier) are strongly immunoreactive (Fig. 1B up). Moreover, unlike the myelin protein Po (Trapp et al, 1981) whose expression on the membrane is restricted to the myelin sheath, SMP is detected in both the myelin extruding from a fiber and on the membrane surrounding the Schwann cell perikaryon. Unmyelinated fibers characterized by the absence of birefringence and several Schwann cell nuclei along the fibers are also immunoreactive (Fig. 1B down).

Thus SMP can be distinguished from the myelin proteins (Po, MAG, MBP, CNPase) which are only expressed by myelinating Schwann cells and only after a signal from the axon.

Intraganglionnic satellite cells of PNS ganglia surrounding neuronal cell bodies are SMP negative (Fig. 1C). So are also enteric glial cells. Thus SMP is the only molecular marker described so far that distinguished the two types of Schwann cells from other peripheral glial cells.

The other unique characteristic of the SMP antigen is its early synthesis by developing Schwann cells. Schwann cell precursors migrate along growing axons and actively divide during the formation of peripheral nerves. From E5 onward in the quail embryo, Schwann cells lining peripheral nerves start to express the SMP glycoprotein (Dulac et al., 1988) while the earliest myelin markers appears only 5 to 6 days later.

In the central nervous system, SMP is expressed exclusively in the white matter of spinal cord (Fig. 1A) and brain as demonstrated by double labelling with anti-SMP and with an antiserum against MBP. Central neurons are never stained. Double labelling experiments performed on brain and spinal cord cultures with anti-SMP and anti-MBP confirm the strict specificity of SMP expression by oligodendrocytes.

→

Fig. 1 . Anti-SMP immunoreactivity

A - Transverse section of an E12 quail embryo at trunk level. Spinal nerves (n) are stained, while the dorsal root ganglion (d) and the paravertebral sympathetic ganglion (s) are negative. Bar = 20 μm.

B - Teased sciatic nerve from adult quail.

- polarized light (left, up and down)
- phase contrast optics (middle, up and down)
- anti-SMP immunoreactivity (right, up and down).

Myelinated fibers, which are birefrengent under polarized light (*), are positive, and nodes of Ranvier (arrows) are brightly fluorescent.
Unmyelinated fibers which lack birefrengence (open arrow) and show several Schwann cell nuclei (arrow heads) are also positive. Bar = 50 μm.

C - Section of an E12 quail embryo dorsal root ganglion.
The fibers that cross the ganglion are positive while neurons and satellite cells are unstained. Bar = 75 μm.

D - Sciatic nerve cells from newly hatched quail cultured for 3 weeks. Schwann cells still express high levels of SMP in such long term cultures. Note that they lost the typical spindle-shaped morphology. Bar = 50 μm.

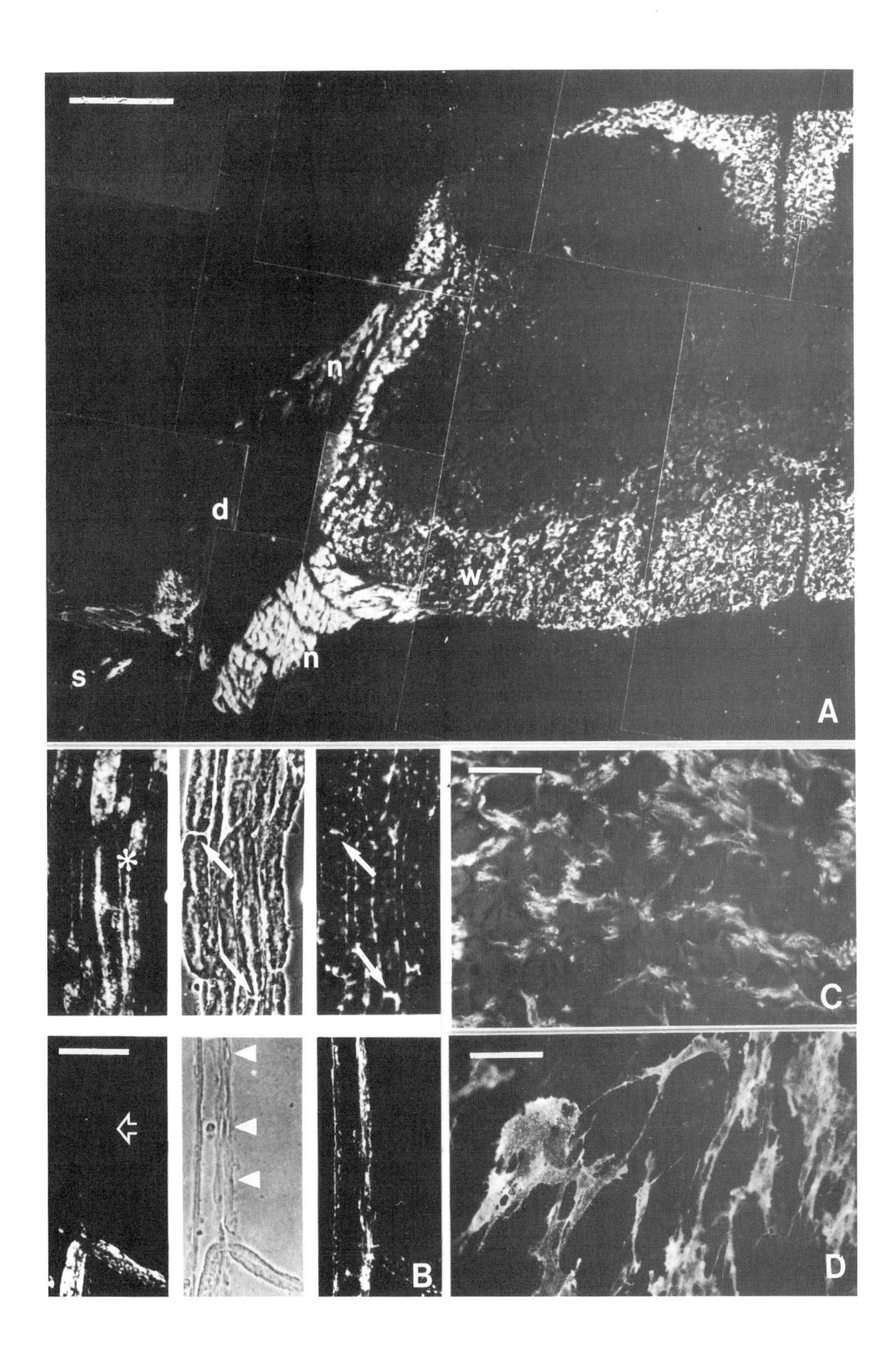
n
d
w
s
n
A
*
C
B
D

Moreover, unlike myelin constituants, SMP synthesis is maintained in cultures from dissociated sciatic nerves or brachial plexuses even after a long period of time (more than 2 weeks) and in the complete absence of neurons (Fig. 1D). We also showed that a subpopulation of SMP-positive cells appears in cultures of neural crest cells after 8 days, revealing the emergence of the Schwann cell lineage (Dulac et al., 1988). Thus this marker allows the segregation between neuronal and the different glial cell lineages to be studied in clonal cultures of neural crest cells.

Glial cell precursors in the neural crest

A culture system based on the use of growth inhibited 3T3 cells (Barrandon and Green, 1975) was adapted in our laboratory to the culture of single neural crest cells. Migratory cephalic crest cells taken from 10 to 13 somite quail embryos can in these conditions expand clonally *in vitro* and differentiate into a variety of cell types (Baroffio et al., 1988, Dupin et al., 1990).

The phenotypes expressed in the colonies were recognized using various markers : melanin which characterizes melanocytes, is identified directly by phase contrast microscopy, cartilage is revealed by metachromasia after toluidine blue staining, neuronal traits are evidenced immunocytochemically by means of several antibodies directed against neurofilament proteins, neuropeptides or the enzyme tyrosine hydroxylase (TH) (a marker for catecholaminergic producing cells). All glial cells are $HNK1^{+}$. The Schwann cell phenotype can be distinguished from the other glial cell types by the presence of both HNK1 and SMP antigens.

Several types of glial precursors were identified among neural crest cells. The large majority of the colonies, 90% that have been analysed contained SMP positive cells ; 40% of the clones including Schwann cells contained also neurons ;13% of the clones yielded a pure population of SMP positive cells. Such clones can therefore be considered as deriving from commited Schwann cell progenitors and this also shows that the first steps of Schwann cell differentiation do not require an influence from neurons. It appears therefore that the migratory neural crest is composed of an heterogeneous population of cells with respect to their state of commitment. A majority of cells are pluripotent and can give rise to both neuronal and glial (including Schwann cells) cell types. Others, in contrast, are more restricted in their potentialities. Fully committed Schwann cell progenitors also exist at these early stages of PNS ongoteny (Dupin et al., 1990).

Molecular characterization of the SMP antigen

SMP is a glycoprotein appearing as a doublet of Mr 75-80 000 in the peripheral nervous system and as a singulet of Mr 80 000 in the central nervous system under non reducing conditions of electrophoresis. The anti-SMP Mab recognizes a conformational polypeptidic epitope as suggested by the fact that immunoreactivity is abolished in western blot under reducing conditions but persists after enzymatic deglycosylation. This glycoprotein bears both the 4B3 and HNK1 carbohydrate epitopes.

Although SMP can be defined as an early marker of the Schwann cell lineage, its role is still unknown. Characterisation of the gene encoding this molecule is now in progress and will hopefully allow to collect further informations about its regulation and its role in gliogenesis.

B. 4B3 : A CARBOHYDRATE EPITOPE RELATED TO EARLY GLIOGENESIS

The 4B3 antigenic determinant, detectable from early stages in development is strictly restricted to neural tissues and more specifically to the glial cells and their precursors.

Cellular distribution in the peripheral nervous system

Immunolocalization *in vivo* showed that nerves (Fig. 2B) and ganglia of the peripheral (Fig. 2A) and enteric nervous system (Fig. 2C) are 4B3 positive, and it was visible on teased nerves, that both myelinated and unmyelinated fibers were labelled. Nodes of Ranvier were evident, meaning that the entire Schwann cell membrane and, not only the portion included in myelin is immunoreactive to the 4B3 Mab. Schwann cells in culture in non-myelinating conditions and in the absence of neurons still express 4B3, thus revealing the same constitutive expression of the 4B3 epitope that was demonstrated for SMP.

In peripheral ganglia the pattern of 4B3 Mab reactivity is different from that of SMP Mab since the membrane sheath of satellite cells enclosing each neuronal cell bodies is immunoreactive as well as Schwann cells lining intraganglionic fibers (Fig. 2A). Cultured peripheral ganglia exhibit many 4B3 positive non neuronal cells, characterized by the absence of tetanus toxin binding sites. Most of these cells are satellite cells since they express the HNK1 epitope and are SMP negative. None of these markers exist on fibroblasts, the third main cell type present in the culture. We can therefore conclude that in the PNS, 4B3 is restricted to the glial cells, irrespective of their implication in the myelination process: the function of some of the molecules carrying this carbohydrate is not likely to be related to that of myelin.

Immunoreactivity in the central nervous system

Distribution of 4B3 immunoreactivity in the central nervous system (CNS), is more difficult to identify precisely, namely because of the complex tissue architecture found in brain and spinal cord.

In the white matter 4B3 Mab yields a diffuse staining pattern (Fig. 2A) ; however stained myelin rings are evident in transversly cut longitudinal fasciculi.

In the gray matter, the staining pattern is also diffuse while the neurons are clearly 4B3 negative (Fig. 2A). Identification of neurons in culture, by tetanus toxin binding or neurofilament staining, confirmed that they do not express 4B3 carrying molecules.

Astrocytes identified in culture as GFAP positive were not double stained by anti-4B3. In general, a down regulation of 4B3 was noticed in cultures of CNS tissues (Cameron-Curry et al., 1990).

Early appearence of the 4B3 epitope

Immunostaining was first detected on peripheral ganglia, in dorsal roots of the spinal nerves and in the ventro-lateral parts of the spinal cord in E2.5 in quail embryos.

Migrating mesencephalic neural crest cells were not stained in vivo, but some labelled cells could be detected after one day in culture, i.e. at a stage when the first neural crest derived structures become 4B3 positive *in vivo*

Biochemical nature of the 4B3 immunoreactivity

The anti-4B3 Mab was generated by immunizing mice with glycoproteins from peripheral myelin and biochemical studies showed that 4B3 carrying molecules co-purify with myelin. Only one band of Mr 100,000

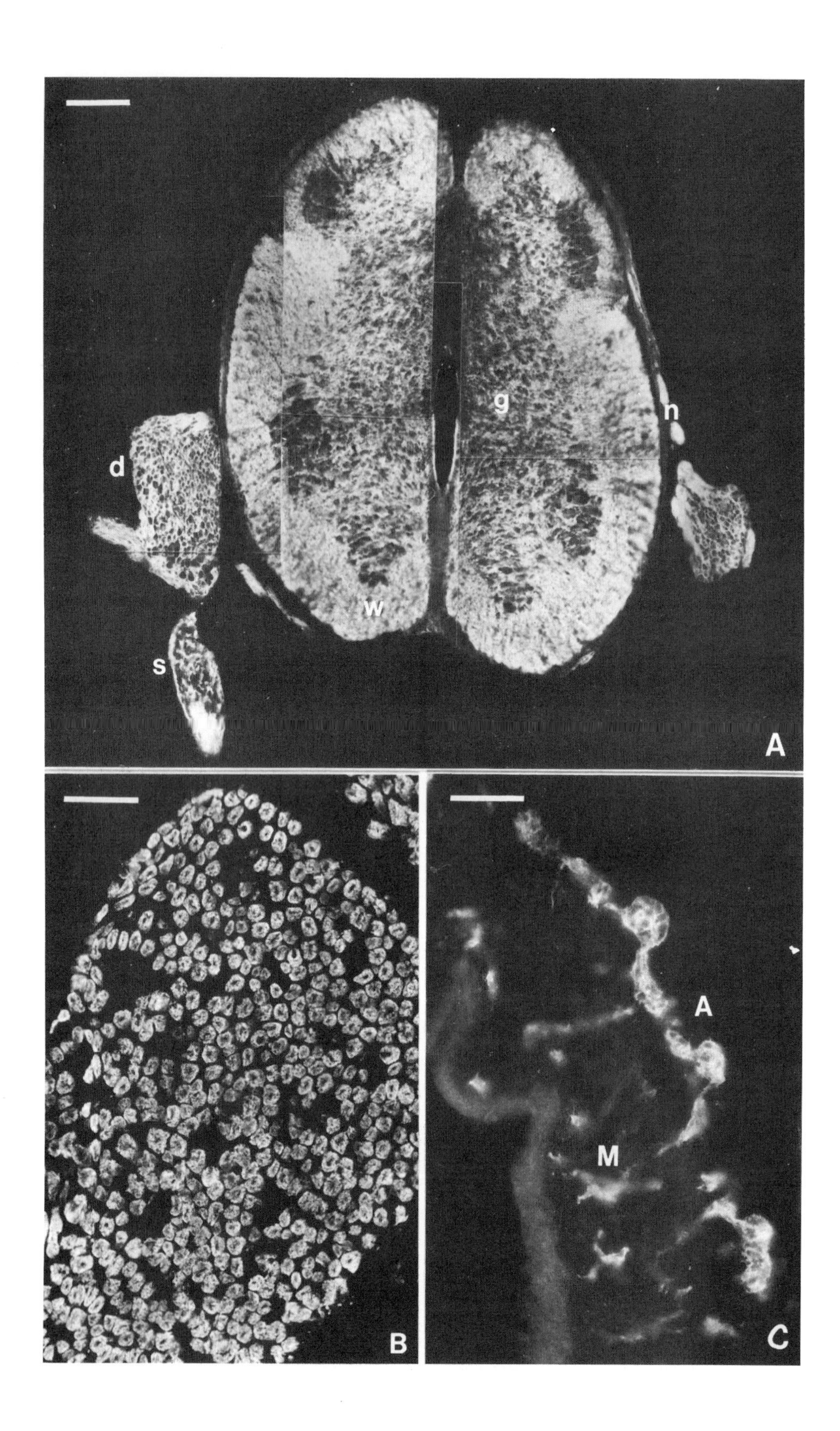
d
g
n
w
s
A
B
A
M
C

is detected by Western blot under reducing conditions that is apparently the same molecular mass that was previously defined for SMP. Moreover, we could demonstrate that, immunopurified SMP from both PNS and CNS reacts with the anti-4B3 Mab (Fig. 3 left). Immunofluorescence studies on

Fig. 2 Anti-4B3 immunoreactivity

A - Transverse section of and E8 quail embryo at trunk level. Spinal cord white (w) and grey (g) matter, dorsal root ganglia (d), sympathetic chain (s) and spinal nerves are stained. Bar = 150 µm.

B - Transverse section of adult quail sciatic nerve.
The whole myelin rings are strongly positive. Bar = 25 µm.

C - Transverse section of the anterior gut of an E8 chick embryo.
Meissner's (M) and Auerbarch's (A) plexuses are both positive. Bar = 25 µm.

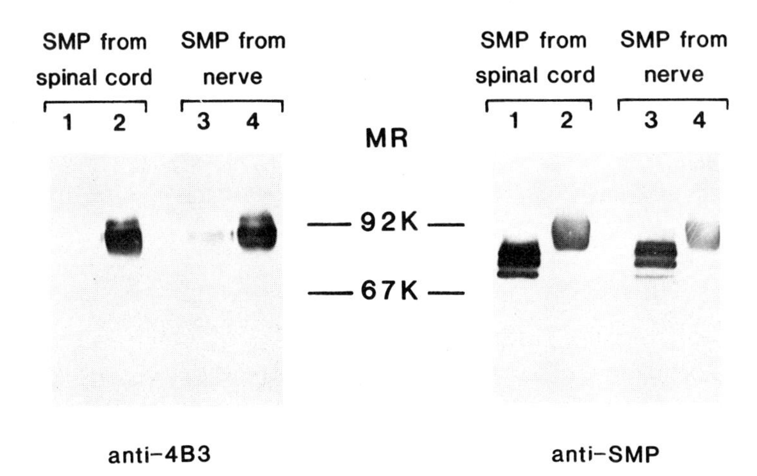

Fig. 3 Molecular characterisation of the 4B3 epitope

SMP purified from spinal cord and peripheral nerve (3,4) was analysed in Western blot under non reducing conditions.
Native (2,4) or glycopeptidase F deglycosylated protein was reacted with anti-4B3 (left) and anti-SMP (right).
Anti-4B3 reacts with native SMP (2,4) but immunoreactivity is abolished by deglycosylation (1,3), while anti-SMP antibody, which is directed against a protein epitope, still recognizes several bands of partly to totally deglycosylated SMP molecules.

transverse nerve sections showed that all the thickness of the myelin ring is stained by anti-4B3 (Fig. 2B), while SMP appears to be mainly concentrated on inner and outer mesaxons. This fact together with the cellular staining pattern in CNS and PNS suggests that besides SMP, at least one more molecule other than SMP bears the 4B3 epitope in peripheral nerves. Whether the same molecule(s) is responsible for the immunostaining of ganglionic satellite cells, of enteric plexuses and of the gray matter in the CNS is presently unknown.

C. BEN : A WIDELY DISTRIBUTED MOLECULE DEVELOPMENTALLY REGULATED ON PERIPHERALLY PROJECTING NEURONS

By mean of a monoclonal antibody, raised against the epithelium of the bursa of Fabricius, a membrane glycoprotein of Mr 95-100.000 named BEN was identified. Besides several types of epithelia (including that of the bursa) BEN is found in the nervous system, and on a few other cell types such as certain hematopoietic cells and chondrogenic cells. Its expression on the cell surface appears in most cases to be developmentally regulated (Pourquié et al., 1990).

Transient and restricted expression in the nervous system

We have studied the expression pattern of the BEN antigen in early neurogenesis in the chick embryo (up to 10 days) by immunocytochemical methods. It appears that it is expressed only by the peripherally projecting neurons, i.e. : by motoneurons of the spinal cord and cranial nuclei, by sensory neurons of the dorsal root ganglia (DRG) and cranial sensory ganglia, and by sympathetic and enteric neurons.

The expression on motor and sensory neurons is transient and occurs soon after their birthdate on the cell body and on the growing axon (Fig. 4A). A maximal expression is detected in the chick by stage 23 (E4), corresponding to the step at which motor and sensory neurons are massively produced (Fig. 4B). Immunoreactivity is detected first in the cell body, then in the growth cone, and the early peripheral nerves are for a while heavily reactive. By stage 25 and later, BEN expression is downregulated and decreases first at the level of the cell bodies and then on their fibers (Fig. 4C). By stage 37 (E10), expression is not detected any more at the level of the ventral horn and ventral root of the spinal cord and in the DRG. Therefore, expression of the BEN molecule seems to be associated to the phase of axonal growth and projection to the periphery. The down regulation observed is concommittant to the beginning of synaptogenesis in the periphery, and also to the proliferation phase of the satellite cells, Schwann cells and astrocytes, in the region of BEN expression.

→

Fig. 4 . Immunoperoxidase staining with BEN Mab of transverse sections of chick embryos.
A - Stage 19 embryo.; B - Stage 23 ; C - Stage 34-35.
Staining can be detected very early at the level of the motoneurons (MN) and the floor plate. It then appears on the dorsal root ganglion (DRG) and the sympathetic ganglion (SG) and on the spinal nerves (SN) and the dorsal funiculus (DF). By stage 34-35, BEN expression starts to be downregulated and progressively disappears at the level of the motoneurons and the DRG neurons.

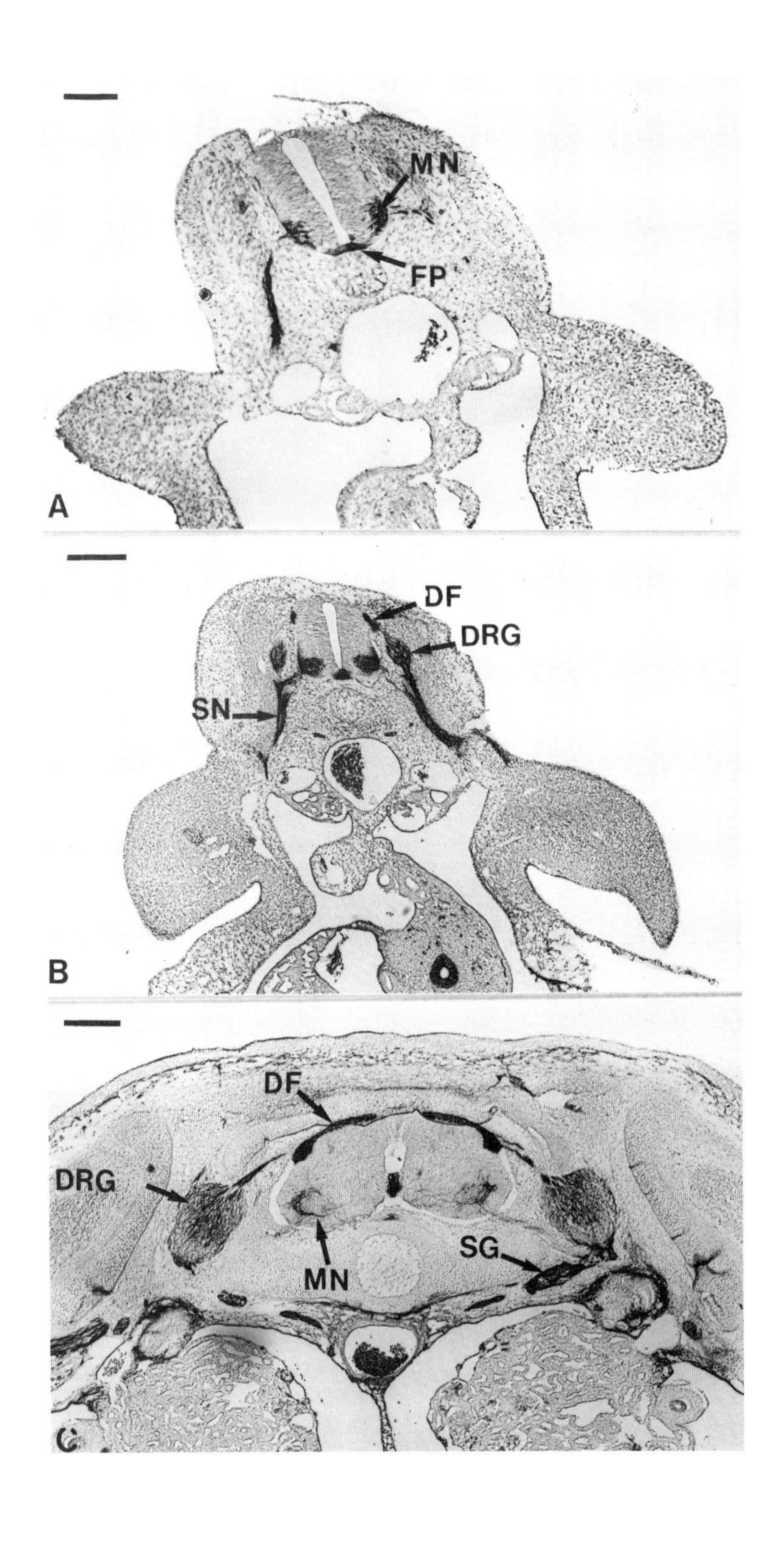
MN
FP
A
DF
DRG
SN
B
DF
DRG
MN
SG
C

In the sympathetic system, BEN expression appears to be down regulated much later in embryonic life ; in the adult, no more immunoreactivity is detected on sympathetic neurons. In the parasympathetic system, namely in the Remak and ciliary ganglia, the timing of expression of BEN is much like in the DRG. In the enteric nervous system, the expression occurs very early (E3) and persists in adult life.

Biochemical analysis of the BEN molecule

The Mab directed against the BEN molecule recognizes in Western blot two molecular forms. The first form of 95 000 Mr is essentially present in neural tissues, and the second of about 100 000 has been identified in the epithelium of the bursa of Fabricius and on hematopoietic cells. We have affinity purified the molecule in order to perform a molecular analysis of the recognized epitope. It appeared from this study that the apparent molecular mass is unaffected by reduction, and that this molecule bears the HNK1-epitope, wich has been identified only on molecules involved in cell/cell like N-CAM, L1, F11... (Edelman, 1987) or cell/matrix (integrin, tenascin...) (Kruse et al., 1984, 1985) adhesion processes. The biochemical features of this molecule do not correspond to those of any previously described neural molecule.

Acknowledgements

This work was supported by the Centre National de la Recherche Scientifique, the Institut National de la Santé et de la Recherche Médicale and by grants from the following fondations : the Association pour la Recherche sur le Cancer, the Fondation pour la Recherche Médicale Française, the Ligue Nationale Française contre le Cancer. The authors acknowledge the skillful help of Bernard Henri and Evelyne Bourson for preparing the manuscript.

REFERENCES

Abo, T., and Balch, C. M., 1981, A differentiation antigen of human NK and K cells identified by a monoclonal antibody (HNK-1). J. Immunol. 127:1024.

Barbu, M., Ziller, C., Rong, P. M., and Le Douarin, N. M., 1986, Heterogeneity in migrating neural crest cells revealed by a monoclonal antibody. J. Neurosci. 6:2215.

Baroffio, A., Dupin, E., and Le Douarin, N. M., 1988, Clone-forming ability and differentiation potential of migratory neural crest cells. Proc. Natl. Acad. Sci. USA, 85:5325.

Cameron-Curry, P., Dulac, C., and Le Douarin, N.M., A monoclonal antibody defining a carbohydrate epitope restricted to glial cells. Europ. J. Neuroscience, in press.

Dulac, C., Cameron-Curry, P., Ziller, C., and Le Douarin, N.M., 1988, A surface protein expressed by avian myelinating and non myelinating Schwann cells but not by satellite or enteric glial cells. Neuron 1:211.

Dupin, E., Baroffio, A., Dulac, C., Cameron-Curry, P., and Le Douarin, N., 1990, Schwann cell differentiation in clonal cultures of the neural crest, as evidenced by the anti-Schwann cell Myelin Protein monoclonal antibody. Proc. Natl. Acad. Sci. USA, 87:1119.

Edelman, G. M., 1987, CAMs and Igs : Cell adhesion and the evolutionary origins of immunity. Immunol. Rev., 100:11.

Kruse, J., Mailhammer, R., Wernecke, H., Faissner, A., Sommer, L., Goridis, C., and Schachner, M., 1984, Neural cell adhesion molecules and myelin-associated glycoprotein share a common carbohydrate epitope recognized by monoclonal antibodies L2 and HNK-1. Nature, 311:153.

Kruse, J., Keihauer, G., Faissner, A., Timpl, R. T., and Schachner, M., 1985, The J1 glycoprotein - a novel nervous system cell adhesion molecule of the L2/HNK-1 family. Nature, 316:146.

Pourquié, O., Coltey, M., Thomas, J.L., and Le Douarin, N.M., 1990, A widely distributed antigen developmentally regulated in the nervous system. Development ,109:743.

Trapp, B. D., Itoyama, Y., Sternberger, N. H., Quarles, R. H., and Webster, H. de K., 1981, Immunocytochemical localization of Po protein in Golgi complex membranes and myelin of developing rat Schwann cells. J. Cell Biol., 90:1.

Tucker, G.C., Aoyama, H., Lipinski, M., Turz, T., and Thiery, J.P., 1984, Identical reactivity of monoclonal antibodies HNK-1 and NC-1 : conservation in vertebrates on cells derived from neural primordium and on some leukocytes. Cell Diff., 14:223.

Vincent, M., Duband, J.L., and Thiery, J.P., 1983, A cell surface determinant expressed early on migrating avian neural crest cells. Dev. Brain Res., 9:235.

DNA CONTENT REVEALED BY CYTOPHOTOMETRY IN NEURONS:

VARIABILITY RELATED TO NEUROPLASTICITY

Stefano Geuna, Alessandro Poncino and
Maria G. Giacobini Robecchi

Dipartimento di Anatomia e Fisiologia Umana
Sezione di Neuroanatomia e Neuroembriologia
Università di Torino, C.so M. D'Azeglio 52
10126 Torino, ITALY

INTRODUCTION

Cytophotometry after Feulgen staining provided a new tool for the determination of nuclear DNA content. Controversial data on the DNA content of neurons in different species of animals, as well as in different regions of the central and peripheral nervous system, have been reported.

Cytophotometric studies of many species, during the 1960's, showed that, in physiological conditions, large neurons such as cerebellar Purkinje cells, pyramidal neurons of the hippocampus, spinal cord motorneurons, etc., contain a tetraploid* amount of DNA (review in: Herman and Lapham, 1969; Mares et al., 1973; Cohen et al., 1973). Some workers (Bregnard et al., 1975; Kuenzle et al., 1978) have even suggested that all rat cerebral cortex neurons may posses an extra-DNA amounting to 3.5 times the haploid content. Further investigations, however, has failed to corroborate many of these findings (McIlwain and Capps-Covey, 1976; Mann et al., 1978; Hobi et al., 1984; Mares et al., 1985). In the case of cerebellar Purkinje cells, several authors have demonstrated that some are endowed with a DNA content exceeding the diploid level (hyperdiploid cells)(see for references Brodsky et al., 1979), whereas other results indicate that it is diploid (Morselt et al., 1972; Fujita, 1974; Mann et al., 1978). The latest reports agree that a small percentage of true hyperdiploid Purkinje cells are normally present, at least in some species (Mares et al., 1980; Marshak et al., 1985; Bernocchi et al., 1986b).

Cytophotometry after Feulgen staining has also been applied, often in association with autoradiography and biochemical methods, to detect changes in both nuclear DNA content and degree of chromatin condensation, in normal and pathological conditions.

Different techniques, applied to investigation of the vertical and subesophageal lobes of octopus brain (Giuditta et al., 1971; De Marianis et al., 1978-1979) have shown that the latter contain larger neurons with an extra-DNA up to several times the diploid level. This extra-DNA

*The terms "haploid", "diploid", "tetraploid", and "hyperdiploid", are here used solely to refer to the nuclear content of DNA, without implying any information on the chromosomal status.

increases progressively with body weight (i.e. with the increase of the innervation district).

The enormous DNA content (20,000-270,000 the haploid value) of the nuclei of the giant neurons of the abdominal and pleural ganglia of the marine mollusc, Aplysia californica, also seems to be related to its size and that of the nerve cells (Coggeshall et al., 1970; Lasek and Dower, 1971).

Bernocchi et al. (1986b) have obtained very interesting results in the hedgehog, Erinaceus europeanus. In addition to heterogeneity of the Purkinje neuron population due to differences in nuclear degree of chromatin condensation, they observed 16% of Purkinje cells with hyperdiploid DNA values. During hibernation hyperdiploid values were never detected, suggesting that they are influenced by some inputs (afferent connections) that should undergo some changes, at least in intensity, during hibernation, with the need of a lower metabolic rate of the neurons. Similar variations of DNA values of Purkinje cells have been obtained in the frog cerebellum (Barni et al., 1983).

It thus seem that variations in the amount of this extra-DNA, although sometimes greatly different in extent, could be influenced by changes in functional or metabolic activity, and hence represent an aspect of neuroplasticity.

Against this, however, is the unsuccessful attempt to influence DNA content in rat and mouse cerebellum Purkinje cells by metabolic and/or functional stimulation of the cerebellum itself; in fact, the percentage of hyperdiploid cells was unaffected (Marshak et al., 1985).

DNA CONTENT RELATED TO NEUROPLASTICITY IN HYPERTROPHIC NEURON POPULATIONS

We have investigated the possibility that changes in DNA content may occasionally represent an aspect of neuroplasticity by examining two classic models of neuronal hypertrophy induced by augmentation of the innervation district.

a) Auerbach's Plexus Neurons Upstream from Surgical Intestinal Stenosis

This provides an excellent model for study of neuronal behaviour during enhanced activity. Filogamo and Vigliani (1955) have shown that subtotal stenosis leads to both cytoplasmic and nuclear hypertrophy, presumably due to a sudden increase in the innervation district represented by pronounced thickening (3-fold after 10 days; 5-fold or more after 1 month in the rat) of muscle wall.

An autoradiographic study using ^{3}H-thymidine, conducted by Giacobini Robecchi et al. (1985), revealed nuclei labelled at varying degrees of intensity in some Auerbach neurons in hypertrophic intestinal loops. The possibility that this might be the result of cell division was ruled out by the constant absence of mitotic figures in these neurons in situations of this kind (Filogamo and Lievre, 1955a; Gabella and Gaia, 1967).

Since the premiss regarded by Swartz and Bhatnagar (1981) as the best suited for the correct use of cytophotometry for identification and quantification of neuronal DNA synthesis in the adult, namely ^{3}H-thymidine uptake, was thus fulfilled, we decided to extend the investigation to include cytophotometry after Feulgen staining (Giacobini Robecchi et al., 1988). An increased F-DNA content (hyperdiploid values) was noted in some neurons on the 10th day after surgery. The picture was also much the same on the 30th day, apart from a slight increase in the number of hyperdiploid neurons; tetraploidy was never observed (Poncino et al., 1990).

The relation betweeen changes in the innervation district and DNA content was also assessed cytophotometrically on hypertrophic intestinal loops in a state of hypoactivity (surgery described by Filogamo and Lievre, 1955b). A marked decrease in Auerbach neuron hypertrophy was noted, together with the virtual disappearance of hyperdiploid neurons. It would thus appear that, in this model, variations in DNA content are a function of cell activity (Geuna and Poncino, 1988a).

b) Sensitive Neurons from Hypertrophic Spinal Ganglia Innervating the Regenerated Lizard Tail

Amputation of part of a lizard's tail is followed by its regeneration, as described by many workers (Terni, 1920; Huges and New, 1959; Pannese, 1963; Simpson, 1968; Cox, 1969; Turner et al., 1972; Duffy et al., 1990). Innervation of the new tail is provided by the last three remaining pairs of spinal nerves, i.e. those lying in the plane of amputation (Terni, 1920; Huges and New, 1959). The corresponding spinal ganglia increase in volume and their neurons undergo both cytoplasmic and nuclear hypertrophy (Terni, 1920; Pannese, 1963).

A preliminary study revealed heterogeneity of the F-DNA content in these neurons (Geuna et al. 1988b). A following cytophotometric analysis (Geuna et al., submitted for publication), disclosed a shift to hyperdiploid values compared with the controls. The fact that such values are not only present in cells with a hypertrophic nucleus suggest that this shift may be, at least partly, the result of a true increase in DNA content as opposed to chromatin changes. Tetraploid values were extremely rare. Once again, therefore, tetraploidy does not appear to be involved.

DISCUSSION AND CONCLUSIONS

Our findings suggest that changes in the DNA content of neurons, in these two models, are linked to alteration in the local functional requirement.

A common feature of these findings and those of other workers (Lasek and Dower, 1971; Giuditta et al., 1971; De Marianis et al., 1978-1979; Bernocchi et al., 1986b) is a link between increased DNA and enhanced neuron activity. The existence of a connection between neuron DNA synthesis and neural function can also be derived from the involvement of DNA in learning (Reinis, 1972; Reinis and Lamble, 1972) and after electroshock (Giuditta et al., 1978).

It may thus be supposed that changes in DNA content are sometimes an aspect of neuroplasticity, even in the adult.

Bernocchi and Scherini (1986a) have advanced the fascinating suggestion that the number of hyperdiploid Purkinje cells is inversely related to the complexity of circuits achieved during phylogenesis. A higher DNA content could thus stem from the fact that, in primitive cerebellum, Purkinje cells receive more input from afferent connections of heterogeneous origin. This would be in keeping with the proposed connection between the extent of stimulation and DNA variability.

This connection, however, lacks general validity, since it is not observed in some conditions where its presence would be expected (Marshak et al., 1985; Chetverukhin et al., 1989).

In cerebellar Purkinje cells and other neuron populations, variations in the degree of chromatin condensation are mainly responsible for nuclear heterogeneity and they may be correlated with different metabolic states (Bernocchi and Scherini, 1986a). Nevertheless a true increase in DNA content is accepted as responsible for at least a small percentage of hyperdiploid nuclei (Mares et al., 1980; Marshak et al., 1985; Bernocchi and Scherini, 1986a; Bernocchi et al. 1986b). Differences in the degree of chromatin condensation may also explain part of our hyperdiploid values. However, the magnitude of the frequency distribution shift, the absence of a strict link between nuclear sizes and F-DNA values, and the parallel incorporation of ^{3}H-thymidine suggest that these hyperdiploid values at least partly reflect a real increase in nuclear DNA quantity.

The significance of these and other observations of hyperdiploid neurons is less clear, since only quantitative values are supplied by cytophotometry.

It is clear that the increased DNA content observed in our models cannot be ascribed to tetraploidy (Giacobini Robecchi et al., 1988; Poncino et al., 1990; Geuna et al., submitted for publication).

Bernocchi et al. (1986b) have suggested that the extra-DNA found in Purkinje cells might be mostly single-stranded.

Brodsky et al. (1979) have put forward the view that increases may reflect amplification of some discrete DNA sequences, i.e. gene amplification. Indeed, amplification of a select genome portion, of specific value to the neuron, has also been proposed as a possible interpretation of the extra-DNA of Aplysia's giant neurons (Lasek and Dower 1971). Furthermore, gene amplification has been described in cells from a murine neuroblastoma line and some human CNS tumors (Brennard et al., 1982; Schwab et al., 1983-1984; Rouah et al., 1989). Schimke et al. (1987) suggest that gene amplification "results from the accumulation of an increased capacity for initiation of DNA synthesis in cells in which DNA synthesis is partially inhibited, but where synthesis of RNA and (especially) protein synthesis can continue". Neurons are typically in a situation of this kind and could hence undergo gene amplification, when their capacity for initiation of DNA synthesis is stimulated in some way.

Amplification of some DNA sequences may thus be proposed as the mechanism responsible for the extra-DNA noted in our models. Characterization of this extra-DNA through the techniques offered by molecular biology may show whether this view is correct, and appropiate experiments are now being devised (Borrione et al., in preparation).

REFERENCES

Barni, S., Bernocchi, G., and Biggioggera, M., 1983, Chromatin organization in frog Purkinje neurons during annual cycle: cytochemical and ultrastructural studies, Basic. Appl. Histochem., 27:129-140.

Bernocchi, G., and Scherini, E., 1986a, DNA content in neurons, in: "Role of RNA and DNA in Brain Function", A. Giuditta, B.B. Kaplan, C. Zomzely-Neurath, eds., Martinus Nijhoff Publishing, Boston.

Bernocchi, G., Barni, S., and Scherini, E., 1986b, The annual cycle of Erinaceus europeanus L. as a model for a further study of cytochemical heterogenity in Purkinje neuron nuclei, Neuroscience, 17:427-433.

Borrione, P., Cervella, P., Geuna, S., Giacobini Robecchi, M.G., Poncino, A., and Silengo, L., Selective DNA amplification in adult spinal ganglion neurons: a neuroplasticity mechanism in lizard tail regeneration, in preparation.

Bregnard, A., Knusel, A., and Kuenzle, C.C., 1975, Are all the neuronal nuclei polyploid?, Histochemistry, 43:59-61.

Brennard, J., Chinault, A.C., Konecki, D.S., Melton, D.V., and Caskey, C.T., 1982, Cloned c-DNA sequences of the hypoxanthine/guanine phosphoribosyltransferase gene from a mouse neuroblastoma cell line found to have amplified genomic sequences, Proc. Natl. Acad. Sci. USA, 59:233-248.

Brodsky, V.J., Marshak, T.L., Mares, V., Lodin, Z., Fulop, Z., and Lebedev, E.A., 1979, Constancy and variability in the content of DNA in cerebellar Purkinje cell nuclei, Histochemistry, 59:233-248.

Chetverukhin, V.K., Salivanova, G.V., Onischenko, L.S., Vlasova, T.D., and Polenov, A.L., 1989, A cytophotometric analysis of the structure of hypotalamic cell populations in the frog, Rana temporaria (L.), with special reference to seasonal changes in the chromatin status, Histochemistry, 88:629-636.
Coggeshall, R.E., Yaksta, B.A., and Swartz, F.J., 1970, A cytophotometric analysis of the DNA in the nucleus of the giant cell, R-2, in Aplysia, Chromosoma, 32:205-212.
Cohen, J., Mares, V., and Lodin, Z., 1973, DNA content of purified preparations of mouse Purkinje neurons isolated by a velocity sedimentation technique, J. Neurochem., 20:651-657.
Cox, P.G., 1969, Some aspects of tail regeneration in the lizard, Anolis carolinensis, J. Exp. Zool., 171:127-150.
De Marianis, B., and Giuditta, A., 1978, Separation of nuclei with different DNA content from the subesoephageal lobe of octopus brain, Brain Res., 154:134-136.
De Marianis, B., Olmo, E., and Giuditta, A., 1979, Excess DNA in the nuclei of the subesophageal region of octopus brain, J. Comp. Neurol., 186:293-300.
Duffy, M.T., Simpson, S.B.JR., Liebich, D.R., and Davis, B.M., 1990, Origin of spinal cord axons in the lizard regenerated tail: supernormal projections from local spinal neurons, J. Comp. Neurol., 293:208-222.
Filogamo, G., and Vigliani, F., 1954, Ricerche sperimentali sulla correlazione tra estensione del territorio di innervazione e grandezza e numero delle cellule gangliari del plesso mienterico (di Auerbach) nel cane, Riv. Patol. Nerv. Ment., 75:1-32.
Filogamo, G., and Lievre, C., 1955a, Aumento di numero delle cellule nervose e moltiplicazione cellulare nei gangli del plesso di Auerbach, in condizioni di esaltata attività funzionale, Boll. Soc. It. Biol. Sper., 31:717-719.
Filogamo, G., and Lievre, C., 1955b, Comportamento delle cellule nervose del plesso mienterico, dell'intestino normale ed ipertrofico, nell'ansa alla Thiry-Vella, Boll. Soc. Piem. Chir., 25:1-3.
Filogamo, G., 1986, Neuronal modulation by number and by volume: a review and critical analysis, in:" Model Systems of Development and Aging of the Nervous System", A. Privat, P.S. Timiras, E. Giacobini, J. Lauder, eds., Martinus Nijhoff Publishing, Boston.
Fujita, S., 1974, DNA costancy in neurons of the human cerebellum and spinal cord as revealed by Feulgen cytophotometry and cytofluorometry, J. Comp. Neurol., 155:195-202.
Gabella, G., and Gaia, E., 1967, La proliferazione cellulare nel plesso di Auerbach di ratto in accrescimento ed in condizioni sperimentali, Boll. Soc. It. Biol. Sper., 43:1584-1586.
Geuna, S., and Poncino, A., 1988a, Analisi citofotometrica del contenuto di DNA di neuroni del plesso mienterico di Auerbach, Boll. Soc. It. Biol. Sper., 64:775-777.
Geuna, S., Poncino, A., and Robecchi, M.G., 1988b, Nuclei of neurons in relation to innervation territory: gangliar neurons innervating the regenerated tail of sauri, Neurosci. Lett., Suppl.33:98 (abstr.).
Geuna, S., Giacobini Robecchi, M.G., and Poncino, A., Nuclear hypertrophy and DNA content changes in sensory neurons during reinnervation of the regenerated lizard tail, submitted for publication.
Giacobini Robecchi, M.G., Cannas, M., and Filogamo, G., 1985, Increase in number and volume of myenteric neurons in the adult rat, Int. J. Devl. Neurosci., 3:673-676.
Giacobini Robecchi, M.G., Poncino, A., Geuna, S., Giaconetti, S., and

Filogamo, G., 1988, DNA content in neurons of Auerbach's plexus under experimental conditions in adult rats, Int. J. Devl. Neurosci., 6:109-115.
Giuditta, A., Libonati, M., Packard, A., and Prozzo, N., 1970, Nuclear counts in the brain lobes of Octopus vulgaris as a function of body size, Brain Res., 25:55-62.
Giuditta, A., Abrescia, P., and Rutigliano, B., 1978, Effect of electroshock on thymidine incorporation into rat brain DNA, J. Neurochem., 31:983-987.
Herman, C.J., and Lapham, L.W., 1969, Neuronal polyploidy and nuclear volumes in the cat Central Nervous System, Brain Res., 15:35-48.
Hobi, R., Studer, M., Ruch, F., and Kuenzle, C.C., 1984, The DNA content of cerebral cortex neurons. Determination by cytophotometry and high performance liquid cromatography, Brain Res., 305:209-219.
Huges, M., and New, D., 1959, Tail regeneration in geckonid lizard, Sphearodactylus, J. Embryol. Exp. Morph., 7:281-302.
Kuenzle, C.C., Bregnard, A., Hubscher, U., and Ruch, F., 1978, Extra DNA in forebrain cortical neurons, Exp. Cell. Res., 113:151-160.
Lasek, R.J., and Dower, W.J., 1971, Aplysia californica: analysis of nuclear DNA in individual nuclei of giant neurons, Science, 17:278-280.
Mann, D.M.A., Yates, P.O., and Barton, C.M., 1978, The DNA content of Purkinje cells in mammals, J. Comp. Neurol., 180:345-348.
Mares, V., Lodin, Z., and Sacha, J., 1973, A cytochemical and autoradiographic study of nuclear DNA in mouse Purkinje cells, Brain Res., 53:273-289.
Mares, V., and van der Ploeg, M., 1980, Cytophotometric re-investigation of DNA content in Purkinje cells of the rat cerebellum, Histochemistry, 69:161-167.
Mares, V., Crkovska, J., Marshak, T.L., and Stipek, S., 1985, DNA content in nerve cells nucleus. A biochemical and cytophotometric study of the rat cerebrum, Neuroscience, 16:45-47.
Marshak, T., Mares, V., and Brodsky, V., 1985, An attempt to influence the DNA content in postmitotic Purkinje cells of the cerebellum, Acta Histochem., 76:193-200.
McIlwain, D.L., and Capps-Covey, P., 1976, The nuclear DNA content of large ventral spinal neurons, J. Neurochem., 27:109-112
Morselt, A.F., Braakman, D.J., and James, J., 1972, Feulgen-DNA and fast green histone estimations in individual cell nuclei of the cerebellum of young and old rats, Acta Histochem., 43:281-286.
Pannese, E., 1963, Investigations on the ultrastructural changes of the spinal ganglion neurons in the course of axon regeneration and cell hypertrophy, Zeitschrift fur Zellforschung, 61:561-586.
Poncino, A., Geuna, S., Scherini, E., Giacobini Robecchi, M.G., and Filogamo, G., 1990, DNA synthesis experimentally induced in neurons: tetraploidy or hyperdiploid?, Int. J. Devl. Neurosci., 8:621-623.
Reinis, S., 1972, Autoradiographic study of ^{3}H-thymidine incorporation into brain DNA during learning, Physiol. Chem. Phys., 4:391-397.
Reinis, S., and Lamble, R.W., 1972, Labeling of brain DNA by ^{3}H-thymidine during learning, Physiol. Chem. Phys., 4:335-338.
Rouah, E., Wilson, D.R., Amstrong, D., and Darlington, G.J., 1989, N-myc amplification and neuronal differentiation in human primitive neuroectodermal tumors of the central nervous system, Cancer Res., 49: 1797-1801.
Schimke, R.T., Roos, D.S., and Brown, P.C., 1987, Amplification of genes in somatic mammalian cells, Methods in Enzymology, 151:85-104.
Schwab, M., Alitalo, K., Klempnauer, K.H., Varmus, H.E., Bishop, J.M., Gilbert, F., Brodeur, G., Goldstein, M., and Trent, J., 1983,

Amplified DNA with limited homology to myc cellular oncogene is shared by human neuroblastoma cell lines and neuroblastoma tumors, Nature, 305:245-248.
Schwab, M., Ellison, J., Busch, M., Rosenau, V., Varmus, H.E., and Bishop, J.M., 1984, Enhanced expression of the human gene c-myc consequent to amplification of DNA may contribute to malignat progression of neuroblastoma, Proc. Natl. Acad. Sci. USA, 81:4940-4944.
Simpson, S.B.JR, 1968, Morphology of the regenerated spinal cord in the lizard, Anolis carolinensis, J. Comp. Neurol., 134:193-210.
Swartz, F.J., and Bhatnagar, K.P., 1981, Are CNS neurons polyploid? A critical analysis based upon cytophotometric study of the DNA content of cerebellar and olfactory bulbar neurons of the rat, Brain Res., 208:267-281.
Terni, T., 1920, Sulla correlazione fra ampiezza del territorio di innervazione e grandezza delle cellule gangliari. II. Ricerche sui gangli spinali che innervano la coda rigenerata, nei sauri (Gongylus ocellatus), Arch. It. Anat. Embriol., 17:507-543.
Turner, J.E., and Singer, M., 1972, Some morphological and ultrastructural changes in the ependyma of the amputation stump during early regeneration of the tail in the lizard, Anolis carolinensis, J. Morph., 140:257-270.

PRENATAL DEVELOPMENT OF THE RAT AMYGDALOID COMPLEX: An Electron Microscopic Study

Anna Manolova and Stephan Manolov

Regeneration Research Laboratory
Bulgarian Academy of Sciences
1431 Sofia, Zdrave str. 2, Bulgaria

ABSTRACT

The prenatal development of the rat amygdaloid complex was studied ultrastructurally day by day starting from the 12th embryonic day (ED12) until birth (ED22). In the earlier stages of embryogenesis, the nerve cells were observed to be of small size. A thin cytoplasmic ring rich in free ribosomes surrounded the oval nuclei. The nucleoli were prominent their number being two or more in a nucleus. The intercellular spaces were wide. A significant advance in the development of the nerve cells and the neuropil was detected in the period between ED15 and ED18. This advance was represented by an increase in the cytoplasmic volume, appearance and development Golgi zones, grouping of the free ribosomes into rosettes, appearance of single synaptic contacts, and a ramification of large-calibre nerve processes that continued during the later stages of the embryogenesis (ED19 - ED22). By the end of the prenatal period the number of the synaptic contacts increased and well-developed synaptic apparatuses were observed. The development of the organelles in the perikarya was also advanced. Nissl bodies were found in some neurons. The number of enlarged Golgi zones and mitochondria increased.

INTRODUCTION

The amygdaloid complex is a brain area lying between the internal capsule and the hypothalamus. This complex is an important component of the limbic system characterized by numerous connections with other parts of the central nervous system (CNS). It has attracted particular interest from the apparent variety of functions it subsumes. The studies dealing with the ontogenetic development of the subnuclei of this complex are quite insufficient. The results from the light microscopic investigations following ^{3}H-thymidine administration show controversial data about the period (embryonic days - ED) of neuronal appearance. The period is different for the various nuclei, but in general, neurons appear in the hamster between ED10 - ED15 (Mc Connel, 1975; Mc Connel and Angevine, 1983; Sidman and Angevine, 1962), at ED12 (Ten Donkelaar *et al.*, 1979), and in the rat between ED13 - ED17 (Bayer, 1980). All

Plasticity and Regeneration of the Nervous System
Edited by P.S. Timiras *et al.*, Plenum Press, New York, 1991

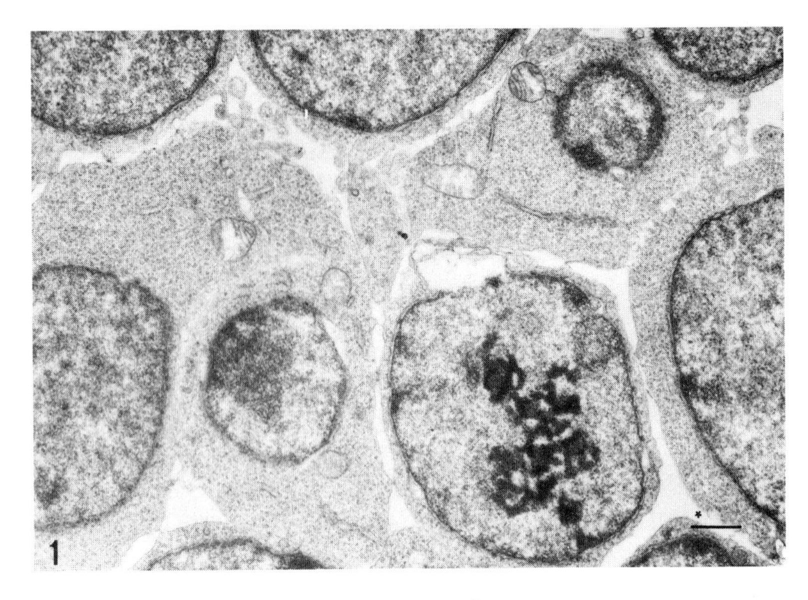

Figure 1. Amygdaloid complex ED13. *All calibration bars = 1μm.

authors cited recognize the existence of a definite rostrocaudal gradient in neuronal appearance and proliferation. There are just a few studies on the embryonic development of the amygdaloid complex in man (Humphrey, 1968; Humphrey, 1972; Nicolic and Kostovic, 1985). Humphrey (1968, 1972) has established that the development of the basolateral group of nuclei is delayed when compared with the corticomedial group - a fact which is synchronous with the phylogenetic development of the amygdaloid components.

To date, no ultrastructural studies on the ontogenetic development of the amygdaloid nuclei could be found in the literature. The establishment of the dynamics in the appearance and in the differentiation of the amygdaloid neurons undoubtedly is an important prerequisite for understanding the complex general organization and the polymorphism of the neurons that comprise the amygdala. In this study, the results from electron - microscopic observations of the prenatal development of rat's amygda-

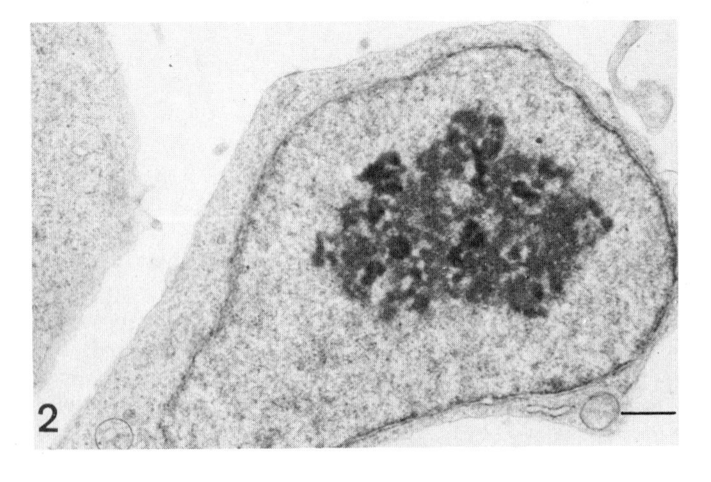

Figure 2. Amygdaloid complex ED12.5.

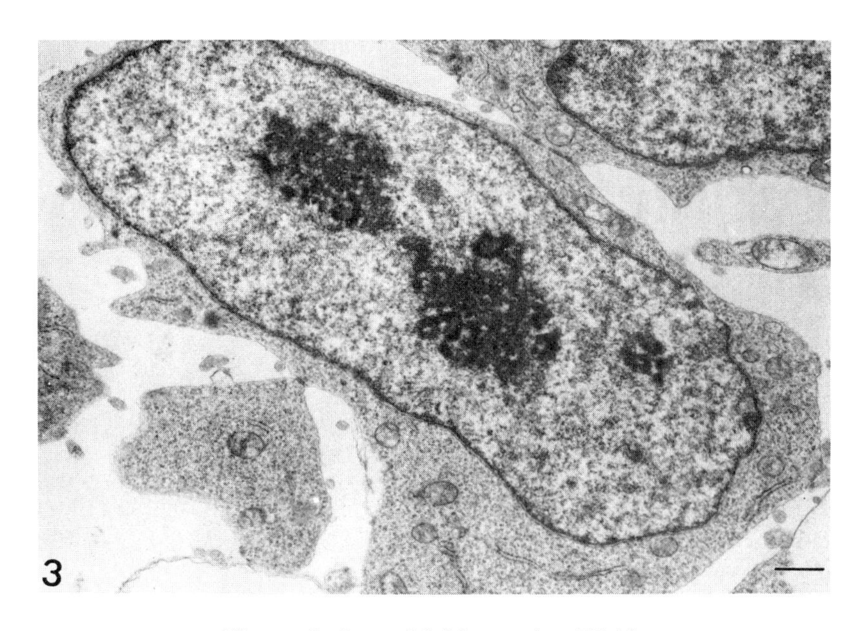

Figure 3. Amygdaloid complex ED13.

loid complex are described, with special attention being paid to the characteristic features of the nerve cells.

MATERIAL AND METHODS

White Wistar rats were used in this study. Under deep ether narcosis, the pregnant animals were laparotomized and their embryos excised. Their embryonic age and number was as follows: ED12 - 8 embryos, ED12.5 - 6 embryos, ED13 - 6 embryos and for each embryonic day from ED14 until ED22 - 3 to 4 embryos on the average. Next, the brains of the embryos younger than ED16 were fixed by immersion in 5.5% solution of glutaraldehyde in 0.05M phosphate buffer, pH 7.4. The embryos

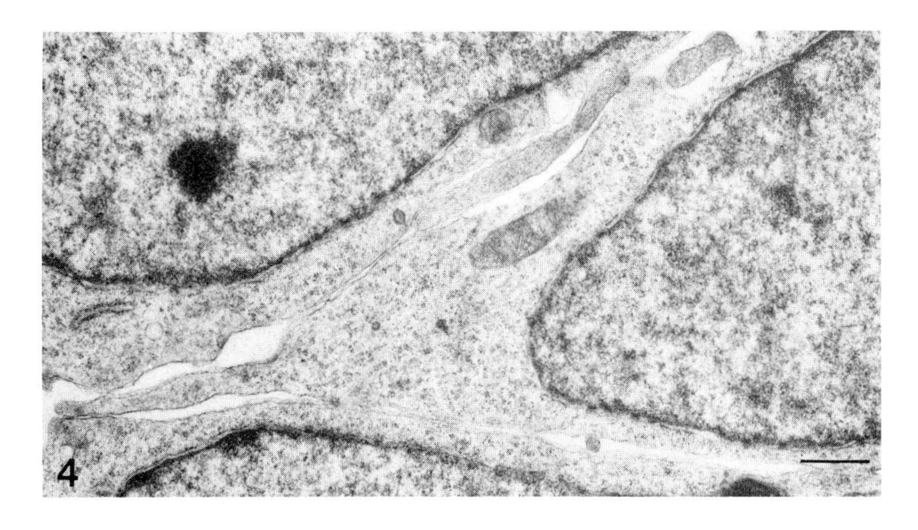

Figure 4. Amygdaloid complex ED12.5.

above this age were perfused via the ascending aorta with the same fixative. Samples from all embryos were sectioned, postfixed in OsO_4, dehydrated in ethanol and embedded in Durkopan (Fluka). Counterstained with uranylacetate and lead citrate, ultrathin sections were viewed and photographed on a Jeol (JEM 100B) electron microscope.

RESULTS

During the earlier stages of embryogenesis (ED12 - ED14), the nerve cells are oval or round in shape and of a small size. The intercellular spaces are wide. A thin ring of cytoplasm surrounds their large nuclei which are oval or round in shape (Fig. 1). The nucleoli are in general large and with various shape (Fig. 2). Quite often, two or even more nucleoli are found (Fig. 3). By the end of this period, invaginated nuclei are viewed. The perikarya are extremely rich in free ribosomes. Single short sacs of the granular endoplasmic reticulum (gER) and light-matrixed mitochondria with few cristas are found among the ribosomes. Occasionally longer sacs of gER are also observed (Fig. 1). The perikarya of some neurons contain microtubules. Single neuronal processes sprout from them (Fig. 4). By the end of this period, the cytoplasmic ring around the nuclei widens and the number of the neuronal processes increases.
The development of the nerve cells considerably advances in the period between ED15 and ED18. the cytoplasm enlarge, and the neuronal processes elarge also. The quantity of free ribosomes diminishes alongside with their grouping in rosettes. Simple Golgi zones appear and they are formed by several saccules and few vacuoles. The mitochondria posses darker matrix and more cristae when compared with earlier stage. The neural processes are enriched with organelles (Fig. 5).

The quantity of microtubules and microfilaments increases (Fig. 6). Later, the Golgi zones enlarge, their saccules become elongated, and the number of associated vesicles increases. The quantity and the length of the profiles of gER also increase (Fig. 7). Synaptic contacts appear. Though the axonal endings are poor in synaptic vesicles, the synaptic apparatus is well represented, with the asymmetry in the thickness of the synaptic membranes being clearly detectable. Significant changes are

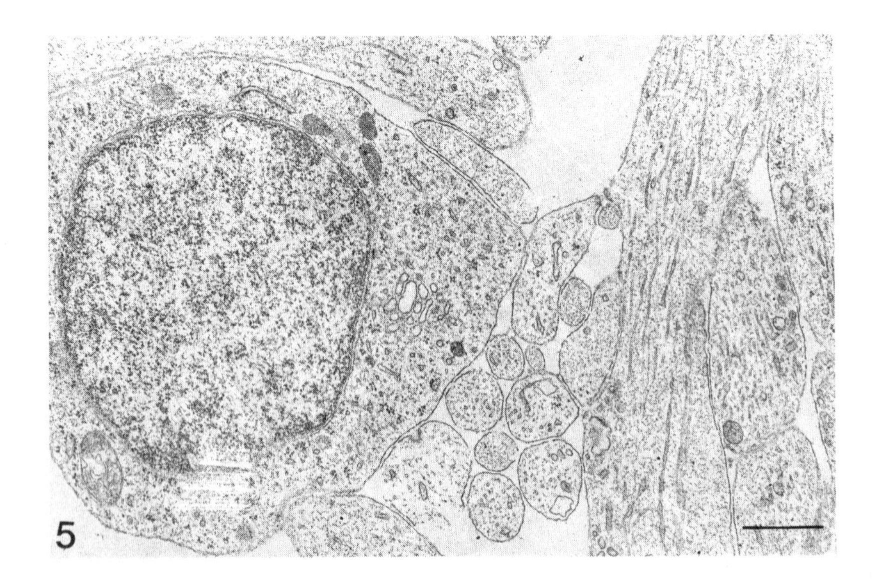

Figure 5. Amygdaloid complex ED15.

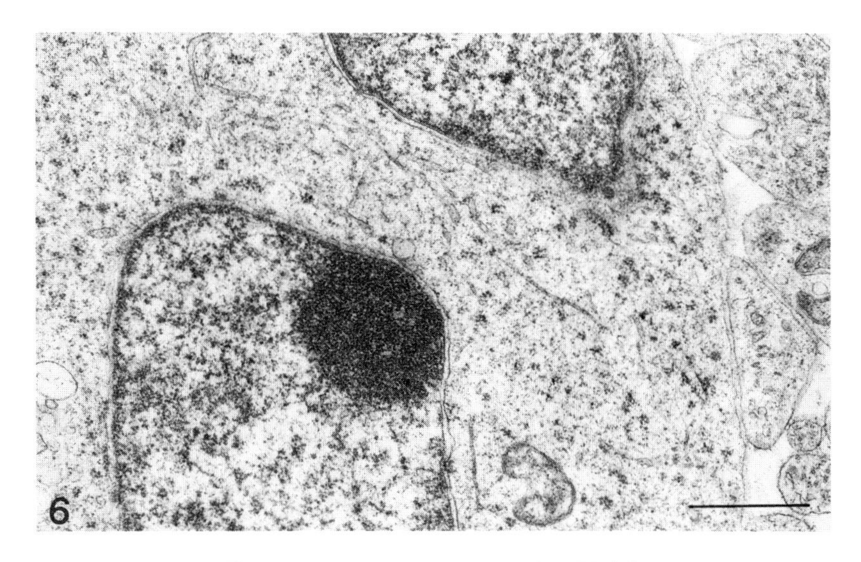

Figure 6. Amygdaloid complex ED16.

observed in the neuropil; the large-calibre neuronal processes typical of the earlier stages are accompanied by small processes; this is due to the abundant ramification of the large processes after ED17 - ED18.

Starting from ED19, large zones of the neuropil are occupied by closely packed small-calibre processes and single oligodendrocytes (Fig. 8). The advancing development is characterized by a decrease in the electron density of the neuropil. The number of the synapses increases. Some axonal endings are still poor in synaptic vesicles; however, the active zones are clearly visible. By the end of the prenatal period, in some cases more mature synaptic contacts are viewed; their axonal endings already contain a lot of synaptic vesicles and well-developed synaptic appara-

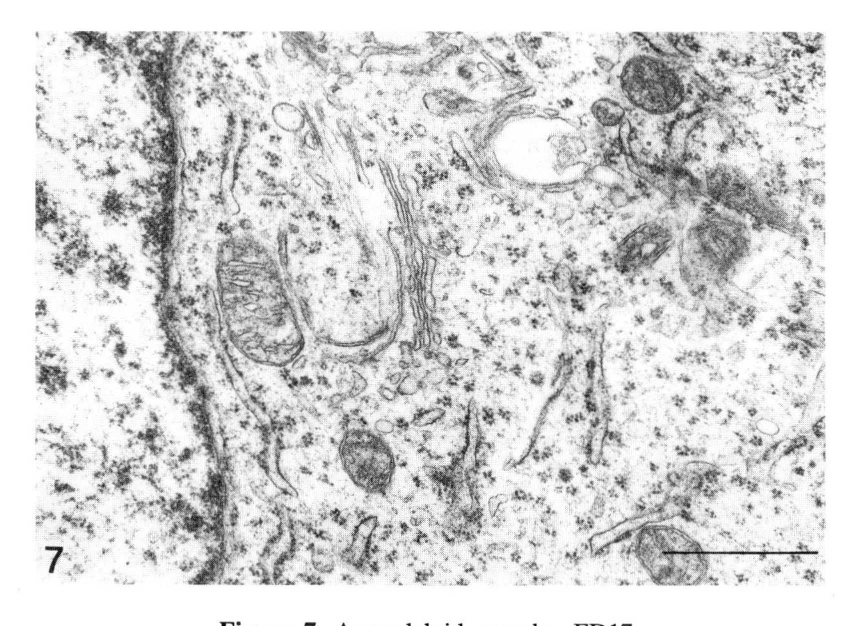

Figure 7. Amygdaloid complex ED17.

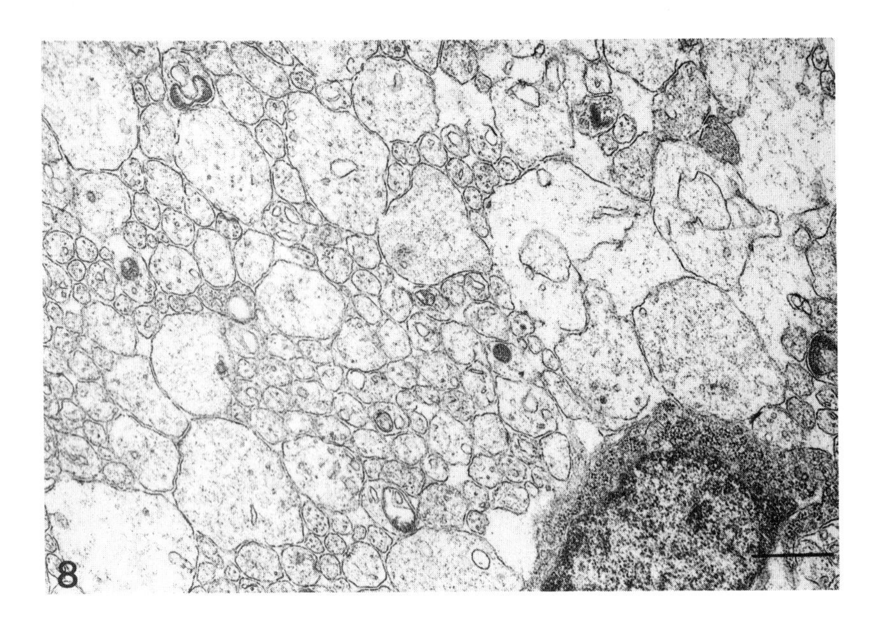

Figure 8. Amygdaloid complex ED19.

tus (Fig. 9). The development of the organelles in the perikaryon is also advanced. The quantity and the length of the profiles of gER increase further; in some neurons they acquire parallel arrangement and form Nissl bodies (Fig. 10). The Golgi zones are already large; they consist of numerous saccules and associated vesicles. The quantity and dimensions of the mitochondria increase (Fig. 11).

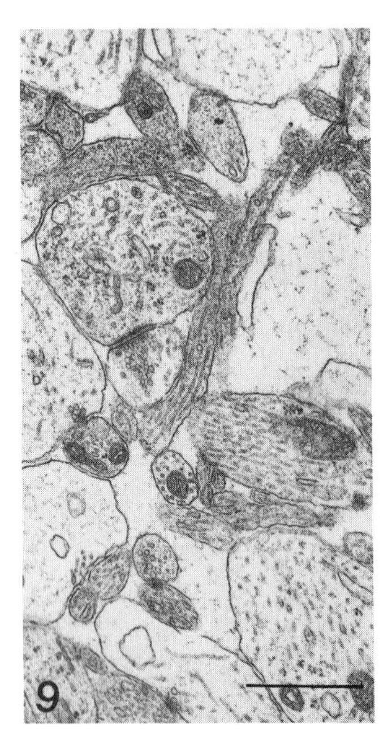

Figure 9. Amygdaloid complex ED22.

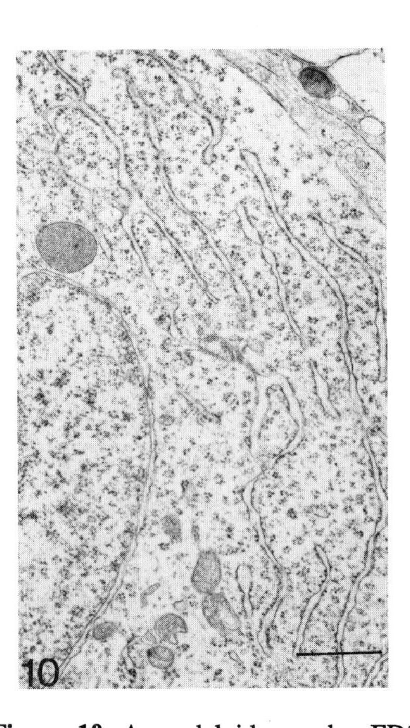

Figure 10. Amygdaloid complex ED19

DISCUSSION

The present study is the first ultrastructural assessment on the prenatal development of the amygdaloid complex in the rat. Also, electronmicroscopic data are available in the literature about the development of this complex in other animal species. That is why, the following discussion will compare the present results with those obtained at the light-microscopic level. It should be recognized however, that since during embryonic development the amygdaloid subnuclei are not definitely outlined, an exact description of the changes occurring in the different nuclei is impossible.

The results from the present study showed that embryonic development of the structures that comprise the amygdaloid complex could be divided into three stages: stage one lasts from ED12 until ED14; stage two lasts from ED15 until ED18 to 19; and stage three lasts from ED19 until ED22.

It was established that there could be observed some nerve cells in the amygdaloid complex at ED12. This observation is in disagreement with the results of Bayer (1980) who, when using light microscopy, stated that the first neurons appear between ED13 and ED17. On the other hand, the presence of nerve cells at ED12 is synchronous with the data concerning the development of the amygdala in mice and hamsters (Mc Connell, 1975; Mc Connell and Angevine, 1983; Sidman and Angevine, 1962; Ten Donkelaar *et al.*, 1979).

The first period is characterized with an irregular transformation of some neuroblasts into primitive neurons. Judging from the formation of single short processes, it may be concluded that these cells undergo a very early differentiation towards primitive nerve cells.

Despite the morphological irregularity occurring in the course of histogenesis of the amygdala, the second period in general, is the time of a rapid and significant proliferation of the neurons accompanied by their further differentiation (till ED19). A characteristic feature of this period is also the appearance of the first synaptic contacts. It is beyond the task and abilities of the present study to state whether these contacts are transitory or definite, as well as to express an opinion about their functional importance.

The third period can be regarded as the stage of further differentiation and maturation of the amygdaloid structures. This supposition is backed by the appearance

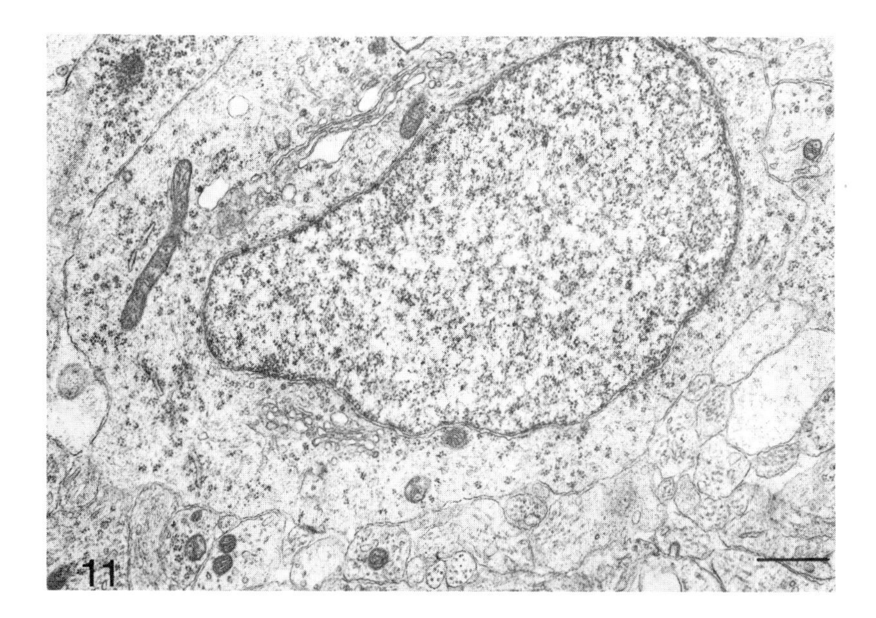

Figure 11. Amygdaloid complex ED21.

of the first cells resembling more mature neurons, the significant increase in the number of synaptic contacts and the differentiation of the neuropil. It should be emphasized that at the end of the period, the ultrastructural maturity of the amygdaloid complex is still far from the image observed in adult animals. So, it is evident that the definite differentiation of the components of the amygdaloid complex takes place during their postnatal development.

REFERENCES

Bayer, S.A. 1980, Quantitative ^{3}H-thymidine radiographic analyses of neurogenesis in the rat amygdala. J. Comp. Neurol., 194:845.

Humphrey, T., 1968, The development of the human amygdala during early embryonic life. J. Comp. Neurol., 132:135.

Humphrey, T., 1972, The development of the human amygdaloid complex, in: "The Neurobiology of Amygdala," B.E. Eleftherion, ed., Plenum Press, New York.

Mc Connell, J.A., 1975, Time of origin of amygdaloid neurons in the mouse. An autoradiographic study. Anat. Rec., 181:113.

Mc Connell, J.A., Angevine, B. Jr., 1983, Time of neuron origin in the amygdaloid complex of the mouse. Brain Res., 272:150.

Nicolic, I., and Kostovic, I., 1985, Development of the lateral amygdaloid nucleus in the human fetus: transient presence of discrete cytoarchitectonic units. Anat. Embryol., 174:355.

Sidman, R.L., and Angevine, B.Jr., 1962, Autoradiographic analysis of time of origin of nuclear versus cortical components of mouse telencephalon. Anat. Rec., 142:326.

Ten Donkelaar, H.J. *et al.*, 1979, Neurogenesis in the amygdaloid nuclear complex in a rodent (the chinese hamster). Brain Res., 165:348.

RECENT FINDINGS ON THE REGULATION OF AXONAL CALIBRE

Ennio Pannese

Institute of Histology, Embryology and Neurocytology
University of Milan, Via Mangiagalli, 14
I-20133 Milan, Italy

The morphological features and functional properties of the neuron are closely related. Concerning the axon in particular, this interrelationship between morphology and function is illustrated, for example, by the fact that conduction is faster in thick axons than in thin ones. Since the transverse diameter of the axon affects the velocity of action potential propagation, the investigation of the factors which control axonal calibre is particularly important.

Usually axonal calibre is positively correlated with nerve cell body volume (Ramón y Cajal, 1909; Marinesco, 1909). Based on this fact, it was thought for many years that axonal calibre was controlled by factors intrinsic to the neuron. Recent morphological studies, carried out principally on large myelinated fibres, have provided data consistent with this old view. These studies have shown that the calibre of an axon is closely correlated with the number of neurofilaments present within it (Friede and Samorajski, 1970; Weiss and Mayr, 1971; Hoffman et al., 1984). These observations suggest that the neurofilament gene expression is a primary determinant of the axonal calibre (see Lasek, 1988 for a review).

Other observations, however, have made it evident that external factors also play an important role in the control of axonal calibre. For example, when a rat neuron is disconnected from its target cells, a reduction in axonal diameter takes place, suggesting that interactions between neurons and their target cells play a role in the control of axon calibre (Hoffman et al., 1988). Moreover, evidence has now accumulated that also interactions between an axon and its supporting cells play a role in this regulation. In the axon of a given motoneuron of the rat, the peripheral portion which is in the ventral root and is myelinated by Schwann cells is significantly thicker than the central portion which is situated in the spinal cord and is myelinated by oligodendrocytes (Fraher, 1978). When a segment of the sciatic nerve from a Trembler mouse is grafted into a normal mouse nerve, the regrowing normal axons become surrounded by Trembler Schwann cells in the grafted segment. The axonal portions situated in the graft not only lack myelin or are hypomyelinated

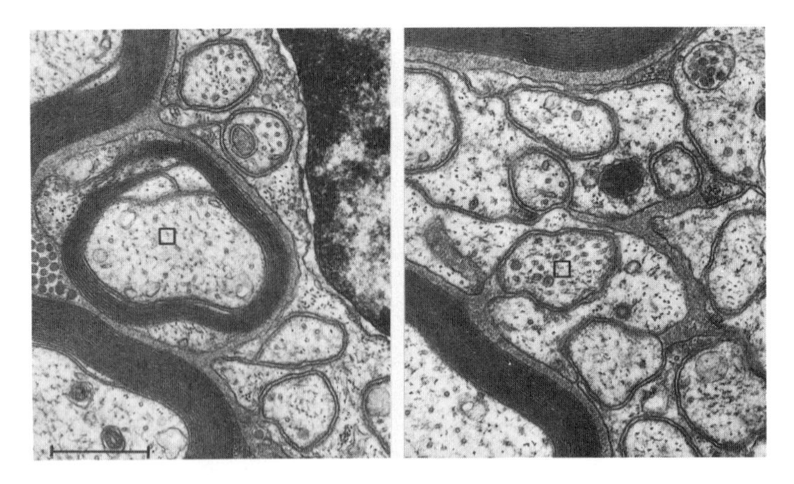

Fig. 1. Electron micrographs showing two sections of a single axon (□) with both myelinated and unmyelinated portions. The section on the left is from a myelinated portion, the section on the right is from an unmyelinated portion. These two sections are about 20 μm from each other. It can be seen that the axonal calibre is much greater in the myelinated portion than in the unmyelinated portion. Dorsal spinal root of a lizard. Bar: 0.5 μm.

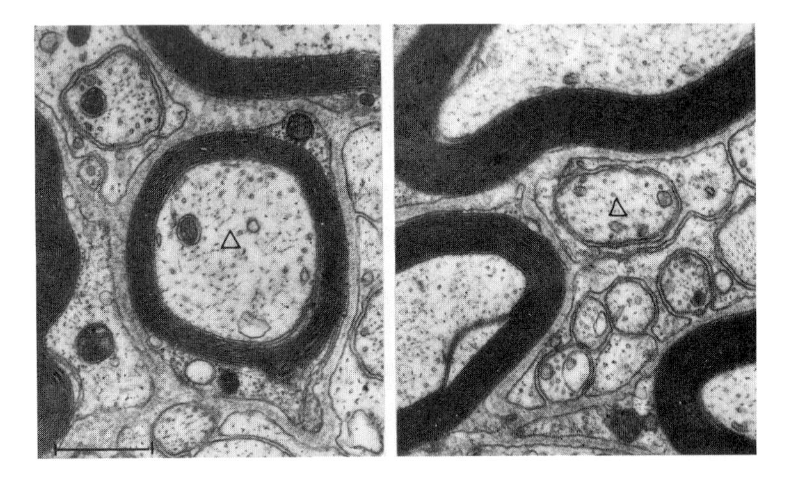

Fig. 2. Electron micrographs showing two sections of another axon (△) having both myelinated and unmyelinated portions. The section on the left is from a myelinated portion, the section on the right is from an unmyelinated portion. These two sections are about 18.5 μm from each other. Again it can be seen that the axonal calibre is much greater in the myelinated portion than in the unmyelinated portion. Dorsal spinal root of a lizard. Bar: 0.5 μm.

but also have reduced transverse diameters, whereas their calibre returns to normal in the distal nerve stump where they are surrounded by normal Schwann cells (Aguayo et al., 1979). In cultures of rat spinal ganglion neurons, myelinated axons have calibres significantly larger than bare axons and axons myelinated by Schwann cells are significantly thicker than axons myelinated by oligodendrocytes (Windebank et al., 1985). Axons with both myelinated and unmyelinated portions have been found in the dorsal spinal roots of normal lizards (Pannese et al., 1988); the calibre of the same axon is significantly greater in the myelinated than in the unmyelinated portion (Figs. 1 and 2). In human demyelinating neuropathies, a marked reduction in axon calibre occurs locally in the regions of segmental demyelination (peripheral nervous system: Raine et al., 1969; Prineas and McLeod, 1976; central nervous system: Prineas and Connell, 1978).

Taken together, the above mentioned observations show that the whole question of the regulation of axon calibre is much more complex than previously believed. They have made it clear that interactions between the axon and its related cells (target and enveloping cells) play an important role in regulating the shape and hence some functional properties of the axon. It is therefore evident that the calibre of the axon is dependent on a variety of factors, both intrinsic and external to the neuron. Very little is known, however, about the mechanisms by which external factors affect the transverse diameter of the axon. It has been suggested that the enveloping cells affect axonal calibre by generating both signals which would act at the level of the nerve cell body and other signals which would act locally at the level of the axon. The first of these signals would travel, by retrograde transport, from the axon to the nerve cell body, where they would promote changes in neurofilament synthesis; the other signals would exert a local control on the rate of the axonal transport of neurofilaments (Lasek, 1988; de Waegh and Brady, 1990). The precise nature of these signals remains unknown at present.

Before closing, it is important to note that the interactions between axon and enveloping cells are much more complicated and intimate than previously thought. It was originally believed, for example, that these interactions consisted exclusively of influences of the axon on the surrounding cells, i.e. that these interactions were unidirectional. It is now evident, however, that enveloping cells can influence the morphological and functional characteristics of the axon, namely there are also interactions in the reverse direction.

REFERENCES

Aguayo, A. J., Bray, G. M., Perkins, C. S., and Duncan, I. D., 1979, Axon-sheath cell interactions in peripheral and central nervous system transplants, in: "Aspects of Developmental Neurobiology," J. A. Ferrendelli, ed., Society for Neuroscience Symposia, 4:361, Society for Neuroscience, Bethesda.

de Waegh, S., and Brady, S. T., 1990, Altered slow axonal transport and regeneration in a myelin-deficient mutant mouse: the trembler as an in vivo model for Schwann cell-axon interactions, J. Neurosci., 10:1855.

Fraher, J. P., 1978, Quantitative studies on the maturation of central and peripheral parts of individual ventral motoneuron axons. I. Myelin sheath and axon calibre, J. Anat., 126:509.

Friede, R. L., and Samorajski, T., 1970, Axon caliber related to neurofilaments and microtubules in sciatic nerve fibers of rats and mice, Anat. Rec., 167:379.

Hoffman, P. N., Griffin, J. W., and Price, D. L., 1984, Control of axonal caliber by neurofilament transport, J. Cell Biol., 99:705.

Hoffman, P. N., Koo, E. H., Muma, N. A., Griffin, J. W., and Price, D. L., 1988, Role of neurofilaments in the control of axonal caliber in myelinated nerve fibers, in: "Intrinsic Determinants of Neuronal Form and Function," R. J. Lasek and M. M. Black, eds., p. 389, Alan R. Liss Inc., New York.

Lasek, R. J., 1988, Studying the intrinsic determinants of neuronal form and function, in: "Intrinsic Determinants of Neuronal Form and Function," R. J. Lasek and M.M. Black, eds., p. 3, Alan R. Liss Inc., New York.

Marinesco, G., 1909, "La cellule nerveuse," tome 1er, O. Doin et Fils Editeurs, Paris.

Pannese, E., Ledda, M., and Matsuda, S., 1988, Nerve fibres with myelinated and unmyelinated portions in dorsal spinal roots, J. Neurocytol., 17:693.

Prineas, J. W., and McLeod, J. G., 1976, Chronic relapsing polyneuritis, J. Neurol. Sci., 27:427.

Prineas, J. W., and Connell, F., 1978, The fine structure of chronically active multiple sclerosis plaques, Neurology, 28:68.

Raine, C. S., Wisniewski, H., and Prineas, J., 1969, An ultrastructural study of experimental demyelination and remyelination: II. Chronic experimental allergic encephalomyelitis in the peripheral nervous system, Lab. Invest., 21:316.

Ramon y Cajal, S., 1909, "Histologie du système nerveux de l'homme et des vertébrés," vol. 1, Maloine, Paris.

Weiss, P. A., and Mayr, R., 1971, Organelles in neuroplasmic ("axonal") flow: neurofilaments, Proc. Nat. Acad. Sci. USA, 68:846.

Windebank, A. J., Wood, P., Bunge, R. P., and Dyck, P. J., 1985, Myelination determines the caliber of dorsal root ganglion neurons in culture, J. Neurosci., 5:1563.

THE BIOGENIC MONOAMINES AS REGULATORS OF EARLY (PRE-NERVOUS) EMBRYOGENESIS : NEW DATA

Gennady A. Buznikov

N.K. Koltzov Institute of Developmental Biology
USSR Academy of Sciences, Moscow

INTRODUCTION

Several neurotransmitters such as acetylcholine, catecholamines, indolylalkylamines and neuropeptides are multifunctional controlling substances. They are involved not only in synaptic transmission and other nervous events but also in various regulatory processes not directly linked to the nervous system. Neurobiologists are interested in these non-nervous neurotransmitter functions as well.

Although synaptic functions of transmitters are better studied, various techniques and approaches of modern neurobiology are used to study pre-nervous and other non-nervous neurotransmitter activities. Results of such new studies may be compared to corresponding data from studies on classical neurotransmitters. The study of non-trivial transmitter events clarifies also developmental neurobiology. One striking instance is represented by the data on the role of non-synaptic biogenic monoamines during neurogenesis (Kater and Haydon, 1987 ; Lauder, 1988 ; Lauder et al., 1988.).

Non-nervous neurotransmitter functions emerge very early during evolution and appear again and again during ontogenesis ahead of classical synaptic functions. In other words, the non-synaptic functions are phylogenetic and ontogenetic precursors of synaptic functions, a fact to remember when studying early developmental onset of synaptic and non-synaptic functions. And "the very onset" precedes fertilization and early embryonic development ; it is preembryonic development (gametogenesis).

CHANGES OF NEUROTRANSMITTER FUNCTIONS DURING ONTOGENESIS

Earlier studies have suggested that neurotransmitters act firstly as intracellular substances controlling cell division, then as local hormones on embryonic motility and lastly as synaptic messengers (Buznikov, 1967, 1984,

Plasticity and Regeneration of the Nervous System
Edited by P.S. Timiras *et al.*, Plenum Press, New York, 1991

1990). This scheme looks now oversimplified. It is apparent that neurotransmitters are multifunctional controlling substances not only in late embryos and adult organisms but also at all other developmental stages. Some neutrotransmitter functions may disappear during development. Other functions can be retained under modified forms or can coexist with synaptic functions without considerable changes. The preparation of a new timetable, valuable as it may be, is not as yet possible. We can only list certain pre-nervous and non-nervous transmitter functions and identify roughly the periods when such functions can be detected (Table 1).

To prepare this table, we have used both our results and data from other authors. Most results were obtained in experiments on echinoderms and vertebrates. Obviously, this list is far from being complete. It cannot be ruled out that some transmitter functions listed individually in the table, belong in fact to the same function.

Periods of appearance and disappearance are known for non-synaptic functions. The biogenic monoamines and acetylcholine act as intracellular triggers of cytokinesis immediately after fertilization. This function disappears during gastrulation - suddently or little by little, we do not know. And it reappears more than once in connection with local bursts of cell proliferation during postgastrulation development. Such function may also reappear in connection with the malignant transformation of differentiated cells (Buznikov, 1984, 1989, 1990). It may also persist throughout life as is the case of serotonin which functions in mollusks as a local hormone of cilia activity : this function persists from the time of ciliated cells appearance (i.e. from trochophora or late blastula stage, respectively) to the end of life.

In other cases, we have obtained only fragmentary information. Acetylcholine and biogenic monoamines would participate in the control of genome activity during sea urchin gastrulation (Gustafson and Toneby, 1971 ; Buznikov, 1971, 1990 ; Gustafson, 1975). However, it is still unclear how these substances act before and after fertilization.

All known non-nervous transmitter functions cannot be considered in this report. I shall try to describe some new data on the role of biogenic monoamines in the control of cleavage division, i.e. first cell divisions after fertilization.

THE BIOGENIC MONOAMINES AS TRIGGERS AND REGULATORS OF CLEAVAGE DIVISIONS

From data on changes of neurotransmitter concentrations during the first cell cycles, we suggested that biogenic monoamines and acetylcholine may control cleavage divisions (Buznikov, 1967). We demonstrated subsequently that biogenic monoamines are, indeed, required for early development of sea urchins. Various drugs, such as lipophilic analogs of

Table 1 Non-nervous Transmitter Functions

Functions	Receptor sites	Periods of existence
- Intracellular triggering and cell division control	Intra-cellular	- Cleavage divisions (up to end blastulation).Cell proliferation bursts during postgastrulation development. Tumor cells (?)
- Non-nervous cell-cell interactions.	Intra-cellular or surface	- Oocytes and follicle cells interactions. Cleavage divisions (from two-cell stage). Embryonic induction and other morphogenetic interactions.
- Control of intracellular metabolite transport	Intra-cellular	- Cleavage divisions (no other information).
- Control of macromolecular syntheses at (a) transcription and (b) post-transcription	Intra-cellular	-(a) Gastrulation(no other information). b. Late postfertilization responses. Cleavage divisions (no other information)
- Control of yolk mobilization	Intra-cellular	- From cleavage divisions to yolk disappearance; sharp rise from gastrulation.
- Control of cytoplasmic Ca level (in addition-to cell division)	Surface	- Full-grown oocytes. Cleavage divisions (no other information)
- Morphogenetic cell movements.	?	- Gastrulation and postgastrulation.
- Non-nervous motility (especially beating of cilia).	Surface	- Spermatozoids (?) Whole ontogenesis from formation of ciliated cells.
- Control of cell sensitivity to other regulatory substances.	Intra-cellular or surface	- Full-grown oocytes. Differentiated neurons (no other information).
- Cytoplasmic influences on cell nuclei.	Intra-cellular	- Full-grown oocytes (?). Cleavage divisions.
- Control of ganglioside synthesis, transport and surface expression.	Intra-cellular	- Cleavage divisions. Tumor cells.

neurotransmitters, inhibit or block cleavage divisions. The corresponding transmitters protect embryos from this block. Later, Renaud et al. (1983) have obtained similar results on sea urchins. We demonstrated also that biogenic monoamines are involved in cleavage divisions in other groups of invertebrates and vertebrates (Buznikov, 1990).

Transport of lipophilic monoamine analogs into cytoplasm is an indispensable condition for their specific cytostatic action on early embryos of sea urchins and amphibians. Early embryos would possess some intracellular receptive structures which interact with these molecules (Buznikov, 1984). It is quite possible that these structures represent only functional equivalents of true monoaminergic or cholinergic receptors. However, great caution must be exerted in our interpretation.

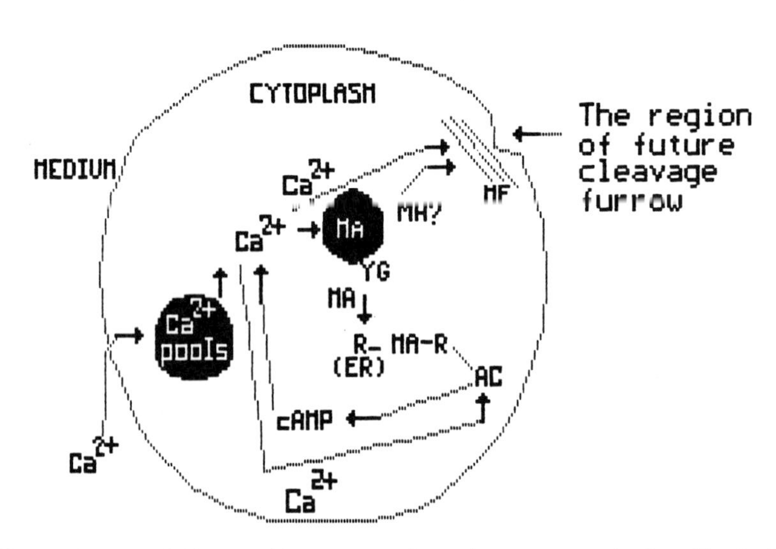

Fig. 1 The scheme of the cyclic monoaminergic process during first cleavage divisions. AC : adenylate cyclase ; ER : endoplasmic reticulum ; BA : biogenic monoamines ; MF : microfilaments ; YG : yolk granules.

A Schematic Representation of Biogenic Monoamines Function during Cleavage Divisions

A proposed scheme points to indolylalkylamines and catecholamines as triggers and controls of sea urchin cleavage divisions (Fig. 1).

Pre-nervous biogenic monoamines are synthesized in yolk granules. These granules represent the main site for non-receptor binding of biogenic monoamines and their lipophilic cytostatic antagonists. The increase of cytoplasmic Ca2+ level triggered by fertilization stimulates monoamine synthesis and also releases monoamines from the yolk granules. Monoamines then reach receptor binding sites which are probably located on the membranes of the endoplasmic reticulum.

It cannot be ruled out that the pre-nervous monoaminergic receptors are directly connected with the effectors of cytokinesis, i.e. the microfilaments of the cortical cytoplasmic layer. The monoamines would act immediately on microfilament contractility without the participation of special receptor sites (a kind of short circuit). Thus, the presence of 3H-serotonin following incubation of early polychaeta embryos or chick embryos with a radiolabeled precursor (5-hydroxytryptophan) was seen to be associated to yolk granules and cytoskeletal elements (Emanuelsson, 1974 ; Emanuelsson et al., 1988). The interaction of monoamines with pre-nervous receptors affects the cAMP level and this, in turn, affects (via this changed level or otherwise) Ca2+ level in the cortical cytoplasm. An increase in the level of Ca2+ provokes microfilament constriction in the region of future cleavage furrow and also induces other changes of the cytoskeleton that induce cytokinesis. In addition, increased Ca2+ level further stimulates monoamine synthesis, in preparation for new cytokinesis. The "relaxation" of cortical microfilaments after cytokinesis is probably also controlled by biogenic monoamines. This cyclic process continues until the end of cleavage.

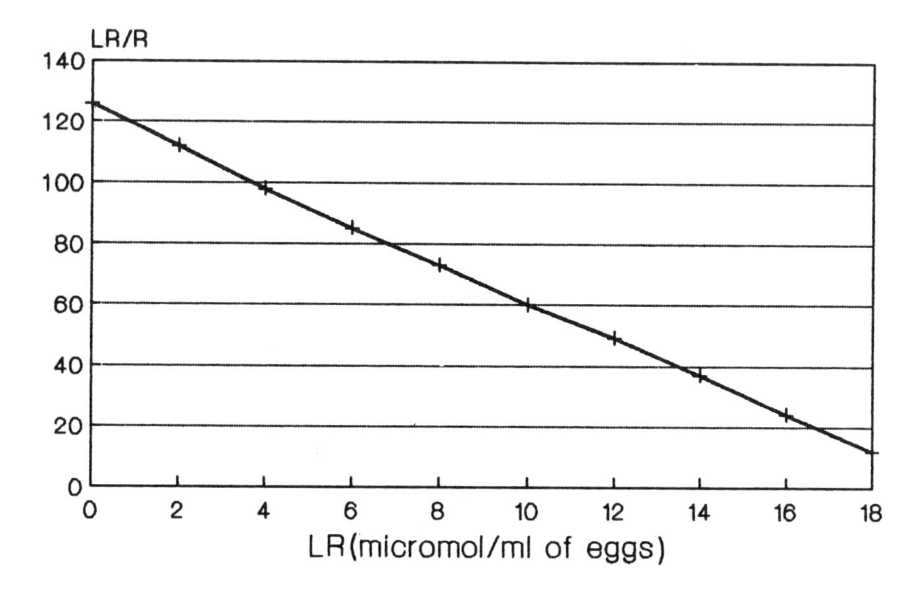

Fig. 2 The binding of 3H-imipramine by the whole embryos of sea urchin Arbacia lixula (two-cell stage) (Scatchard plot).

The Binding of 3H-Serotonin and its Antagonists by Sea Urchin Embryos

There is some evidence in favour of this binding. It has been demonstrated in experiments with radioligands that cytostatic serotonin antagonists belong to two intercellular binding pools. The first has a very high capacity and surely is the non-receptor pool (Fig. 2). The dissociation constants (KD) for this pool exceed by one-two orders of magnitude the physiological active concentrations of these antagonists.

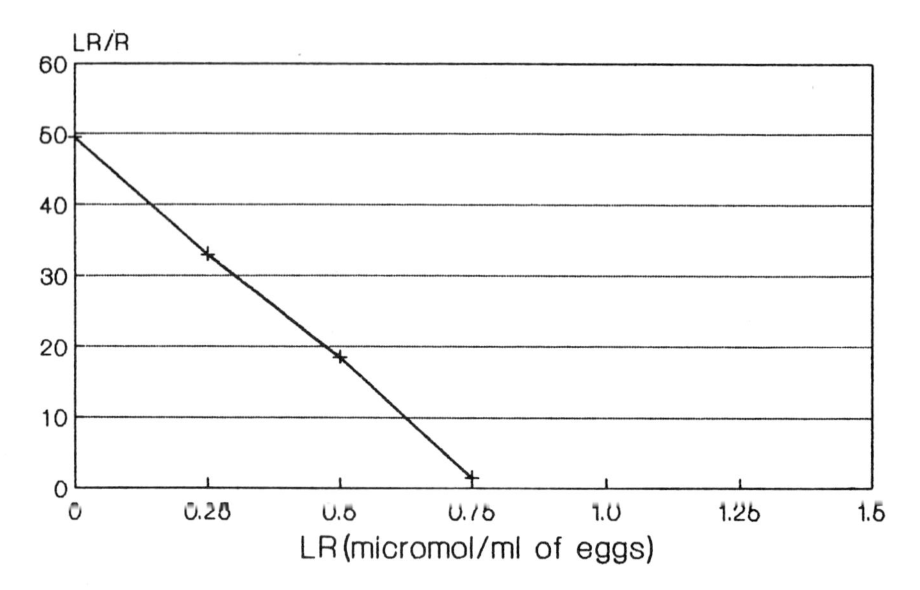

Fig. 3 The binding of 3H-imipramine by embryos-fragments (upper halves) of sea urchin A. lixula (two-cell stage) (Scatchard plot).

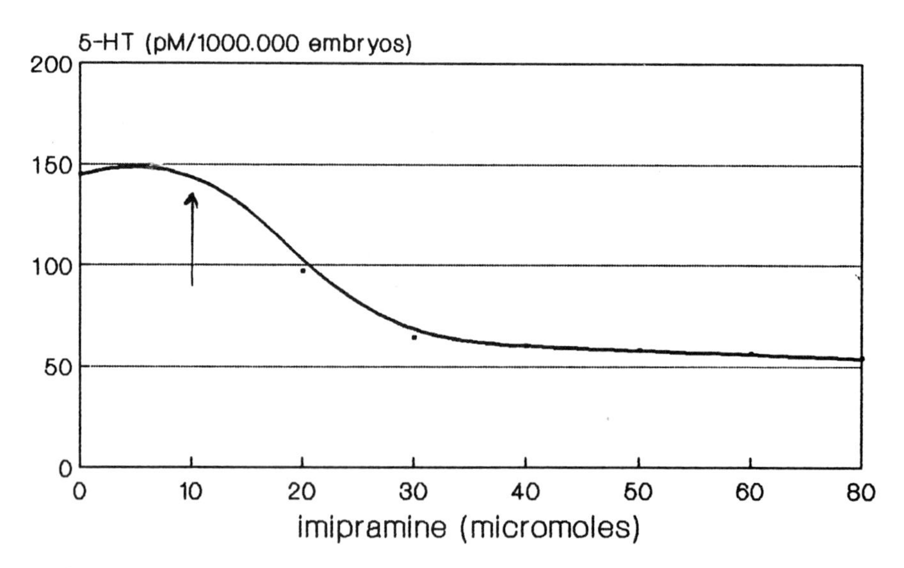

Fig. 4 Imipramine action on binding of 3H-serotonin by the one-cell embryos of A. lixula. (arrow: beginning of cytostatic action of imipramine).

The second pool is very small and probably remains usually unnoticed. However, this pool is unmasked in embryos-fragments (upper halves) characterized by low yolk content (Fig. 3). As can be judged from the dissociation constant, it is the receptor pool. The corresponding dissociation constant for imipramine is equal to 15-20 uM. This concentration range of imipramine corresponds to its specific cytostatic action. With similar concentrations, imipramine can inhibit the binding of 3H-Serotonin (Fig. 4).

The non-receptor pool is probably associated with yolk granules. Direct evidence for this statement emerges from experiments with the binding of 3H-Serotonin and 3H-imipramine by isolated yolk granules (Fig. 5). In these granules there is only one binding pool with high capacity and dissociation constant very similar to those for non-receptor binding by intact embryos. The other pool with high affinity and low capacity is absent here.

Radioligand binding appears to be irreversible in the case of whole (non-fragmented) embryos because the capacity of the non-receptor pool is very high (Fig. 6). The inhibition of this binding takes place only at high concentrations of the unlabeled ligand. This may explain why Brown and Shaver (1989) reached the conclusion that, intracellular binding of 3H-Serotonin in early sea urchin embryos is non-specific. The binding of this radioligand is reversible and specific as clearly demonstrated in experiments with the embryos-upper halves.

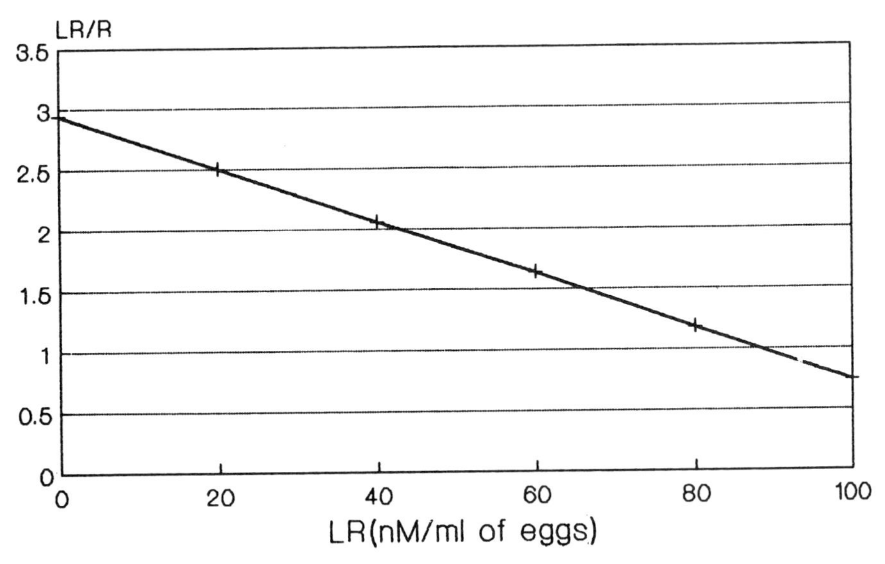

Fig. 5 The binding of 3H-imipramine by the yolk granules isolated from 4-8-cell embryos of sea urchin Paracentrotus lividus (Scatchard plot).

The fragmentation of embryos can be used for weakening or eliminating of non-receptor binding. A good method for inhibiting receptor binding is to increase embryo concentration in the incubation medium. Intracellular receptors are masked because of this increase. They become inaccessible for related exogenous ligands. The non-receptor binding is not practically changed in this situation but the cytostatic activity of the used ligand drops to zero. Thus, the concentration of the cytostatic serotonergic or adrenergic blocking drug in the cytoplasm would be hundred times larger than the minimal concentration capable of blocking cleavage, development is quite normal and sensitivity to this blocking drug does not increase.

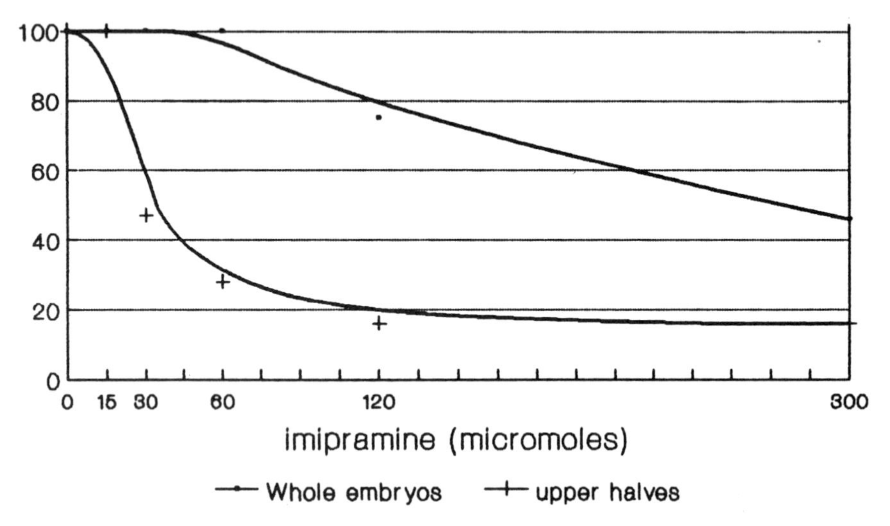

Fig. 6 "Cold" imipramine action on the binding of 3H-imipramine by whole embryos and embryos-upper halves of sea urchin A. lixula.

The gangliosides of early embryos play an important but unclear role in masking intracellular monoaminergic receptors (Prokazova et al., 1981 ; Zvezdina et al., 1989). Cells of transplantable tumors such as ascite Ehrlich carcinoma or ascite hepatoma 22a are sensitive to lipophilic cytostatic analogs of biogenic monoamines acting on the intracellular level. This sensitivity is also decreased by rising the cell concentration of these substances and this decrease is connected with important but unclear ganglioside effects (Shaposhnikova et al., 1984). Because of these results we suggested the reappearance of the cytokinetic monoamine function in tumor cells (Buznikov, 1990).

The Biogenic Monoamines of Cleaving Embryos Coupled to Second Messenger Systems

In experiments with early embryos of sea urchins and amphibians we also obtained other results predicted on the basis of our scheme. We found that adenylate cyclase of cleaving sea urchin embryos was mainly located in the membranes of the endoplasmic reticulum. Forskolin can activate the intracellular adenylate cyclase. It has been reported elsewhere that the adenylate cyclase of sea urchin embryos is activated by biogenic monoamines, especially by dopamine, a major pre-nervous catecholamine (Capasso et al., 1987, 1988). Conversely, some antagonists of pre-nervous monoamines decrease cAMP level in early embryo cells (Renaud et al., 1983). Influences leading to increased cAMP or $Ca2+$ cytoplasmic levels evoke a reliable decrease of the embryo sensitivity to the monoaminergic blocking drugs (Buznikov,1990).

We were first to obtain results on intracellular functional coupling of pre-nervous biogenic monoamines to the other group of second messengers, to diacylglycerols (Buznikov, Grigoryev, unpublished). Some other results relate the coupling of biogenic monoamines to various second messengers in cleaving embryos. If this coupling really exists, then the early embryos should possess the whole system for intracellular transduction of transmitter signals into messenger ones.

The Peculiarity of Pre-nervous Intracellular Monoaminergic Receptors

The intracellular functional equivalents of monoaminergic receptors are, as far as we know, not identical with the classical receptors. For example, l-alprenolol, a beta-adrenergic blocking drug, and its inactive stereoisomer d-alprenolol have practically the same specific cytostatic activity in experiments with cleaving sea urchin embryos (Fig. 7). If, in such experiments, we measure the radioligand binding, we should not hesitate to state that beta-adrenoceptors are absent from such embryos (see Laduron, 1984). I am referring here to the specific responses of the embryonic cells to antagonists of pre-nervous catecholamines, but not to binding. In this experiment, both d- and l-alprenolol are such antagonists. Accepting the old physiological definition of a receptor, we can conclude that we are dealing here with real receptors but very special ones. Perhaps, a similar conclusion can be made for pre-nervous serotoninoceptors.

I propose that the special nature of such receptors may be linked to very specific aspects of pre-nervous cytokinetic signalling. Perhaps, the situation is more complex than it appears. A new experimental model is badly needed for further progress. We must combine, if possible, quantitating the ligand binding and simultaneously measuring the physiological response to the same ligand. Unfortunately, neither whole cells nor subcellular fractions can serve as such a model. In the first case, correct binding quantitation is impossible ; we measure not this binding but transport of the ligand into the cytoplasm. In the second case, the physiological response to the cytokinetic signal is absent.

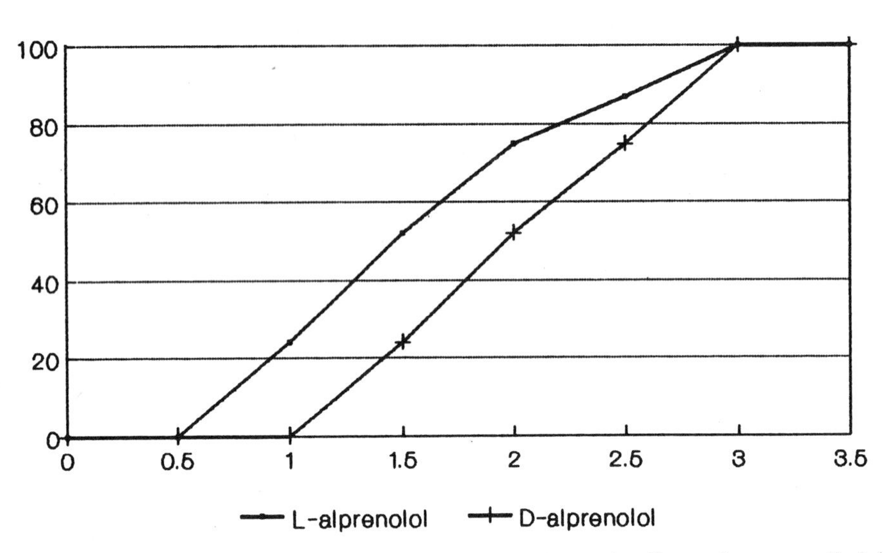

Fig. 7 The blocking action of l- and d-alprenolol on the first cleavage division of sea urchin A. lixula

The Action of the Biogenic Monoamines and their Antagonists on the Cortical Cytoplasmic Layer

A possible way to solve this situation is the use of of "living" subcellular structures which allow us to partially simulate the cytokinetic response and estimate the ligand binding. It is quite feasible. The cortical cytoplasmic layer can be used, as a model. In experiments with the cortical cytoplasmic layer, the effects of monoamines and their antagonists are studied on contractility of the cortical layer in one-cell sea urchin embryos permeabilized shortly before the first cleavage division (see Vacquier and Moy, 1980). Under such conditions, normal cytokinesis is impossible. Cytokinetic signal inducing contraction of the cortical microfilaments may be simulated by adding exogenous ATP. The resulting contraction ruptures the cell membrane and extrudes jets of cytoplasm from some embryos (Fig. 8). Exogenous serotonin added to the medium 1-4 min before ATP does not induce any contraction of cortices, but stimulates the action of ATP (Fig. 9). In this case, all cortices contract much more strongly than those on the previous picture. This action of serotonin is specific. It is completely eliminated by cyproheptadine and other serotonergic-blocking drugs. Moreover, these drugs also block the cortex contraction induced by ATP (Fig.

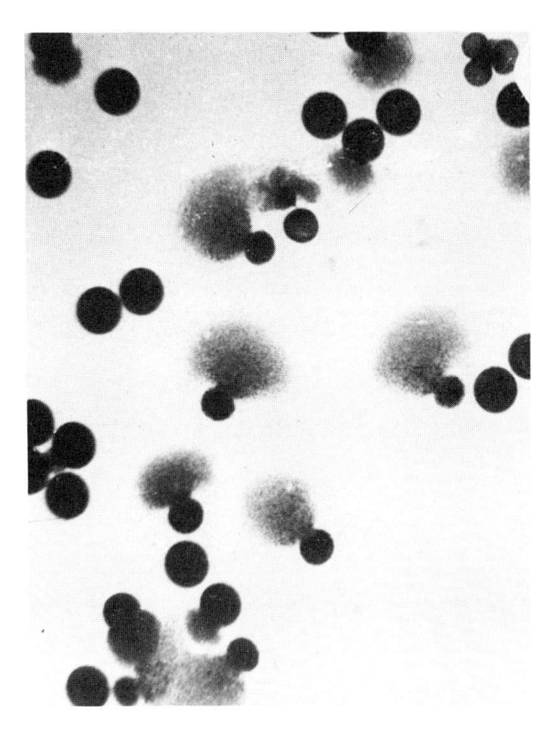

Fig. 8 ATP (1 uM) induces the contraction of cortical microfilaments and extrusion of cytoplasm from permeabilized one-cell embryos of sea urchin A. lixula (according to Vacquier and Moy, 1980).

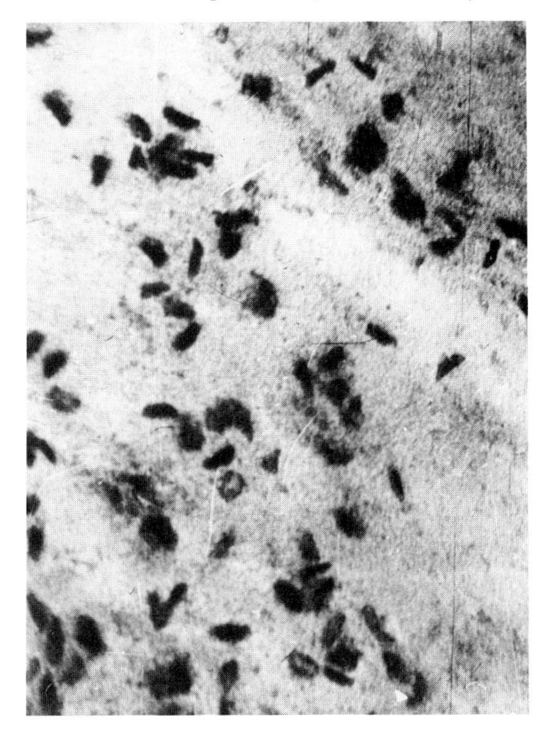

Fig. 9 Serotonin (100 uM) stimulates the ATP-induced contraction of cortical microfilaments.

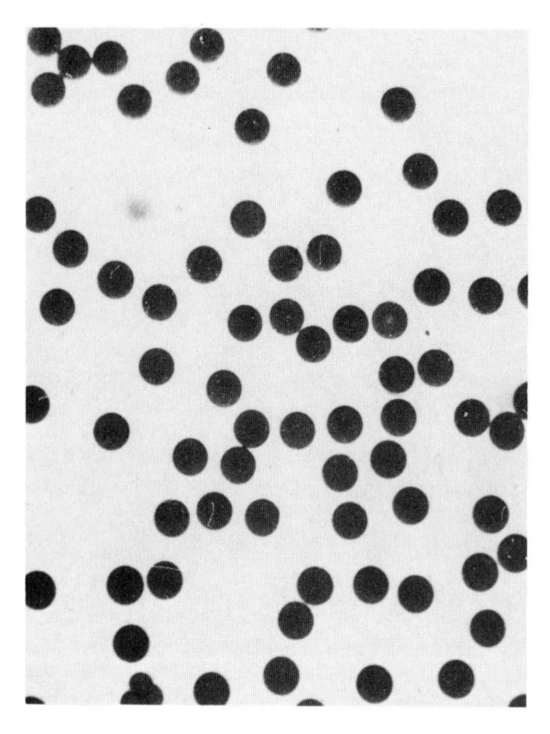

Fig. 10 Cyproheptadine (80 uM) blocks the ATP-induced contraction of cortical microfilaments.

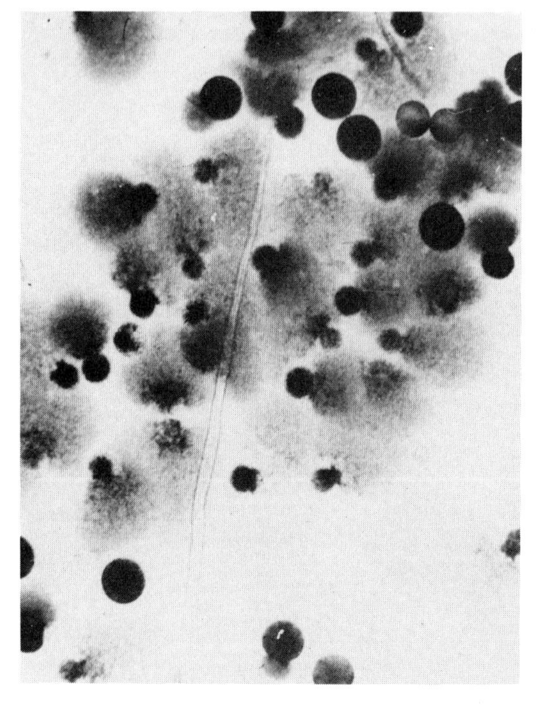

Fig. 11 l-Alprenolol (2 uM stimulates the ATP-induced contraction of cortical microfilaments.

10). Beta-adrenergic blocking drugs (in this case it was alprenolol) enhance the cortex contraction (Fig. 11), while adrenaline inhibits it (Buznikov and Grigoryev, 1990).

Thus, with the use of this model we were able to observe fast physiological responses to monoamines and their antagonists in the absence of any requirement for transport and non-receptor binding of these substances. We obtained first data on the direct and opposite specific effects of serotonin and adrenaline on the functional state of cortical microfilaments.

On the Possible Presence of Monoaminergic Receptors on the Cell Surface of Cleaving Embryos

We can avoid some serious difficulties in experiments with isolated cortices of early embryos although new problems may be associated with this system. One difficulty is associated with the possible presence of receptor binding sites on the outer surface of the embryonic cells. It is known that classical transmitter receptors exist on the oocytes and eggs of all animals studied. These surface receptors seem to disappear after fertilization (Buznikov, 1990). It should be noted that the surface transmitter receptors do not play any role in cleavage divisions. We could ignore these receptors in the case of intact cell models. However, in experiments with isolated cortices, these receptors may still be important. Let me give one instance.

Serotonin potentiates the action of threshold concentrations of 1-methyladenine, the hormone reinitiating meiosis of full-grown oocytes of the starfishes. Many serotonergic-blocking drugs have the opposite action, i.e. they decrease the sensitivity of the oocytes to 1-methyladenine (Buznikov et al., 1990 a,b). These effects are mediated via cell surface receptors which look like 5-HT2-receptors of differentiated cells. The serotoninoreceptors of oocytes play a certain role in the control of Ca^{2+} transport through the cell surface. The serotonergic-blocking drugs appear to inhibit this transport and, moreover, they protect the oocytes against the lethal doses of ionomycine, a calcium ionophore (Fig. 12). To our great surprise, these drugs protect also the cleaving echinoderm embryos against calcium shock. Therefore, surface serotonino-reactive structures are retained during the cleavage divisions.

The functional significance of these structures remains unknown. It cannot be excluded that they take part in early cell-cell interactions. The surface receptors seems be retained in isolated cortices. This should be kept in mind in binding experiments.

In conclusion, let me say that our task is made easy by the use in our work of sea urchins. Eggs and early embryos of sea urchins were and still are one of the main experimental models for the investigation of pre-nervous actions of transmitters. It is no coincidence that Alberto Monroy, a famous embryologist, referred to the centennial debt of developmental biology to the sea urchins (1986).

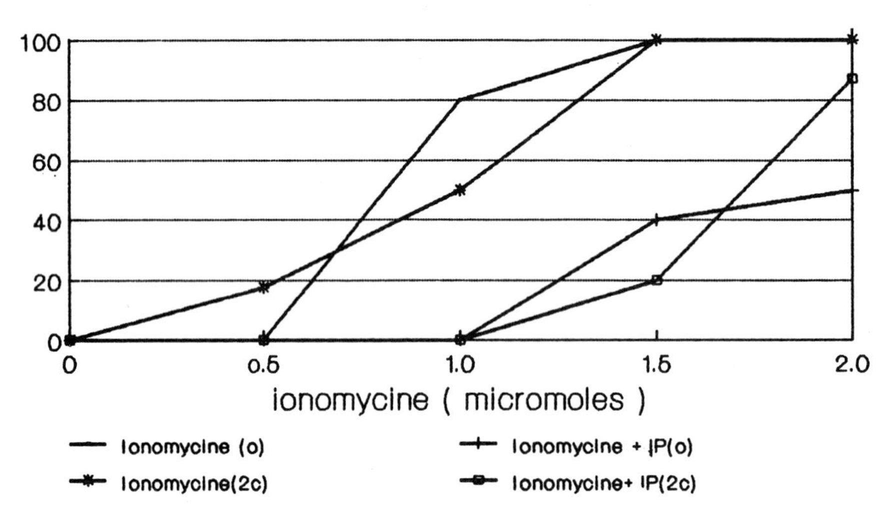

Fig. 12 The action of serotonergic blocking drug IP on the sensitivity of eggs (0) and two-cell embryos (2c) of sea urchin A. lixula to ionomycine.

CONCLUSION

It is accepted now that the biogenic monoamines are necessary for cleavage divisions. The functional response of the cortical cytoskeleton to the cytokinetic signal is impossible without pre-nervous transmitters. We begin to work with a new experimental model which permits the analysis of specific monoaminergic mechanisms related to cytokinesis. However, progress in this area of developmental neurobiology seems to be slow. I hope that this report will promote an increase of interest in pre-nervous transmitter functions.

REFERENCES

Brown, K. B., and Shaver, J. R. 1989, 3-H-serotonin binding to blastula, gastrula, prism, and pluteus sea urchin embryo cells. Comp. Biochem. Physiol., 93C: 281.

Buznikov, G. A., 1967, "The Low-moleculary Weight Regulators of Embryonic Development", Nauka: Moscow (In Russian).

Buznikov, G. A., 1971, The role of neurotransmitters in development. Ontogenez, 2:5 (In Russian).

Buznikov, G. A., 1984, The action of neurotransmitters and related substances on early embryogenesis. In: "Developmental pharmacology", J. Gy. Papp, ed., Pergamon: London, pp. 23-59.

Buznikov, G. A., 1989, Transmitters in early embryogenesis. Ontogenez, 20:637 (In Russian).
Buznikov, G. A., 1990, "Neurotransmitters in Embryogenesis. Harwood Academic Publ.: Chur.
Buznikov, G. A., and Grigoryev, N. G., 1990, The action of biogenic monoamines and their antagonists on the cortical cytoplasmic layer in the early sea urchin embryos. Zh. evol. Biokhim. Fisiol., 26:000 (In Russian)..
Buznikov, G. A., Malchenko, L. A., Nikitina, L. A., Galanov, A. Yu., and Emanov, V. S., 1990a, The action of neurotransmitters and their antagonists on the oocyte maturation. 1. The influence of serotonin and serotonergic blocking drugs on the sensitivity of starfish oocytes to 1-methyladenine. Ontogenez, 21:000 (In Russian).
Buznikov, G. A., Malchenko, L. A., Nikitina, L. A., Galanov, A. Yu., Pogosyan, S. A., and Papayan, G. L., 1990, The action of neurotransmitters and their antagonists on the oocyte maturation. 2. The influence of serotonin and serotonergic blocking drugs on the sensitivity of starfish oocytes to forskolin and ionomycine. Ontogenez, 21:000 (In Russian).
Capasso, A., Creti, P., De Petrocellis, B., De Prisco, P., and Parisi, E., 1988, Role of dopamine and indolamine derivatives in the regulation of the sea urchin adenylate cyclase. Biochem. Biophys. Res. Commu., 154:758.
Capasso, A., Parisi, E., De Prisco, P., and De Petrocellis, B., 1987, Catecholamine secretion and adenylate cyclase activation in sea urchin eggs. Cell Biol. Int., 11:457.
Emanuelsson, H., 1974, Localization of serotonin in cleavage embryos of Ophryotrocha labronica. W. Roux Arch. Entwicklungsmech., 175:253.
Emanuelsson, H., Carlberg, M., and Lowkvist, B., 1988, Presence of serotonin in early chick embryos. Cell Differ., 24:191.
Gustafson, T., 1975, Cellular behavior and cytochemistry in early stages of development. In: "The Sea Urchin Embryo. Biochemistry and Morphogenesis", G. Czihak, ed., Springer: Berlin, pp. 233-266.
Gustafson, T., and Toneby, M., 1971, How genes control morphogenesis: The role of serotonin and acetylcholine in morphogenesis. Amer. Sci., 59:452.
Kater, S. B., and Haydon, P. G., 1987, Multifunctional role of neurotransmitter: the regulation of neurite outgrowth, growth-cone motility and synaptogenesis. In: "Model Systems of Development and Aging of the Nervous System",. A. Vernadakis et al., ed., Martinus Nijhoff: Boston, pp. 239-255.
Laduron, P. M., 1984, Criteria for receptor sites in binding studies. Biochem. Pharmacol., 33:833.
Lauder, J. M., 1988, Neurotransmitters as morphogens. Progr. Brain Res., 73:365.
Lauder, J. M., Tamir, H., and Sadler, T. W., 1988, Serotonin and morphogenesis. I. Sites of serotonin uptake and binding protein immunoreactivity in the midgestation mouse embryo. Development, 102: 709.
Monroy, A., 1986, A centennial debt of developmental biology to the sea urchin. Biol. Bull., 171:509.
Prokazova, N. V., Mikhailov, A. T., Kocharov, S. L., Malchenko, L. A., Zvezdina, N. D., Buznikov, G. A., and Bergelson, L. D., 1981, Unusual gangliosides of eggs and embryos of the sea urchin Strongylocentrotus intermedius. Structure and density dependence of surface localization. Eur. J. Biochem., 115:671.

Renaud, F., Parisi, E., Capasso, A., and De Prisco, E. P., 1983, On the role of serotonin and 5-methoxytryptamine in the regulation of cell division in sea urchin eggs. Develop.Biol., 98:37.

Shaposhnikova, G. I., Prokazova, N. V., Buznikov,G.A., Zvezdina, N. D., Teplitz, N. A., and Bergelson, L. D., 1984, Shedding of gangliosides from tumor cells depends on cell density. Eur.J.Biochem., 140:567.

Vacquier, V. D., and Moy, G. V., 1980, The cytolytic isolation of the cortex of the sea urchin egg. Develop.Biol., 77:178.

Zvezdina, N. D., Sadykova, K. A., Martynova, L. E., Prokazova, N.V., Mikhailov,A.T., Buznikov, G. A., and Bergelson, L. D., 1989, Gangliosides of sea urchin embryos: their localization and participation in early development. Eur.J.Biochem., 186:189.

HORMONE-DEPENDENT PLASTICITY OF THE MOTONEURONS OF THE ISCHIOCAVERNOSUS MUSCLE: AN ULTRASTRUCTURAL STUDY

C.Cracco and A.Vercelli*

Department of Human Anatomy and Physiology
corso M.d'Azeglio 52, I-10126 Torino, Italy

INTRODUCTION

The motoneurons innervating the perineal muscles of the rat are located in the lumbosacral spinal cord within the nucleus of Onuf, made up by a dorsolateral column and a dorsomedial one (Schroeder, 1980).

In the adult animal the motoneurons innervating he bulbocavernosus, the levator ani and the ischiocavernosus muscles display a marked sexual dimorphism (Breedlove and Arnold, 1980; Sengelaub and Arnold, 1989). Moreover, experimental alterations in testosterone serum levels, if performed during the perinatal period, modulate their number (Breedlove and Arnold, 1983a, b); alterations in the circulating levels of androgen at puberty and in adulthood influence their size (Breedlove and Arnold, 1981; Vercelli and Cracco, 1990). Therefore, these motoneurons provide an interesting model of hormone-dependent plasticity.

In a previous study (Vercelli and Cracco, 1990) we have explored the effects of prepubertal castration on the motoneurons of the ischiocavernosus muscle. Since after castration their somatic and nuclear areas display a significant reduction, we wanted to exploit this feature to find out whether some cytological parameters vary as a function of the neuronal size. Furthermore, it is of interest to investigate the fine structure of these motoneurons in control and in castrated adult male rats, in order to establish to what extent their ultrastructural features are influenced by the fluctuating levels of androgen, and eventually reflect different functional properties.

MATERIALS AND METHODS

Six Wistar rats (male, 20 weeks of age, 350 grams body weight) were used. Three of them had been previously orchiectomized at the age of puberty (4 weeks of age). All rats were kept in a room at constant temperature (22-24°C) with constant dark-light period and were allowed to drink and eat ad libitum.

*present address: Institut d'Anatomie, Rue du Bugnon 9, 1005 Lausanne, Switzerland.

Approximately 48 hours before sacrifice, the animals were anaesthetized with diethylether and their left ischiocavernosus muscle was injected with 10 μl 30% saline solution of horseradish peroxidase (HRP, type VI, Sigma). The procedure is described in detail elsewhere (Vercelli and Cracco, 1990). At the selected time, the animals were again anaesthetized with diethylether and perfused transcardially with the aid of a peristaltic pump at a pressure of about 80 mm Hg, first with 0.9% NaCl containing heparin (250 units/100 ml) and subsequently with a mixture of 1.25% glutaraldehyde and 1% paraformaldehyde in 0.1M phosphate buffer, ph 7.4 (PB). The perfusion lasted about 10 minutes.

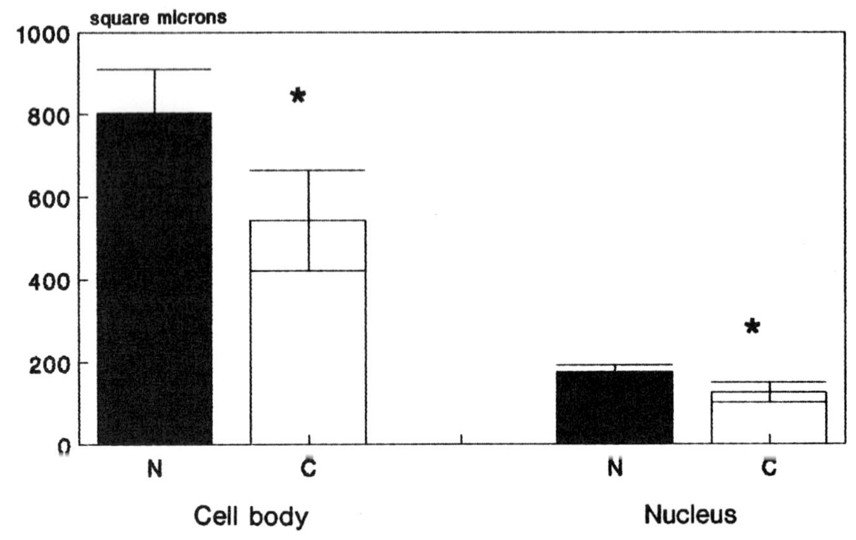

Fig. 1. Histogram related to the mean areas of cell body and nucleus of the HRP-labeled motoneurons of normal (intact) control (N) and castrated (C) rats. Number of rats: 15 per group. Bar: standard error of means.

The lumbosacral spinal cord was dissected out and immersed in the same fixative; two hours later, 50 μm-thick sections were transversally cut with an Oxford Vibratome and washed in PB for 30 minutes. The free-floating sections were then incubated in 5% $CoCl_2$ in PB for 15 minutes, washed again in PB, incubated for 20 minutes in PB containing 0.05% 3,3'-diaminobenzidine (DAB, Sigma) at 37°C in the dark and finally for 20 minutes in PB containing 0.05% DAB and 0.06% H_2O_2 at room temperature in the dark. The sections containing HRP-positive motoneurons were selected observing them in a Olympus light microscope, immersed in fixative again (2.5% glutaraldehyde in PB, 30 minutes), washed in PB, osmicated (2% osmium tetroxide in PB, 1 hour), dehydrated in ethanols and epoxypropane, infiltrated with Araldite and flat embedded between two acetate films.

Only motoneurons from the dorsolateral column were used for electron microscopy: the cell bodies of retrogradely labelled motoneurons were easily identifiable in both 50 μm-thick Araldite embedded slices and semithin sections, being filled by the brown granules of the reaction product. Sections of the HRP containing motoneurons, 2 μm thick, were cut on LKB ultramicrotomes with dry glass knives, collected on slides, stained with toluidine blue, mounted in oil and observed in a Olympus light microscope, in order to reach a midnuclear plane (in

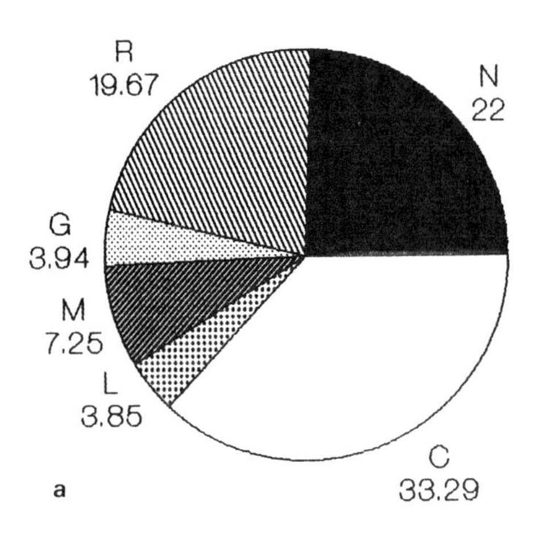

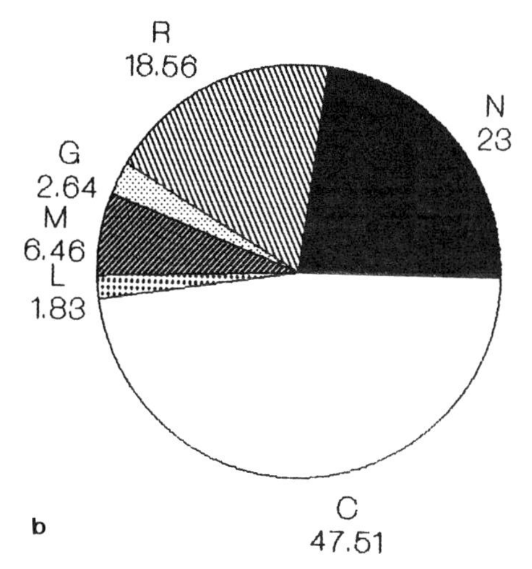

Fig.2. Percentage (%) of somatic cross-sectional area occupied by mitochondria (M), rough endoplasmic reticulum (RER), Golgi apparatus (G) and lysosomes (L), in control motoneurons (a) and in castrated ones (b). C= cytoplasm.

correspondence of the nucleolus). Thin sections (from gold to silver) were then cut with a diamond knife, collected on copper grids, stained with uranyl acetate and lead citrate and examined with a Siemens 102 electron microscope.

The electron microscopic quantitative work was carried out by photographing (at x3000 and at x10000) entire nucleated profiles of HRP-filled motoneurons and was performed on assembled montages, printed at a final magnification of x7500 and x25000. On every motoneuron, somatic, nuclear and nucleolar areas and perimeters were calculated by means of a digitizing tablet and a morphometric program run on a Apple IIe computer. The percentage surface occupied by the cytoplasmic organelles (mitochondria, Golgi apparatus, rough endoplasmic reticulum and lysosomes) was calculated by overlying to every print montage of the entire neuronal profiles a graticule of 5 mm-spaced test points, drawn on a transparent acetate sheet. The absolute number of axosomatic synapses and their frequency (intended as number of synapses/100 μm of membrane length) were calculated. A synapse was identified by its presynaptic terminal with accumulation of vesicles near the presynaptic membrane, and by its characteristic postsynaptic density. Statistical analysis of the data obtained was performed using the student T-test.

RESULTS

All HRP-positive motoneurons considered were located within the dorsolateral column and were, as expected, ipsilateral to the injected ischiocavernosus muscle, extending between segments L6 and S1 of the spinal cord.

The method followed to reveal the presence of HRP in the retrogradely labelled motoneurons resulted fully suitable for electron microscopy. The cytoplasm of cell bodies and proximal dendrites was full of roundish and electron-dense reaction product complexes, and allowed, at the same time, a good preservation of ultrastructural features.

On 15 neuronal profiles from each experimental group, all of them displaying a round nucleus with a prominent nucleolus, the following parameters were preliminarily measured: somatic perimeter, somatic, nuclear and nucleolar areas. Mean values (fig.1) served for the subsequent quantitative analysis and confirmed that castration significantly reduces the size of the motoneurons of the ischiocavernosus muscle. The ratio nuclear area/somatic area remained the same (about 22%, fig.2).

The general ultrastructural organization of motoneuronal cytoplasm was similar in the two groups, being rough endoplasmic reticulum, Golgi apparatus, mitochondria and lysosomes the most prominent organelles (fig.3 and 4). Nevertheless, some morphological differences could be detected. The larger motoneurons of the control group were characterized by a regularly stacked rough endoplasmic reticulum, assembled in wide Nissl bodies, and by numerous Golgi apparatuses, consisting of several units linked to one another and distributed through the pericarion. In the castrated group, the cisternae of both rough endoplasmic reticulum and of Golgi complexes appeared loosely assembled and presented a more dispersed pattern. Moreover, neurofilaments and neurotubules were less conspicuous and the cytoplasm appeared more electron-lucent (fig.5 and 6).

The quantitative analyis of the cytoplasmic organelles was carried out on 15 entire neuronal profiles from each experimental group. Percentages were obtained for somatic cross-sectional areas occupied by mitochondria, rough endoplasmic reticulum, respectively (6% and 18% in control and in castrated group). There was of course a certain range of variability for the values obtained,

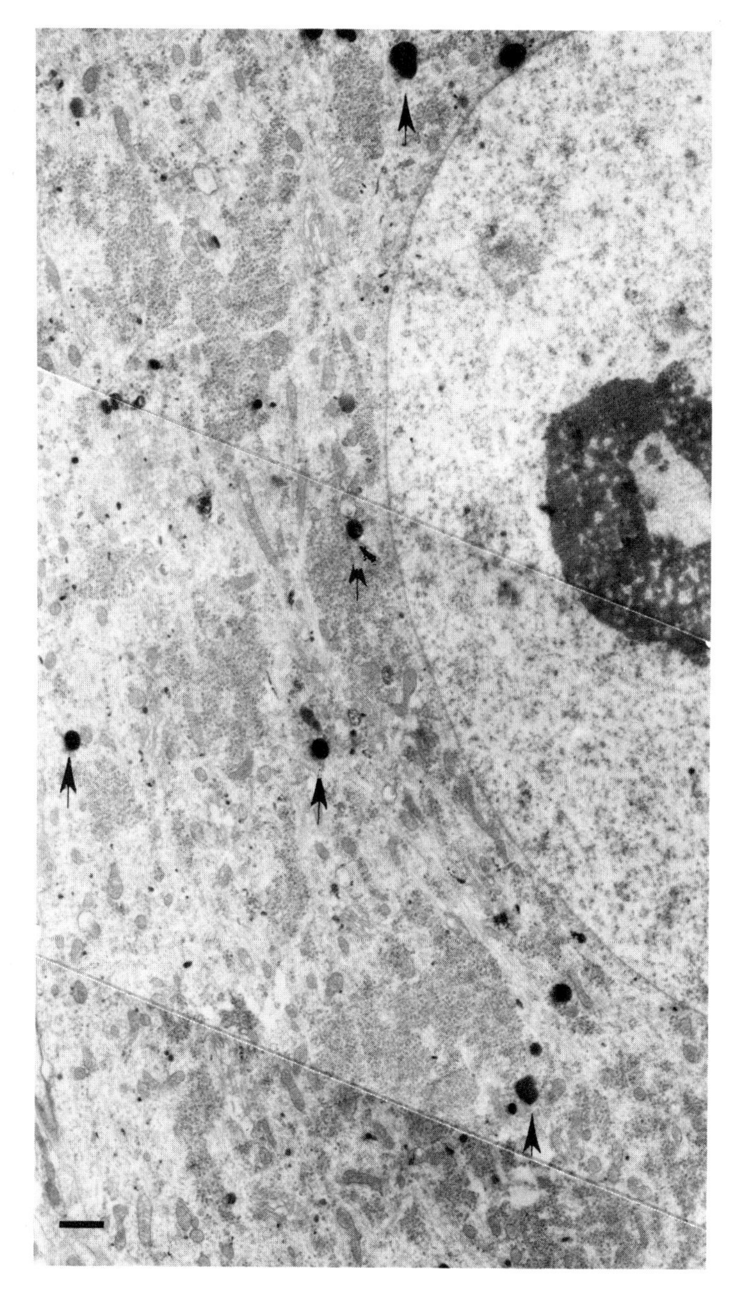

Fig.3. HRP-labeled motoneuron of the ischiocavernosus muscle, control rat. Arrows = reaction product complexes revealing the retrogradely transported HRP. Calibration bar = 1 μm.

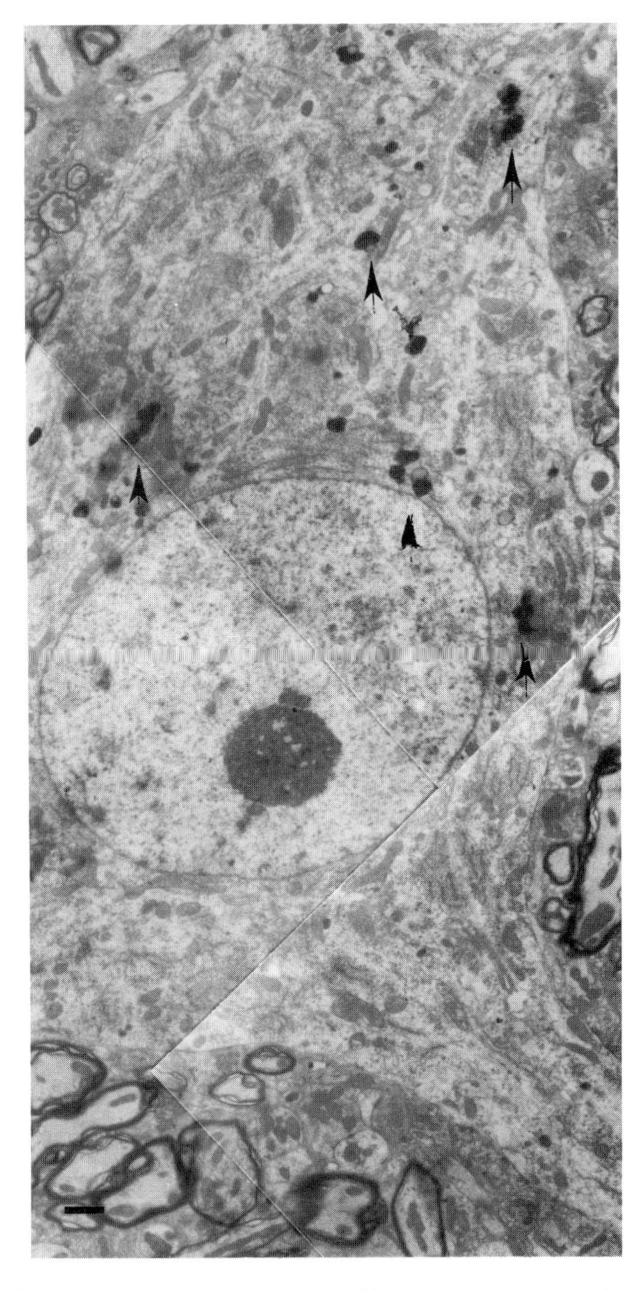

Fig.4. HRP-labeled motoneuron of the ischiocavernosus muscle, castrated rat. Arrows = reaction product complexes. Calibration bar = 1 μm.

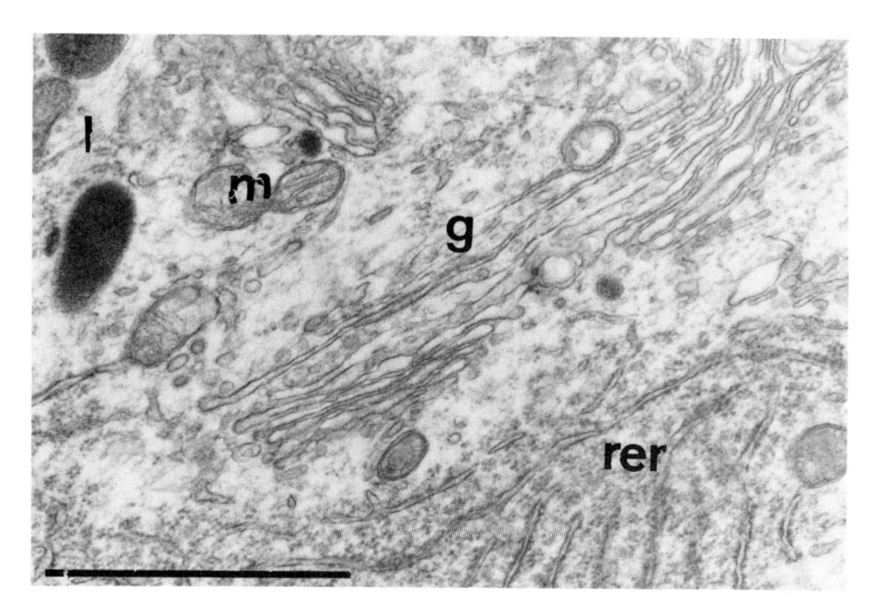

Fig.5. Cytoplasm of a control motoneuron. rer= rough endoplasmic reticulum; g = Golgi apparatus ; m = mitochondria; l = lysosomes. Calibration bar= 1 μm

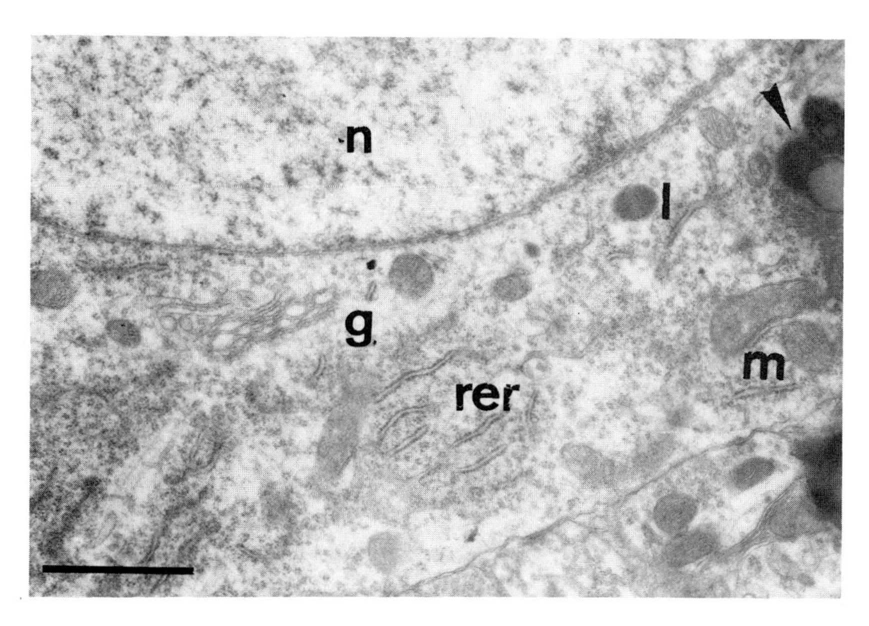

Fig.6. Cytoplasm of a castrated motoneuron. n = nucleus; rer, g, m and l as in Fig 5. Arrow = HRP; Calibration bar = 1 μm

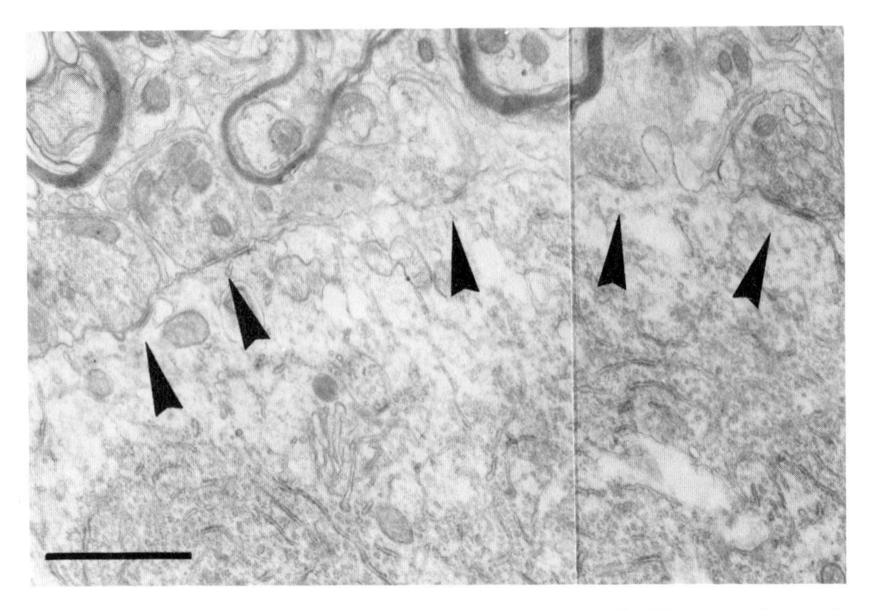

Fig.7 . Synapses of a control motoneuron (arrows) . Calibration Bar = 1 µm

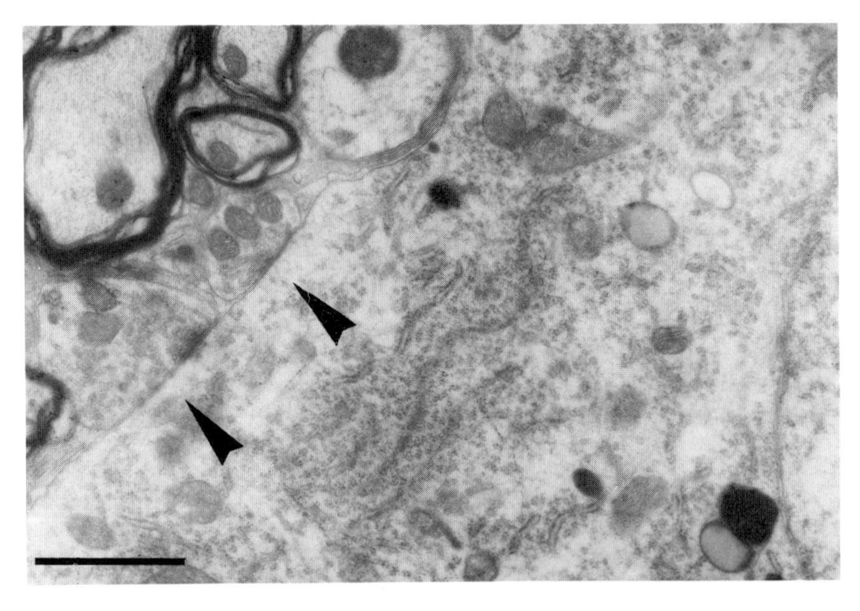

Fig. 8 . Synapses of a castrated motoneuron (arrows). Calibration bar = 1 µm

but it was not correlated with the group considered or with the neuronal size. With regard to the Golgi complexes, the percentage of volume they occupied was about 4% in the control motoneurons, nearly twice as much than the value obtained for the castrated ones. The same happened for the percentage volume occupied by the lysosomes (about 4% in the control motoneurons, about 2% in the castrated ones). Both these differences were statistically significant ($p<0.01$ for the Golgi complexes; $p<0.001$ for the lysosomes).

Axosomatic and axodendritic synapses were common, displaying tight clusters of round clear vesicles (about 50 nm in diameter) and mitochondria, and showing a certain variability in size and shape. Occasionally, some neuronal endings were found to contain a small number of large, dense-core granular vesicles (about 100 nm in diameter). Desmosome-like contacts could be observed between somata and dendrites. Apparently castration reduced the synaptic covering of the motoneuronal membrane to a larger extent then the somatic perimeter, covered in many sites by glial processes which isolated them from any contact (fig.7 and 8).

The absolute number of synapses per cell soma in the castrated group was strongly reduced; their number was three times higher on the control motoneurons. Moreover, their frequency (number of synapses/100 μm membrane) was also definitively lower (nearly half) than in controls.

DISCUSSION

In addition to increasing evidence for sexual dimorphism in the central nervous system (Arnold and Gorski, 1984), other parameters influenced by the hormonal state have been reported, such as neuronal number, size, ultrastructure and synaptic organization (Breedlove and Arnold, 1980; Matsumoto and Arai, 1986). With regard to the mechanism of steroid action on nervous cells, there are still considerable controversies; beside the classical activation of intracellular receptors modulating gene transcription, recent evidence supports rapid membrane effects, modifying the neuronal firing frequency (Schumacher, 1990).

Our results indicate that the motoneurons of the ischiocavernosus muscle, after prepubertal castration, are characterized by: 1) reduction of somatic and nuclear areas; 2) preservation of a constant density of cytoplasmic organelles; 3) reduction both in the absolute number of synapses per cell soma and in the number of synapses per unit membrane length.

The reduction in somatic and nuclear areas induced by prepubertal castration (40% and 30%, respectively) confirms the quantitative data already obtained in light microscopy (Vercelli and Cracco, 1990). Adult hormone manipulation strongly influences the size of these motoneurons (Breedlove and Arnold, 1981). The significance of cell size has been well investigated in motoneurons, particularly in relationship with their electrical properties. Smaller motoneurons have smaller axons, with a slower conduction velocity than the larger motoneurons, and innervate smaller motor units, with a lower threshold (Henneman and Mendell, 1981). Moreover, they present larger values of input resistance, causing them to develop larger synaptic potentials in response to a given input (Mendell et al., 1990). Therefore, the population of motoneurons present after prepubertal castration might have a modified electrical activity.

Currently morphological, histochemical and cytological studies have failed to find clear-cut differences between small and large neurons of a homogeneous population, even if quantitative enzymatic differences have sometimes been reported (Burke, 1981; Henneman and Mendell, 1981; Gabella et al., 1988; Edgerton et al., 1990). Our quantitative ultrastructural data apparently complement these statements: in fact, with regard to the percentage volume of the soma occupied by the cytoplasmic organelles (rough endoplasmic reticulum, Golgi apparatus, mitochondria and lysosomes, about 30% comprehensively), we could not detect substantial changes between the control group and the castrated one. This may be explained by a similar reduction in density of the cytoplasmic organelles as that involving the somatic area.

In this study, we carried out the cytologic quantitation only on the somatic portion of the motoneurons studied, even if we know that it represents only a very small precentage of their cytoplasm. In fact, the most consistent part of it is in the dendrites, which display a reduced length after castration (Kurz et al., 1986). Therefore, we may assume that dendritic tree and motoneuronal area are exposed to the same hormonal influence, thus expecting similar variations in the cytological parameters.

The ability of gonadal steroids to influence the fine structure of cells in the nervous system is already well known, for instance for the ventromedial nucleus of the hypothalamus (Meisel and Pfaff, 1988). Our results are the first, however, to describe the effects of testosterone deprivation on the motoneuronal ultrastructure of the Onuf's nucleus. Moreover, the morphological changes of the rough endoplasmic reticulum we observed between control and castrated motoneurons match with those described in the ventromedial hypothalamic neurons of ovariectomized female rats treated with no hormone or with estradiol only, and of those treated with estradiol and progesterone (Meisel and Pfaff, 1988).

Our results indicate that prepubertal castration reduces the absolute number of synapses per cell soma as a consequence of the reduced neuronal perimeter; in addition, there is also a reduction in their number per unit length of cellular membrane, thereby decreasing over all the synapses beyond the expected rate. These data, similar to those obtained by Leedy et al. (1987) and Matsumoto et al. (1988) on motoneurons of the bulbocavernosus muscle, indicate a great plasticity of the synaptic inputs to the motoneurons of the ischiocavernosus muscle, depending on the hormonal state. Such an influence of steroid hormones on development and maintenance of synapses has been previously reported in other sexually dimorphic nuclei of the central nervous system: suprachiasmatic nucleus (Guldner, 1982), arcuate nucleus (Olmos et al., 1989) and ventromedial nucleus (Matsumoto and Arai, 1986) of the rat.

We find it important to speculate on the functional consequences of this androgen-remodeling of neuronal circuits. The distribution of terminals to motoneurons plays a prominent role in the determination of their different thresholds. The number and arrangement of endings on neurons must be precisely regulated, in order to ensure adherence to the size principle under a wide variety of inputs (Mendell et al., 1990). This androgen-regulated remodeling might be the consequence of rapid membrane effects (Schumacher, 1990), possibly influencing the ionic conductivity and therefore the availability of the somatic membrane to synaptic contacts.

In conclusion, prepubertal castration reduces the size of the motoneurons of the ischiocavernosus muscle and induces cytological and synaptic changes, possibly reflecting modified metabolic and electric activities. A parallel hormone-sensitivity affects the ischiocavernosus muscle itself and modifies the morphology of its striated fibers, the size of its neuromuscular endplates (Vercelli and Cracco, 1989) and its contractile properties (Hughes et al., 1982). Currently, it is impossible to determine whether testosterone acts primarily on motoneurons or directly on the muscle; it is also unknown whether all the changes observed must be considered an effect or a consequence of the modified physiological properties.

ACKNOWLEDGEMENTS

This study was supported by grants from the Italian Ministry of Public Education (M.P.I., to prof.G.Filogamo). We are grateful to prof. G.Filogamo for encouragement and critical discussion of the data and to prof. G.Gabella for his helpful suggestions. We also thank Dr. R.Bertone for technical assistance.

REFERENCES

Arnold, A.P., and Gorski, R.A., 1984, Gonadal steroid induction of structural sex differences in the central nervous system, Ann.Rev.Neurosci., 7:413-442.

Breedlove, S.M., and Arnold, A.P., 1980, Hormone accumulation in a sexually dimorphic nucleus of the rat spinal cord, Science, 210:564-566.

Breedlove, S.M., and Arnold, A.P., 1981, Sexually dimorphic nucleus in the rat lumbar spinal cord: response to adult hormone manipulation, absence in androgen-insensitive rats, Brain Res., 225:297-307.

Breedlove, S.M., and Arnold, A.P., 1983a, Hormonal control of a developing neuromuscular system: I. Complete demasculinization of the male rat spinal nucleus of the bulbocavernosus using the anti-androgen flutamide, J.Neurosci., 3:417-423.

Breedlove, S.M., and Arnold, A.P., 1983b, Hormonal control of a developing neuromuscular system: II.Sensitive periods for the androgen-induced masculinization of the rat spinal nucleus of the bulbocavernosus, J.Neurosci., 3:424-432.

Burke, R.E., 1981, Motor units: anatomy, physiology and functional organization, in: "Handbook of Physiology -The nervous system", vol II, part 1, V.B.Brooks, ed., American Physiological Society, Bethesda, MD, pp.345-422.

Edgerton, V.R., Roy, R.R., and Chalmers, G.R., 1990, Does the size principle give insight into the energy requirements of motoneurons?, in: "The segmental motor system", M.D.Binder and L.M.Mendell, eds., Oxford University Press, New York, pp.150-164.

Gabella, G., Trigg, P., and McPhail, H., 1988, Quantitative cytology of ganglion neurons and satellite glial cells in the superior cervical ganglion of the sheep. Relationship with ganglion neuron size, J.Neurocytol., 17:753-769.

Guldner, F.-H., 1982, Sexual dimorphism of axo-spine synapses and postsynaptic density material in the suprachiasmatic nucleus of the rat, Neurosci.Lett., 28:145-150.

Henneman, E., and Mendell, L.M., 1981, Functional organization of motoneuron pool and its inputs, in: "Handbook of Physiology -The nervous system", vol.II, part 1, V.B.Brooks, ed., American Physiological Society, Bethesda, M.D., pp. 423-507.

Hughes, B.J., Bowen, J.M., Campion, D.R., and Bradley, W.E., 1982, Effect of prepubertal castration on porcine bulbospongiosus muscle, Anat.Embryol., 168:51-58.

Kurz, E.M., Sengelaub, D.R., and Arnold, A.P., 1986, Androgen regulates the dendritic length of mammalian motoneurons in adulthood, Science, 232:395-398.

Leedy, M.G., Beattie, M.S., and Bresnahan, J.C., 1987, Testosterone-induced plasticity of synaptic inputs to adult mammalian motoneurons, Brain Res., 424:386-390.

Matsumoto, A., and Arai, Y., 1986, Male-female difference in the organization of the ventromedial nucleus of the hypothalamus in the rat, Neuroendocrinology, 42:232-236.

Matsumoto, A., Mycevich, P.E., and Arnold, A.P., 1988, Androgen regulates synaptic input to motoneurons of the rat spinal cord, J.Neurosci., 8:4168-4176.

Meisel, R.L., and Pfaff, D.W., 1988, Progesterone effects on sexual behavior and neuronal ultrastructure of female rats, Brain Res., 43:153-157.

Mendell, L.M., Collins, W.F., and Koerber, H.R., 1990, How are Ia synapses distributed on spinal motoneurons to permit orderly recruitment?, in: "The segmental motor system", M.D.Binder and L.M.Mendell, eds., Oxford University Press, New York, pp. 308-327.

Olmos, G., Naftolin, F., Perez., J., Tranque, P.A., and Garcia-Segura, L.M., 1989, Synaptic remodeling in the rat arcuate nucleus during the estrous cycle, Neuroscience, 32:663-667.

Schroeder, H.D., 1980, Organization of the motoneurons innervating the pelvic muscles of the rat, J.Comp.Neurol., 192:567-587.

Schumacher, M., 1990, Rapid membrane effects of steroid hormones: an emerging concept in neuroendocrinology, TINS, 13:359-362.

Sengelaub, D.R., and Arnold, A.P., 1989, Hormonal control of neuron number in sexually dimorphic nuclei of the rat: I. Testosterone-regulated death in the dorsolateral nucleus, J.Comp.Neurol., 280:622-629.

Vercelli, A., and Cracco, C., 1989, Influence of testosterone on the development of the ischiocavernosus muscle of the rat, Acta Anat., 134:177-183.

Vercelli, A., and Cracco, C., 1990, Effects of prepubertal castration on the spinal motor nucleus of the ischiocavernosus muscle of the rat, Cell Tissue Res., 262:551-557.

REACTIVE SPROUTING (PRUNING EFFECT) IS ALTERED IN THE BRAIN OF RATS PERINATALLY EXPOSED TO MORPHINE

A. Gorio, B. Tenconi, N. Zonta, P. Mantegazza and
A.M. Di Giulio

Dept. of Medical Pharmacology, University of Milan,
School of Medicine, Via Vanvitelli 32, Milano 20129, Italy

INTRODUCTION

Recent publications have suggested that the endogenous opioid system may be involved in the natural process of brain growth and development (1). When opioids such as morphine, heroin, or methadone are supplied exogenously to man and laboratory animals, there is a reduced somatic and neuronal development (1,2,3,4). In addition if opioids are added to the culture media of neurons and cell lines, the process of growth and differentiation is altered (3,4). These effects are apparently mediated by the opioid receptor since the administration of specific antagonists prevents the opioid-induced alterations (3,4). Also opioid antagonists as naltrexone alter brain and body development when administered to pregnant mothers (1). The diffusion of opiates from the mother blood stream to the fetus brain occurs without apparent difficulties (3,4). The "opiate foetal syndrome", encountered in children of mothers addicted to opiates, is apparently becoming widespread with an incidence superior to other neonatal problems such as those caused by alcoholic mothers, or cancer and Down syndrome (3,4). It is unclear to what extent the brain of neonates from opiate addicted mothers is altered, and furthermore no molecular and cellular clues are available for understanding such damage. We challenged such a problem by monitoring brain development in pups perinatally exposed to morphine; in addition, we further examined the brain functions of these animals by stimulating brain reactive processes as the ones caused by specific neurotoxic lesions.

MATERIALS AND METHODS

We utilized female Sprague-Dawley rats (Charles River, Como, Italy) weighing 250-275 grams. Animals were reared in standard conditions with food and water ad libitum for one week. Then, mating was performed overnight utilizing male Sprague-Dawley rats weighing 300-350 grams.

When female rats were found with spermatozoa in the vaginal fluid, morphine treatment began. The opiate was dissolved in the drinking water at the dose of 0.1 mg/ml and supplied to the pregnant mothers. With time the dosage was gradually increased to the maximal level of 0.4 mg/ml, on the 19th day of pregnancy. At 21 day of pup life the dose was dropped to 0.1 mg/ml of water for avoiding an excessivedosage of morphine to the pups now capable of drinking water directly. Animals were sacrificed at the appropriate time by decapitation, brain areas were dissected and frozen in liquid nitrogen. Then samples were stored at -80°C until assayed.

Tissue samples were placed in polypropylene tubes containing ice cold 1N acetic acid and homogenized with polytron (PTA 10TS, Kinematica). Approppriate aliquots were stored at 0-4°C for protein determination. Test tubes containing the homogenate were placed in boiling water for 8-10 minutes, cooled down in ice, and centrifuged at 40,000g for 20 minutes. The resulting supernatant was lyophilized; samples were then reconstituted in distilled water, and subjected to radioimmunoassay (RIA) for Substance P and Met-enkephalin . RIAs were performed using highly specific antisera raised in our laboratory (5,6).

TABLE I

Substance P and Met-Enkephalin Content in the Cerebral Cortex of Control and Morphine Treated Animals

Days	Substance P (ng/mg prot)		Met-Enkephalin (ng/mg prot)	
	Control	Morphine	Control	Morphine
0	0.3 (0.05)	0.3 (0.04)	0.2 (0.01)	0.2 (0.01)
4	0.4 (0.01)	0.5 (0.01)**	0.5 (0.01)	0.7 (0.01)**
8	0.5 (0.01)	0.7 (0.01)***	0.5 (0.01)	0.7 (0.01)***
12	0.5 (0.01)	1.0 (0.1)***	0.7 (0.02)	1.9 (0.1)***
20	1.0 (0.1)	1.0 (0.1)	2.5 (0.01)	3.0 (0.02)***
30	2.1 (0.1)	3.1 (0.2)***	3.5 (0.1)	4.3 (0.02)**

The high-pressure liquid chromatography (HPLC) analysis of monoamines was performed as follows: tissues were homogeneized in ice-cooled 0.4 N perchloric acid; a known amount of 3,4-dihydroxybenzylamine hydrobromide (DHBA, Sigma Chemical Co) was then added to each sample as internal standard. The homogenate was split into two aliquots, one for 5-hydroxyindoles (serotonin and 5-hydroxyindolacetic acid) and the other for catecholamines (noradrenaline and dopamine) quantification. The aliquots for 5-HT and 5-HIAA assay were centrifuged at 40,000 g for 15 min and the resulting supernatant directly injected into the HPLC appararus (7). Catecholamine assay required an intermediate purification step by adsorption into allumina. Catechols were then extracted from allumina by adding 200 ul of 0.4 N perchloric acid and shaking for 8 minutes. Following centrifugation, samples were injected into the HPLC system (Spectra Physics, SP8750) within 24 hours from tissue homogenization. The method used for monoamines extraction, purification, and HPLC/electrochemical detection is a slight modification of the original methods (8,9,10).

The data were evaluated statistically by the Student T test and significant differences between the means (calculated as P values) are shown. The differences were considered statistically significant when the p values were ≤ 05. Standard error is shown in parenthesis.
* ≤0.05
**≤0.001
***≤0.0001

RESULTS

Peptide assays were performed in several brain areas at various time after birth.

TABLE II

Substance P and Met-Enkephalin Content in the Pons-Medulla of Control and Morphine Exposed Rats

Days	Substance P (ng/mg prot)		Met-Enkephalin (ng/mg prot)	
	Control	Morphine	Control	Morphine
0	0.2 (0.01)	0.4 (0.01)***	0.2 (0.01)	0.2 (0.01)
4	0.3 (0.01)	0.8 (0.01)***	0.2 (0.01)	0.3 (0.01)*
8	0.9 (0.1)	1.0 (0.1)	0.2 (0.01)	0.4 (0.01)*
12	1.3 (0.1)	1.4 (0.1)	0.5 (0.08)	0.9 (0.1)***
20	2.0 (0.2)	2.0 (0.1)	0.7 (0.2)	1.0 (0.1)
30	5.0 (0.2)	5.1 (0.2)	0.9 (0.1)	1.8 (0.2)***

The data reported in TABLE I show that in control animals there is a gradual and constant increase in substance P and met-enkephalin content in cerebral cortex. Exposure to morphine induced further increase in this levels. Preliminary data indicate that the two groups of animals show similar peptide levels after 30 days of life.

The results shown in table II indicate that substance P is more abundant in the pons-medulla of morphine exposed animals only in the very early stage of life. Conversely met-enkephalin content is similar in the two groups at birth, at later stages, however, the met-enkephalin content is significantly higher in morphine exposed rats. At 60 days of life also met-enkephalin content is identical in the two groups.

The caudate nucleus of morphine exposed rats displays significantly higher content of met-enkephalin throughout the postnatal period. Such a difference has almost disappeared at 60 days of life as shown by preliminary results (table III).

TABLE III

Substance P and Met-Enkephalin Content in the Nucleus Caudatus of Control and Morphine Exposed Animals

Days	Substance P (ng/mg prot)		Met-Enkephalin (ng/mg prot)	
	Control	Morphine	Control	Morphine
0	0.8 (0.1)	0.8 (0.1)	2.0 (0.1)	2.8 (0.1)**
4	2.1 (0.2)	2.4 (0.3)	1.0 (0.1)	1.9 (0.1)***
8	3.6 (0.4)	3.9 (0.2)	1.0 (0.1)	2.1 (0.3)*
12	3.6 (0.2)	4.3 (0.1)*	3.0 (0.2)	5.0 (0.2)***
20	4.8 (0.3)	5.6 (0.1)*	7.1 (0.3)	12.1 (0.8)***
30	5.8 (0.2)	5.8 (0.3)	9.2 (0.3)	13.2 (0.4)***

Monoamines are remarkably unchanged in morphine exposed animals, only in the nucleus caudatus there is an interesting change. During the period of met-enkephalin hyper-innervation there is a transient significant reduction in content of dopamine. These data are suggesting that the perinatal exposure to morphine has perturbed the normal process of developmental neuroplasticity.

We have challenged this hypothesis by assessing reactive brain plasticity as induced by neonatal treatment with the neurotoxin 5,7-HT. We have chosen to use this toxin specific for the serotoninergic system for 2 reasons. One is that 5-HT distribution in the brain of morphine treated animals is similar to normal. Second that the lesion causes denervation of the distal targets that are gradually reinnervated (reinnervation time: 8 weeks); the reinnervation process is preceded by a transient hyperinnervation of the proximal targets due to a tremendous outgrowth of the serotoninergic shorter collaterals (11,12).

TABLE IV

Serotoninergic Reinnervation of the Spinal Cord
8 Weeks after Neonatal 5,7-HT in Normal and
and Morphine Exposed Animals

	Serotonin (ng/mg prot)			
	Control	Control +5,7-HT	Morphine	Morphine +5,7-HT
Cervical	12.2 (1.1)	14.2 (0.9)	11.6 (0.7)	10.9 (0.8)
Thoracic	13.6 (0.8)	12.8 (0.9)	12.6 (0.5)	6.8 (2.1)*
Lumbar	29.0 (2.9)	24.2 (1.8)*	28.7 (0.7)	6.6 (0.9)***

The data shown in the table suggest that perinatal exposure to morphine profoundly alters the process of serotoninergic reinnervation of the distal targets such as the lumbar segment of the spinal cord. In addition to the reduced regenerative capacity, serotoninergic neurons of morphine treated rats are incapable of reactive sprouting. In healthy rats during the first 4 weeks after 5,7-HT treatment, hyper-outgrowth of the short collaterals following the neurotoxic axotomy of the most distal ones causes the typical " pruning effect"; this process is practically absent in rats perinatally exposed to morphine.

We have further challenged this problem utilizing another neurotoxin. 6-OHDA applied intraperitoneally to rat pups within 6 hours of birth causes nor-adrenergic (NA) denervation in the cortex. The lesioned axons are incapable of regrowing into the area that is reinnervated by collateral sprouting by other neural systems such as the dopaminergic and the serotoninergic. On the other hand the lesioned NA axons, that cannot regrow distally, hyperinnervate greatly the mesencephalon. Recent results are showing that in animals perinatally exposed to morphine both collateral reinnervation in the cortex by serotoninergic and dopaminergic axons and noradrenergic hyperinnervation of mesencephalon are absent.

DISCUSSION AND CONCLUSIONS

It is our opinion that the above outlined results may lead us to understand better the subtle changes occurring in the brain of drug addicts and also to improve our knowledge of the molecular and cellular mechanisms underlying neuronal repair in the brain. In the brain of rats perinatally exposed to morphine we observed very minor or no changes at all in the monoaminergic innervation. However when we challenged these neuronal systems with specific neutoxins we observed that the reactive processes were poorly efficacious or absent when compared to the highly efficacious repair mechanisms present in the brain and spinal cord of healthy animals. In the case of the 5,7-HT experiments we observed that lesioned serotoninergic neurons could regenerate their axons very poorly along the spinal cord so that 8 weeks after lesioning the thoracic and lumbar segments of the spinal cord were poorly reinnervated. Therefore, we can assume that axonal elongation might be a problem for animals exposed perinatally to morphine. In addition, the lesioned serotoninergic axons give rise to a tremendous, though transient, sprouting of the short collaterals innervating the pons medulla "pruning effect"; again morphine exposed rats are incapable of such a reactive process. We could compare such a process to a growing tree with a very poor budding in the spring, that leads to a poor development and growth. However, these abnormalities may not be very obvious by study of the typical parameters for brain development as neurotransmitter levels or morphology. Only a challenge such as the ones we used could reveal the gross abnormalities caused by morphine. In addition, recently we have been using another neonatal neurotoxic lesion utilizing 6-OHDA. The results, just summarized above, are striking. Also the noradrenergic neurons are incapable of giving rise to the "pruning effect" in the mesencephalon, and the serotoninergic and dopaminergic axons do not collaterally reinnervate the frontal cortex as they do in healthy animals after neonatal depletion of noradrenaline. In conclusion, we have two further information from this experiment. One is that the altered plastic properties in the brain of animals perinatally exposed to morphine are not restricted to serotonin only, but also noradrenaline and dopamine systems are affected. The other is that collateral reinnervation by intact axons is fully inhibited in the frontal cortex suggesting that this might be a very generalized process in the brain.

We yet do not know by which mechanisms morphine damages the repair mechanisms in the brain; it is our goal to pursue this problem.

REFERENCES

1. Zagon J.S. and McLaughlin P.J., 1983, Increased brain size and cellular content in infant rats treated with an opiate antagonist, Science, 221: 1179-1180
2. Zagon J.S. and McLaughlin P.J., 1983, Naltrexone modulates growth in infant rats, Life Science, 33: 2449-2454
3. Zagon J.S., McLaughlin P.J., Weaver D.J., and Zagon E., 1982, Opiates, endorphins and the developing organism: a comprehensive bibliography, Neuroscience and biobehavioral reviews, 6:439-479
4. Zagon J.S., McLaughlin P.J., and Zagon E., Opiates, endorphins, and the developing organism: a comprehensive bibliography, 1982-1983, 8:387-403
5. Di Giulio A.M., Mantegazza P., Dona' M, Gorio A., 1985, Peripheral nerve lesions cause simultaneous alterations of substance P and enkephalin levels in the spinal cord. Brain Res. 342:405-408
6. Di Giulio A.M., Borella F., Mantegazza P., Hong J.- S., Panozzo C., Zanoni R., Gorio A., 1985, Structural and biochemical alterations in the dorsal horn of the spinal cord caused by peripheral nerve lesions, Peptides, 6:249-256
7. Di Giulio A.M., Tenconi B., Mannavola A., Mantegazza P., Schiavinato A., Gorio A., 1987, Spinal cord interneuron degenerative atrophy caused by peripheral nerve lesions is prevented by serotonin depletion, J. Neurosc. Res.,18:443-448
8. Keller R., Oke A., Mefford I., Adams R.N., 1976, Liquid chromatographic analysis of catecholamines routine assays for regional brain mapping, Life Science, 19:995-1004
9. Lackovic Z., Parenti M., Neff H., 1981, Simultaneous determination of fentomoles quantities of 5-hydroxytryptophan, serotonin and 5-hydroxyindoleacetic acid in brain using HPLC with electrochemical detection, Eur. J. Pharmacol., 69: 347-352.
10. De Saint Blanquat G., Lamboeuf Y., Fritsch P., 1987, Determination of catecholamines in biological tissues by liquid chromatography with coulometric detection. J. Chromatogr., 415:388-392
11. Jonsson G., Pollare T., Hallman H., and Sachs C., 1978, Developmental plasticity of central serotonin neurons after 5,7-dihydroxytryptamine treatment, 1978, Ann. NY Acad. Sci., 305:329-345
12. Jonsson G, Gorio A., Hallman M., Janigro D., Kojima H., Zanoni R., 1984, Effect of Gm1 ganglioside on neonatally neurotoxin induced degeneration of serotonin neurons in the rat brain, Dev. Brain Res., 16:171-180

EFFECTS OF SEROTONIN ON TYROSINE HYDROXYLASE AND TAU PROTEIN IN A HUMAN NEUROBLASTOMA CELL LINE

N.J. John[*#], G.M. Lew[+], L. Goya[*] and P.S. Timiras[*]

[*]Department of Molecular and Cell Biology, University of California at Berkeley, Berkeley, California 94720 USA; [+]Department of Anatomy, College of Human Medicine, Michigan State University, E. Lansing, Michigan 48824 USA

ABSTRACT

The direct effects of the neurotransmitter serotonin on the catecholaminergic enzyme, tyrosine hydroxylase and the microtubule-associated tau protein were studied in a human neuroblastoma cell line. Undifferentiated LAN-5 cells, cultured in serum supplemented basal medium, or cells induced to differentiate by 6 day exposure to 10 uM retinoic acid were treated for 48 hr with 50 nM and 50 uM serotonin. In undifferentiated cells, serotonin treatment (50 uM) decreased both tyrosine hydroxylase activity and a 50 kD cytoplasmic fraction tau protein while 50 nM serotonin treatment caused this 50 kD protein to increase in the cytoplasmic fraction but decrease in the membrane fraction. While basal tyrosine hydroxylase activity increased in differentiated vs. undifferentiated cells, serotonin treatment had no effect on the enzyme or tau in differentiated LAN-5. This study shows serotonin to have direct effects on the biochemistry and cytoskeleton of undifferentiated cultured human neuroblastoma.

Key words: serotonin, neuroblastoma, tyrosine hydroxylase, microtubules, tau protein, retinoic acid, differentiation.

INTRODUCTION

The best known cytoskeletal functions in neurons include axoplasmic transport, cell contractility and, during development, nerve fiber outgrowth. Some of these actions

[#]current address: Institut fur Anatomie, Universitat Essen, 4300 Essen 1, West Germany

are related or depend on specific neurotransmitters: for example, dopamine (together with Ca^{++} and cAMP) regulates elongation and contraction of retinal rods and cones (Burnside and Dearry, 1986; Burnside, 1988). Microtubules, major cytoskeletal components, are involved in intracellular transport of cell products and possibly in their secretion and absorption as well (Burgess and Kelly, 1987; Pfeffer and Rothman, 1987). It may be expected, then, that changes in microtubule assembly and disassembly, perhaps promoted by specific proteins such as tau proteins, influence metabolism and transport of neurotransmitters. Reciprocally, neurotransmitters may affect microtubular and associated proteins and, thereby, regulate their own transport and secretion and that of other neurotransmitters. These experiments represent a first attempt to explore some neurotransmitter-microtubule interrelations.

Serotonin [5-hydroxytryptamine (5-HT)] effects have been implicated in both biochemical and cytoskeleton regulation. 5-HT-mediated mechanisms may exert an inhibitory control of noradrenaline metabolism in the locus ceruleus (Renaud et al., 1975: Lewis et al., 1976; Crespi et al., 1980; Devau et al., 1987) and in microtubule-dependent induced stimulation of corticosteroid production (Feuilloley et al., 1988). Moreover, the amine appears associated with microfilaments and microtubules, particularly in developing neural cells (Emanuelsson et al., 1988).

Evidence that 5-HT exerts an inhibitory control on tyrosine hydroxylase (TH) (the rate-limiting enzyme in the catecholamine biosynthesis pathway) activity in noradrenergic neurons of the locus ceruleus (LC) is given by the following biochemical studies: (a) Destruction of specific raphe nuclei which send serotoninergic projections to the LC resulted in an increase in TH activity in the LC region (Lewis et al., 1976). (b) Destruction of serotinergic neurons by 5,6-dihydroxy-tryptamine increased TH activity in rat LC (Renaud et al., 1975; Keane et al., 1978). (c) Inhibition of tryptophan hydroxylase, the rate limiting enzyme of 5-HT synthesis, by parachlorophenylalanine administration, produced a significant increase in TH activity in the rat LC (Crespi et al., 1980). (d) Recently, Devau et al. (1987) have directly demonstrated the inhibitory role of 5-HT on TH activity in *in vitro* cultured explants of newborn rat LC.

While microtubules have been implicated in the coupling of corticosteroid secretion in adrenal cells in response to 5-HT (Feuilloley et al., 1988), and may play a role in cell-shape changes and morphogenesis in the early chick embryo (Emanuelsson et al., 1988), there are no reports in the literature on the effects of 5-HT on TH or microtubules in neuroblastomas.

In this investigation we used a human neuroblastoma cell line, LAN-5, which has adrenergic properties and can be induced to differentiate (Sidell et al., 1983; 1984; Cole and Timiras, 1987), as a tissue culture model to study the direct effects of 5-HT on TH activity and tau protein, a component of neural microtubules, which acts *in vivo* chiefly to induce the assembly of tubulin and *in vitro* to promote microtubule polymerization (Drubin & Kirschner, 1986).

EXPERIMENTAL PROCEDURES

Cell Culture, Cell Differentiation and Serotonin (5-HT) Treatment

Human neuroblastoma LAN-5 cells (Sidell et al., 1983) were propagated as undifferentiated monolayers in basal medium consisting of Ham's nutrient mixture F12/Dulbecco's Modified Eagle's medium, 1:1 (Sigma), supplemented with 10% fetal bovine serum and 1% penicillin-streptomycin (100 units/ml) (Cell Culture Facility, University of California, San Francisco). Medium was routinely changed every 3 days. The cells were maintained in a humidified incubator at 37° C with 5% CO_2.

To induce neuroblastoma differentiation, LAN-5 cells were treated with 10 uM retinoic acid (Sigma) for 6 days prior to further treatment (Sidell et al., 1983; Cole and Timiras, 1987). The retinoic acid medium was changed every other day.

For serotonin (5-HT) treatment, cells were plated at an initial density of 5 x 10^5 cells/dish (Falcon 100 x 20 mm) in control or retinoic acid medium. Six days after initial plating and retinoic acid treatment, both the differentiated and undifferentiated cells were treated for 48 hrs with 50 uM and 50 nM 5-HT. (The time course and concentrations of 5-HT used were those found to give maximal effect on the TH regulation with cultures of newborn rat locus ceruleus explants, Devau et al., 1987). Dilutions were made immediately before use from a stock solution of 5-HT containing 1.5 mM HCl and 0.05 M ascorbic acid in distilled water (10 ul was added to each plate with 10 ml of appropriate medium). Vehicle control dishes were treated with equimolar amounts of the diluent.

Cell Proliferation Assay

LAN-5 cells were plated at an initial density of 10^4 cells per 2-cm^2 well (Falcon multi-well plate) in basal or retinoic acid containing medium. At various times through a 10-day time course, triplicate samples of cells were fixed in 70% ethanol, dried in vacuo at room temperature, and 0.4 ml of 10% 3,5-diaminobenzoic acid-dihydrochloride (Aldrich Chemical Co.) was added to each well. After 1 hour at 60^0 C, 2 ml of 1 N HCl was added and the DNA-mediated fluorsecence (Hinegardner, 1971) in each solution was monitored at excitation and emission wavelengths of 405 and 465-540 nm, respectively, in a Model 112 Turner fluorimeter. Under these conditions, a reading of one fluorescence unit equals 10^4 cells.

Tyrosine Hydroxylase (TH: EC1.14.16.2) Assay

Both the differentiated and undifferentiated cells were treated with 5-HT (50 uM and 50 nM) or vehicle for 48 hrs as described above and harvested in phosphate-buffered saline (PBS), pH 7.4, centrifuged at 300 x g for 10 min and then washed once in PBS.

TH was measured according to Waymire et al. (1971) as described previously (Safaei and Timiras, 1985) with the following modification. After harvesting in PBS, the final cell pellet was resuspended in lysis buffer (10 mM Tris, pH 8.3, 0.1 mM phenylmethylsulfonilfluoride, 10 ug/ml leupeptin, 10 ug/ml aprotinin, 10 ug/ml pepstatin A, 5 mM benzamidine, 1 mM EDTA) and stored frozen at -70^{0} C until assay.

SDS-Page and Western Blot Analysis

Cells were harvested, suspended in lysis buffer, and stored frozen at -70° C as described above. The thawed samples were centrifuged at 3,000 x g to separate the membrane fraction (pellet) from the cytoplasmic fraction (supernatant). The pellet was resuspended in 35 ul of lysis buffer and sonicated at 30,000 Hz for 5 secs.

Total protein in both supernatant and pellet fractions was determined according to the method of Bradford (1976). SDS-Page was performed according to a modification of Laemmli (1970). Samples were solubilized in 25 ul SDS gel sample buffer (10% glycerol, 18.4% SDS, 0.5 M Tris, pH 6.8) and 0.5 M dithiothreitol (DTT) at 100° C for 3 min; after a 2-min Eppendorf centrifugation, the supernatants were electrophoresed in SDS polyacrylamide gels containing 12% acrylamide.

Proteins separated by SDS-Page were blotted onto nitrocellulose membranes and incubated in blocking buffer as described previously (Argasinski et al., 1989). The blot was then washed 5 times with Tris/saline (0.01 Tris-base, 0.14 M NaCl, pH 7.6) and incubated (3 hr, 0° C) with a 1:25 dilution of mouse monoclonal antibody to tau, called Alz-50 (generously supplied by Dr. P. Davies, Albert Einstein Medical College, New York). Following the 3 hr incubation the blot was washed 5 times with Tris/saline, pH 7.6, and placed on the shaker for 5 mins in fresh Tris/saline. The blot was incubated 30 min with an alkaline phosphatase-conjugated second antibody (Fisher) diluted 1:1000, washed for 1 hr and subsequently developed with the use of substrates, nitroblue tetrazolium (50 mg/ml in 70% dimethyl formamide) and 5-bromo-4-chloro-3-indolyl phosphate (50 mg/ml in 100% dimethyl formamide) (Sigma) in Tris/saline, pH 8.8.

Intensity of staining on nitrocellulose was measured with a Zeineh Soft Laser Densitometer (model SL-504-XL) and bands were compared to controls within the same gels. The data were subjected to statistical analysis according to the Student's "t" test.

RESULTS

Effect of retinoic acid on cell morphology and proliferation

In serum supplemented basal medium, the human LAN-5 neuroblastoma cells have an immature neuroblast-like morphology with rounded cell bodies and occasional short processes (Fig. 1a). Treatment with 10 uM retinoic acid, the optimal concentration for induction of differentiation of the LAN-5 cells (Cole and Timiras, 1987), resulted in morphologically differentiated cells which were more spread out and polar with processes showing more varicosities (Fig. 1b).

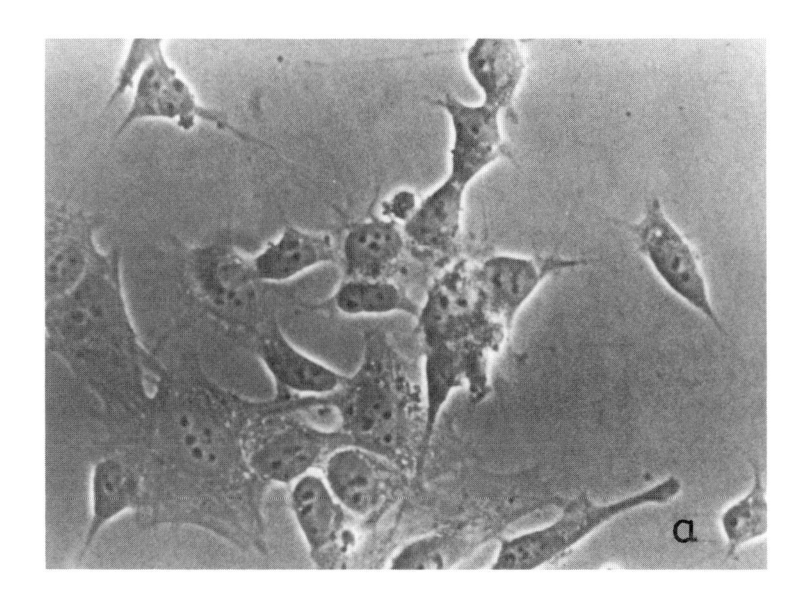

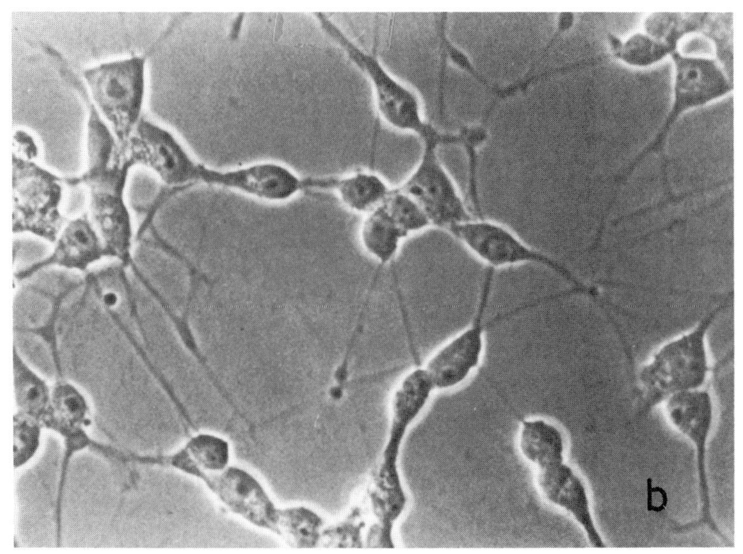

Fig. 1. Effect of retinoic acid treatment on LAN-5 cell morphology. Phase contrast micrographs of neuroblastoma cells grown in serum supplemented basal medium (a), or following 7 day treatment with 10 uM retinoic acid (b). Total magnification 2000 X.

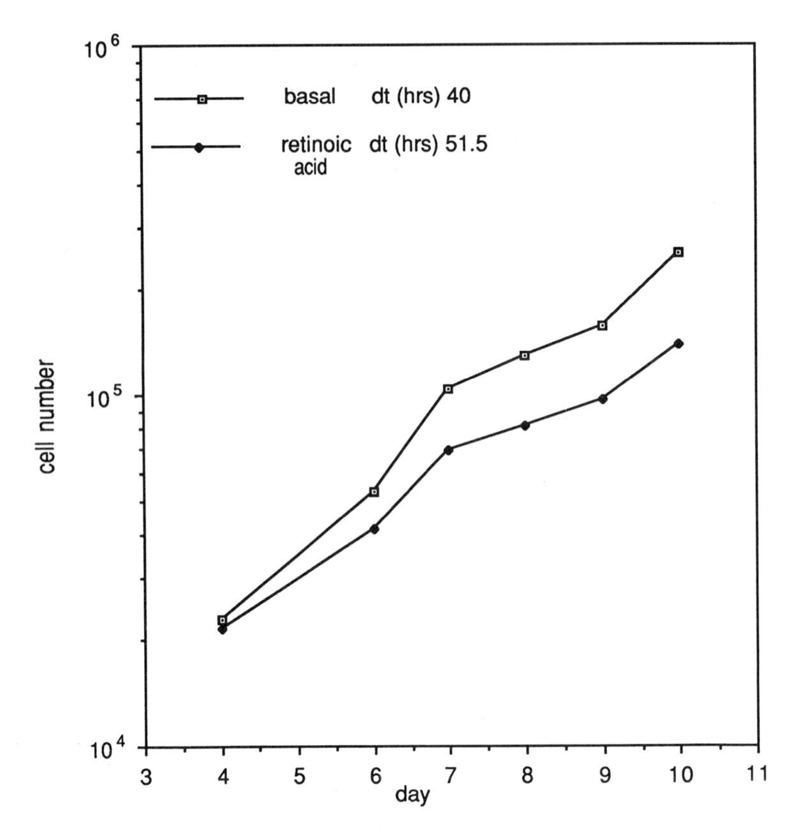

Fig. 2. Effect of retinoic acid on LAN-5 cell growth. LAN-5 cells were plated in the presence or absence of 10 uM retinoic acid and the DNA content was analyzed by a diaminobenzoic acid fluorescence assay through a 10-day time course. Using a standard curve, DNA-mediated fluorescence was monitored at the indicated times and converted to cell number per cm^2.

The morphological differentiation induced by retinoic acid treatment in LAN-5 cells was accompanied by a significant decrease in cell proliferation rate. The doubling time increased from 40 hours (untreated) to 51.5 hours (retinoic acid treatment) (Fig. 2).

<u>Effect of 5-HT on TH activity in undifferentiated and differentiated cells</u>

5-HT treatment (50 uM) significantly decreased TH activity ($p < .05$) in undifferentiated LAN-5 cells. While basal TH activity increased significantly in LAN-5 differentiated by culture in retinoic acid medium compared to undifferentiated control ($p < .0025$), 5-HT treatment had no effect on TH activity in differentiated cells (Table 1).

<u>Effect of 5-HT on Tau protein in undifferentiated and differentiated cells</u>

5-HT treatment (50 nM) resulted in a significant 188% increase ($p < .025$) in the 50 kD band of tau protein in the cytoplasmic fraction of undifferentiated cells (Fig. 3). On

Table 1. Effect of 5-HT on tyrosine hydroxylase activity in LAN-5 cultured in serum supplemented basal or retinoic acid treated medium.

	TH activity (nmoles/hr/mg prot)	
Treatment	Serum supplemented basal	Retinoic acid (10 uM)
Control	1.43 ± 0.10	2.26 ± 0.09[a]
Vehicle	1.25 ± 0.03	2.01 ± 0.05
5-HT 50 nM	1.20 ± 0.10	2.20 ± 0.07
50 uM	1.11 ± 0.07[b]	2.39 ± 0.22

Enzyme activities are expressed as means ± S.E.M. of 5-6 individual cultures. Statistical analysis is by Student's *t* test: [a] for cells cultured in retinoic acid medium compared to serum supplemented basal medium, $p \leq 0.025$. [b] compared to respective vehicle control, $p \leq 0.05$.

the other hand a higher concentration of 5-HT (50 uM) caused a significant decrease (to 70%, $p < .05$) in this same 50 kD band (Fig. 3). In the membrane fraction of undifferentiated cells 50 nM 5-HT significantly decreased (to 33%, $p < .0025$) this 50 kD band (Fig. 3).

No significant changes were seen in tau protein (cytoplasmic or membrane fractions) in differentiated LAN-5 cells after 5-HT treatment (50 nM or 50 uM).

DISCUSSION

Vitamin A and its derivates (retinoids) promote differentiation in human neuroblastoma (Reynolds et al. 1986). Consistent with the findings of Sidell et al., 1983; 1984; and Cole and Timiras, 1987, culture of LAN-5 neuroblastoma in retinoic acid treated medium induces morphological and biochemical changes indicative of differentiation toward more mature neurons. Differentiation is manifested morphologically by the formation of neurite extension, cellular clustering, and significant growth inhibition, and biochemically, by increased basal TH activity. Our results are in agreement with other *in vitro* studies of neuroblastoma in which differentiation by retinoic acid or sodium butyrate (Prasad, 1980) paralled increases in TH activity in a variety of neuroblastomas: A2(1) and E (Safaei & Timiras,1985), SH-SY5Y (Ino et al., 1986), and SMS-KCNR (Pennypaker et al.1989).

That retinoic acid may be involved in differentiation by regulation at the genomic level is evidenced by its direct

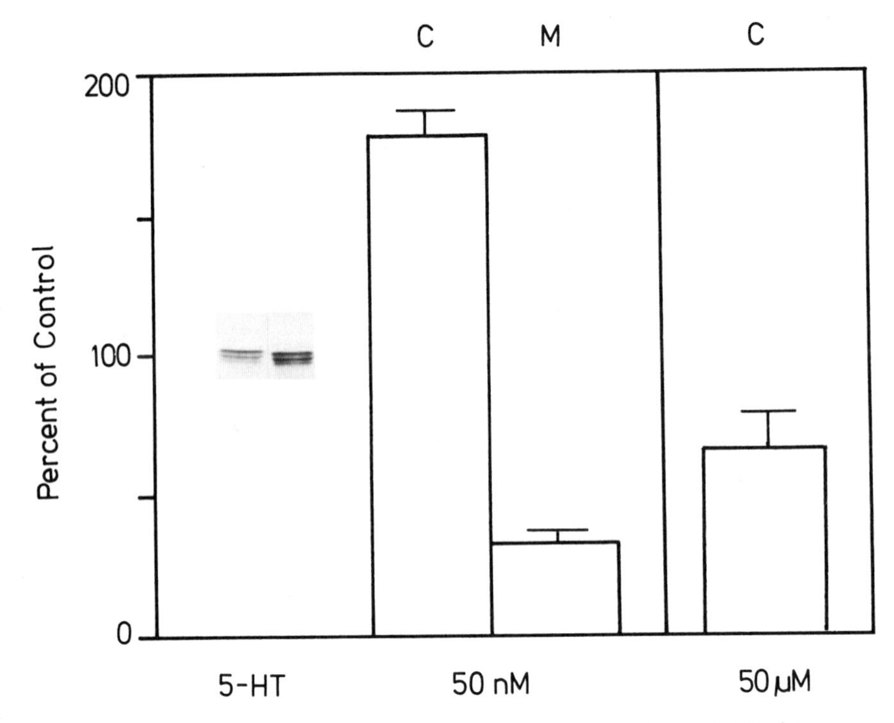

Fig. 3. Effect of 5-HT treatment on levels of 50 kD tau protein associated with the cytoplasmic or membrane fraction in LAN-5 neuroblastoma. Bar graphs: Undifferentiated LAN-5 cells were treated with 50 nM or 50 uM 5-HT for 48 hrs, and levels of 50 kD tau protein associated with the cytoplasmic (C) or membrane (M) fraction were analysed by Western blot using Alz-50 mouse monoclonal antibody and laser densitomerty of the resulting bands. Data points are means of 4 samples with standard errors indicated and are expressed as percent of control = 100. Western blot insert: Undifferentiated LAN-5 cells were treated with or without 50 nM 5-HT for 48 hrs, and cytoplasmic fractions were prepared and blotted onto nitrocellulose as described in the text. 50 kD tau isomers in untreated (left lane) or 50 nM 5-HT treated (right lane) samples were visualized using Alz-50 mouse monoclonal antibody.

affects on transcription and translation in F9 terato-carcinoma cells and macrophages (Griep et al., 1986; Chiocca et al., 1988). It may also change the pattern of protein synthesis in differentiating carcinoma cells (Griep et al., 1986).

The inhibitory effect of 5-HT on TH activity has been observed in vivo, where lesions in the raphe nucleus to prevent 5-HT effect (McRae-Degueurce et al.1982), or destruction of 5-HT terminals resulting from the intracisternal injection of 5,6-dihydroxytryptamine (Renaud et al., 1975), elicited increases in TH activity. The same effect has been proven lately in vitro, in cultures of newborn rat locus ceruleus explants, where the presence of 5-HT in the culture medium for a 24-hr period was followed by an inhibition of TH activity in the explants (Devau et al.1987). This effect lasted several days, with a maximal effect 48 hrs after treatment. More recently, the same inhibitory effect of 5-HT on TH has been described in the nucleus accumbens of the rat (Hetey,1988).

Our results with human neuroblastoma LAN-5 in culture agree with 5-HT inhibitory role on TH activity. In differentiated neuroblastoma cells in vitro we found no effect of 5-HT on the increased TH activity induced by retinoic acid. It is possible that a more inhibitory effect would have been seen with a higher concentration of 5-HT or with a longer incubation time with the amine. Increases in TH activity were seen in more differentiated cells such as those in the brain and sympathetic ganglia of the mature rat after in vivo reserpine treatment which depletes the brain and ganglia of 5-HT and catecholamines (Black et al., 1985). The latter finding is consistent with the reported inhibitory effect of 5-HT in mature neurons (Renaud et al., 1975; Lewis et al., 1976; Keane et al., 1978; Crespi et al., 1980).

No data exist in the literature for comparison with our findings on tau protein in neuroblastoma after 5-HT treatment. Recent evidence strongly suggests that in adrenal cells microtubules are involved in the mechanism of 5-HT action and of ACTH, presumably at the level of their interaction with the receptors (Feuilloley et al., 1986; 1988). Our finding of increased tau protein, which in vitro promotes microtubule polymerization, in the cytoplasm of undifferentiated neuroblastoma after a lower concentration of 5-HT, is consistent with the early association of this amine with microfilaments and microtubules in early chick embryos and especially in developing neural cells (Emanuelsson et al., 1988); the elevated concentration of 5-HT at gastrulation, and the intracellular distribution of the amine during early organization indicated a major role for it in morphogenesis and cell-shape changes in the early chick embryo. It is interesting to note that a higher concentration of 5-HT was found in the more advanced embryos with a better developed microtubular system. Its actions in vivo may therefore be dose-dependent with a higher concentration being more effective in more differentiated cells.

The exact mechanism of 5-HT action on microtubules is unknown. However, it is tempting to speculate that it initiates a process similar to that described by Gustafson (1981) who theorized that 5-HT is responsible for the formation of cAMP and a rise in the intracellular concentration of ionic calcium; the latter changes cellular morphology by breaking down the microtubules in the embryonic cells and activates the actin-myosin complex. More experimentation is necessary not only to elucidate the exact mechanism of action of 5-HT on tau protein and on TH activity in neuroblastoma, but also the overall relationship between cytoskeleton and neurotransmission.

Acknowledgment. The authors are grateful to Dr. D. Hudson for reviewing the manuscript and to Dr. W.B. Quay for helpful advice and suggestions. This work was supported by a grant from the State of California, Department of Health Services.

REFERENCES

Argasinski, A., Sternberg, H., Fingado, B., and Huynh, P., 1989, Doxorubicin affects tau protein metabolism in human neuroblastoma cells, Neurochem. Res., 14:927.

Black, I.B., Chikaraishi, D.M., and Lewis, E.J., 1985, Trans-synaptic increase in RNA coding for tyrosine hydroxylase in a rat sympathetic ganglion, Brain Res., 339:151.

Bradford, M., 1976, A rapid and sensitive method for the quantification of microgram quantities of proteins in utilizing the principle of protein dye and binding. Anal. Biochem., 72:248.

Burgess, L.T., and Kelly, B.R., 1987, Constitutive and regulated secretion of proteins, Ann. Rev. Cell Biol., 3:343.

Burnside, B., 1988, Photoreceptor contraction and elongation. Calcium and cyclic adenosine 3'5'monophosphate regulation of actin- and microtubule-dependent changes in cell shape, in: "Intrinsic Determinants of Neuronal Form and Function," R.J.Lasek, and M.M. Black, eds., Alan R. Liss, New York.

Chiocca, E.A., Davies, P.J., and Stein, J.P., 1988, The molecular basis of retinoic acid action, J. Biol. Chem., 263:11584.

Cole, G.M., and Timiras, P.S., 1987, Aging-related pathology in human neuroblastoma and teratocarcinoma cell lines, in: "Model Systems of Development and Aging of the Nervous System," A. Vernadakis, A. Privat, J.M. Lauder, P.S. Timiras, and E. Giacobini, eds., Martinus Nijhoff Publishing, Boston.

Crespi, F., Buda, M., McRae-Degueurce, A. and Pujol, F., 1980, Alteration of tyrosine hydroxylase activity in the locus coeruleus after administration of p-chlorophenylalanine, Brain Research, 191:501.

Devau, G., Multon, M.F., Pujol, J.P., and Buda, M., 1987, Inhibition of tyrosine hydroxylase activity by serotonin in explants of newborn rat locus ceruleus, J. Neurochem., 49:665.

Drubin, D., and Kirschner, M.W., 1986, Tau protein function in living cells, J. Cell. Biol., 103:2739.

Emanuelsson, H., Carlberg, M., and Lowkvist, B., 1988, Presence of serotonin in early chick embryos, Cell Differentiation, 24:191.

Feuilloley, M., Netchitailo, P., Delarue, C., Leboulenger, F., Benyamina, M., Pelletier, G., and Vaudry, H., 1988, Involvement of the cytoskeleton in the steroidogenic resonse of frog adrenal glands to angiotensin II, acetylcholine and serotonin, J. Endocr., 118:365.

Feuilloley, M., Netchitailo, P., Lihrmann, I., and Vaudry, H., 1986, Effect of vinblastine, a potent anti-microtubular agent, on steroid secretion by perifused frog adrenal glands, J. Steroid Biochem., 25:143.

Griep, A., Kendrick, N.C., and DeLuca, H.F., 1986, Modulation of protein biosynthesis during early stages of differentiation in retinoic acid-treated F9

teratocarcinoma cells, Arch. Biochem. Biophys., 249:180.
Gustafson, T., 1981, The cellular basis of morphogenesis and behavior during sea urchin development, in: "The Biology of Normal Human Growth," M. Ritzen, ed., Raven Press, New York.
Hetey, L., 1988, Putative interaction of presynaptic dopamine and serotonin receptors modulating synaptosomal tyrosine hydroxylase activity in the nucleus accumbens of rats, Biomed. Biochim. Acta, 12:1089.
Hinegardner, R.T., 1971, An improved fluorometric assay for DNA, Anal. Biochem., 39:197.
Ino, M., Cole, G.M., and Timiras, P.S., 1986, Tyrosine hydroxylase and monoamine oxidase-A activity increases in differentiating human neuroblastoma after elimination of dividing cells, Develop. Brain Res., 30:120.
Keane, P., Degueurce, A., Renaud, B., Crespi, F., and Pujol, J.F., 1978, Alteration in tyrosine hydroxylase and dopamine B-hydroxylase activity in the locus coeruleus after 5,6-dihydroxytryptamine, Neurosci. Lett., 8:143.
Laemmli, U.K., 1970, Cleavage of structural protein during assembly of head of bacteriophage T4, Nature, 277:680.
Lewis, B.D., Renaud, B., Buda, M., and Pujol, J.P., 1976, Time-course variations in tyrosine hydroxylase activity in the rat locus coeruleus after electrolytic destruction of the nucleus raphe dorsalis or raphe centralis, Brain Res., 108:339.
Lowry, O.H., Rosebrough, N.J., Farr, A.L., and Randall, R.J., 1951, Protein measurement with the folin phenol reagent, J. Biol. Chem., 193:265.
Mcrae-Degueurce, A., Berod, A., Mermet, A., Keller, A., Chouvet, G., Joh, T.H., and Pujol, J.F., 1982, Alterations in tyrosine hydroxylase activity elicited by raphe huclei lesions in the rat locus coeruleus: Evidence for the involvement of serotonin afferents, Brain Res., 235:285.
Pennypacker, K.R., Kuhn, D.M., and Billingsley, M.L., 1989, Changes in expression of tyrosine hydroxylase immunoreactivity in human SMS-KCNR neuroblastoma following retinoic acid or phorbol ester-induced differentiation, Mol. Brain Res., 5:251.
Pfeffer, S.R., and Rothman, J.F., 1987, Biosynthetic protein transport and sorting by endoplasmic reticulum and Golgi, Ann. Rev. Biochem., 56:829.
Pickel, V., Joh, T.H., and Reis, D.J., 1977, A serotoninergic innervation of noradrenergic neurons in nucleus locus coeruleus: demonstration by immunocytochemical localization of the transmitter specific enzyme tyrosine and tryptophan hydroxylase, Brain Res., 131:197.
Prasad, K.N., 1980, Butyric acid: a small fatty acid with diverse biological functions, Life Sci., 27:1351.
Renaud, B., Buda, M., Douglas Lewis, B., and Pujol, J.-F., 1975, Effects of 5,6-dihydroxytryptamine on tyrosine hydroxylase activity in central catecholaminergic neurons of the rat, Biochem. Pharmacol., 24:1739.
Reynolds, C.P., Biedler, J.L., Spengler, D.A., Reynolds, R.A., Ross, E.P., Frenkel, A., and Smith, R.G., 1986, Characterization of human neuroblastoma cell lines established before and after therapy, J. Natl. Cancer Inst., 76:375.
Safaei, R., and Timiras, P.S., 1985, Thyroid hormone binding and regulation of adrenergic binding in two

neuroblastoma cell lines, J. Neurochem., 45:1405.
Sidell, N., Altman, A., Haussler, M., and Seeger, R.C., 1983, Effects of retinoic acid on the growth and phenotypic expression of several human neuroblastoma cell lines, Exp. Cell Res., 148:21.
Sidell, N., Colin, A.L., and Dreutzberg, G.W., 1984, Regulation of acetylcholinesterase activity by retinoic acid in a human neuroblastoma cell line, Exp. Cell Res., 155:305.
Waymire, J.C., Bjur, R., and Weiner, N., 1971, Assay of tyrosine hydroxylase by coupled decarboxylation of dopa formed from 1-14C-L-tyrosine, Anal. Biochem., 43:588.

CRITICAL PERIODS OF NEUROENDOCRINE DEVELOPMENT: EFFECTS OF PRENATAL XENOBIOTICS

Sumner J. Yaffe, M.D.* and Lorah D. Dorn, R.N., Ph.D.**

* National Institute of Child Health and Human Development, National Institutes of Health
** National Institute of Mental Health
9000 Rockville Pike
Bethesda, MD 20892

INTRODUCTION

It is now generally accepted that the developing fetus may be adversely affected by exposure to drugs and environmental chemicals. The stage of development of the intrauterine host is a major determinant of the resultant drug or chemical action. With rare exception, all foreign compounds are transmitted across the placenta, and depending upon their solubility and chemical structure, achieve varying concentrations in the fetus. Historically, the concept that external agents could adversely effect the fetus was first expounded by Gregg[1] in Australia nearly 50 years ago when he demonstrated that rubella infection in the mother could lead to congenital cataracts in the newborn infant. During the several decades following Gregg's report, most concern regarding drug effects on the fetus had to do with the perinatal period, particularly with the effect of narcotics and analgesics on the ability of the newborn infant to sustain respiration following delivery.

In the early 1960s the devastating effects upon the fetus of thalidomide administered prenatally were reported. It was widely appreciated that drugs and chemicals could affect embryonic development and differentiation and give rise to birth defects. It is important at this point to emphasize that although thalidomide had a distinct cluster of anatomic defects that are virtually pathognomonic for this agent, it required several years of use with the production of many thousands of grossly malformed infants before the cause and relationship between its administration to the mother and its harmful effects were recognized. Furthermore, thalidomide had been evaluated for its safety in several animal species and had been found free of adverse effects, both in mother and her offspring. This example of thalidomide serves to emphasize the difficulties that exist in humans in incriminating drugs and chemicals as harmful when they are administered during pregnancy. Thalidomide also played a significant role in establishing the specialty of teratology and supporting the research endeavors in this area.

Plasticity and Regeneration of the Nervous System
Edited by P.S. Timiras *et al.*, Plenum Press, New York, 1991

In discussing drug effects upon the fetus, it is important that concepts be extended beyond the congenital malformations evident at birth to short- and long-term functional effects upon the fetus.

Concern for the delayed effect of intrauterine drug exposure was first raised following the tragic discovery that female fetuses exposed to diethylstilbestrol (DES) had a high incidence of adenocarcinoma of the vagina. Yet, the malignancy was not discovered until after the exposed individuals reached puberty. Thus, 15 to 20 years were required for this effect following intrauterine exposure to become evident. Ironically, it should be noted that diethylstilbestrol was no more effective in reducing the complications of pregnancy than placebo and need not have been administered at all.

The concept of long-term latency and delayed effect is, in our opinion, extremely important. We will emphasize this upon by citing data from investigations that were conducted in the laboratory at the University of Pennsylvania several years ago.[2] In these studies in the rodent, the effects of the widely used hypnotic-sedative agent phenobarbital were investigated. Phenobarbital was selected for several reasons. First of all, it was an old drug that was widely prescribed. It is quite clear that like most other agents, use of this drug did not abate during pregnancy. Dr. Chhanda Gupta, co-investigator in these studies, had shown in previous experiments that phenobarbital can act upon the neuroendocrine axis to affect reproductive function in adult animals. It, therefore, seemed appropriate to investigate the effects of phenobarbital administration to the fetus (via the mother) in order to correlate the biologic changes observed postnatally with measurements of steroid sex hormones at various stages during postnatal development. Furthermore, since the benefit of prophylactic phenobarbital administration, either prenatally or neonatally, on neonatal-hyperbilirubinemia has been well documented[3], phenobarbital continues to be widely used during this critical period of development. However, the effects of such treatment on the endocrine-related functions of children or adults exposed in the neonatal period have not been studied (see below).

In our early research [Yaffe & Gupta], phenobarbital was administered in a small dose during the latter part of pregnancy and the offspring were carefully observed after birth until adult life. Pregnancy was uneventful and the treated mothers delivered litters of normal size and weight. In the presentation of results, we have separated these into effects upon the female offspring and effects upon the male offspring. Since effects on either sex were different, this mode of presentation will facilitate discussion.

EFFECTS ON FEMALE OFFSPRING

In our initial experiments, phenobarbital was administered in a dose of 40 mg per kg per day subcutaneously from day 12 to day 20 of gestation. Subsequently, the critical period of action was narrowed to day 17-20 of pregnancy, suggesting that phenobarbital was interfering with neuroendocrine differentiation[4].

The first observation in the female offspring was that the animals receiving phenobarbital in utero grew at a slower rate and were significantly smaller than the control animals. This effect was observed shortly before weaning and continued until the end of the maximum growth spurt in adulthood.

There was a significant delay in the time of vaginal opening (a marker for the onset of puberty) in the females offspring that had received phenobarbital in utero. When the animals reached adulthood, vaginal smears were examined to evaluate the menstrual cycle. In the control group, as anticipated, 100% of the animals had a normal cycle, whereas only one-third of the treated animals had a normal cycle. This abnormal cycle was associated with infertility in approximately 60% of the females.

To further investigate the underlying cause of the reproductive failure of the phenobarbital exposed offspring, we assayed estrogen and progesterone concentrations in plasma. Significant increases in plasma concentration of estrogen were found before the onset of puberty and during adulthood in the treated animals. Progesterone concentrations were also higher before puberty, but the difference was no longer significant after puberty.

Although circulating hormone levels are responsible for certain reproductive functions, action at the molecular level is mediated by combination with specific hormone receptors. Concentrations of the receptors for estrogen in the uterus were found to be significantly higher than control values at all periods of development. It is possible that the high levels of estrogen resulted in increased synthesis of uterine receptor protein. Alternatively, the increased receptor level in the cytoplasm may indicate either a defective conversion of cytoplasmic to nuclear receptor, or decrease in the permeability of the nuclear membrane to the cytoplasmic hormone receptor complex. It is also possible that the nuclear receptor is less responsive to the action of the hormone, and as a consequence, higher amounts of hormone and its receptor are synthesized.

The concentration of luteinizing hormone (LH) in the plasma was also determined and found to be approximately 50% of normal concentration. These subnormal values of LH of the animals exposed to phenobarbital in utero, suggest a defect in hypothalamic-hypophyseal regulation of tropic hormone secretion. This phenomenon has been described previously in adult rats given phenobarbital and values returned to normal once the drug was withdrawn. In contrast, the effects that we saw were present throughout the lifespan following prenatal exposure to phenobarbital. (For additional details concerning the experimental findings, the reader is referred to the original studies.[5,6])

EFFECTS ON MALE OFFSPRING

Effects on male offspring differ considerably from those observed in female offspring. They also provide us with some important clues as to the mechanism of action of phenobarbital. As in the female, phenobarbital was administered in a dosage of 40 mg per kg of body weight, subcutaneously, during the latter part of pregnancy (day 17-20). The results of phenobarbital use did not produce any observable effects in the mothers (similar to results in female offspring). Both control and phenobarbital treated mothers delivered on day 21 or 22 of pregnancy. Offspring showed no gross anatomic malformations and litter size and birth weights were similar to those of control offspring.

Anogenital distance, an indication of anatomic masculinization, was significantly shorter in phenobarbital treated pups than in controls. Testicular descent, a marker for the onset of puberty, also was delayed.

Mating studies demonstrated that male offspring who received the drug in utero had decreased fertility (64% versus 100% in controls). The normal females who were mated with the phenobarbital treated male offspring had sperm plugs, but they did not conceive. Reproductive dysfunction was also correlated with endocrine hormone determinations; testosterone and LH were decreased in the plasma of adult male rats exposed to phenobarbital in utero.

The reproductive dysfunctions noted closely resemble the effects of neonatal perinatal androgen deficiency. Therefore, it is reasonable to speculate that phenobarbital administration in utero can decrease fetal or neonatal testosterone concentration and thereby produce defects in masculine sexual differentiation. In support of this hypothesis, we found that testosterone concentrations were decreased significantly in the plasma and brain of perinatal animals exposed to phenobarbital. Furthermore, lower concentrations of testosterone persisted throughout postnatal life into adulthood. In addition, testicular synthesis of testosterone was significantly decreased when measured on day 1 and day 3 of postnatal life. The defect in testicular synthesis of testosterone appears to be permanent since adult animals exposed to phenobarbital in utero had a marked deficit in this function.[7,8]

The precise mechanism of action of phenobarbital remains unknown. It is reasonable to conclude that prenatal phenobarbital exposure causes androgen deprivation during the critical prenatal period of fetal masculine development. As a result, reproductive function in the adult is abnormal. These effects are presumably mediated by alterations in hypothalamic pituitary function and interference with the normal process of masculinization.

The animal data presented, clearly demonstrated the short- and long-term effects of prenatal administration of phenobarbital on the offspring. Researchers using animal models, have the opportunity to carry out a variety of well controlled double blind drug trials which have afforded us invaluable information. For example, the timing of drug administration can be varied in order to show differentiating effects during critical periods of development. Structural changes in the brain also can be demonstrated once the animal is sacrificed; something which cannot, for obvious reasons, be done in humans. Furthermore, long term effects of drugs administered in the prenatal period are more readily examined in the rat as compared with the human, because the lifespan in a rat is completed in about a two year period.

A unique and singular opportunity arose for examining both short- and long-term developmental effects in a human population who were exposed prenatally to phenobarbital for the prevention of neonatal hyperbilirubinemia.

In the original study pregnant women on the Greek Island of Lesbos participated in a controlled randomized double-blind trial of prenatal phenobarbital administration for the prevention of neonatal hyperbilirubinemia.[9,10] The offspring of these mothers were examined as newborns, again between the ages of 5 and 7 years, and most recently as adolescents. We will briefly give an overview of the methodology of the study and report on summary findings during the neonatal period and follow-up examinations which took place in early childhood[10] and adolescence.

THE ORIGINAL STUDY

Women enrolled in the study were seen for prenatal care between October 1968 and November 1971 on the Greek island of Lesbos and from the city of Athens. The methodology of the trial was described in more detail in earlier publications.[9,10] In brief, the treatment group took a 100 mg tablet of phenobarbital (PB) daily beginning at 34 to 36 weeks of gestation and up to the delivery. The first part of the study (Phase I), was randomized double-blind and the control group received a placebo. The treatment was highly effective and there were no immediate untoward effects of phenobarbital noted. Thus, after this determination, PB was administered to all those expectant mothers who were seen 4 to 6 weeks before delivery while the mothers who missed their prenatal appointment during the last 4 to 6 weeks of pregnancy constituted the control group (Phase II). There was no difference between Phase I and II in the entry characteristics and the effect on jaundice. Thus, for analysis, the two phases were combined.

A total of 3075 infants (1522 PB and 1553 control) were studied in the initial trial. A subgroup of mothers ($\underline{n}$ = 212) in the PB group who took less than 10 tablets of PB (< 1.0 gm total) were not statistically different from the control group with respect to mean bilirubin levels or the incidence of marked jaundice and were deleted from most analyses.

Results indicated that prenatal phenobarbital significantly decreased the incidence of neonatal hyperbilirubinemia. In particular, the incidence of marked jaundice and the need for an exchange transfusion was decreased by a factor of six. In a subsample of offspring with ABO incompatibility, marked jaundice was decreased by three times and the need for an exchange transfusion was reduced by ten. In addition, it appeared that the dosage of phenobarbital had no negative effects on the offspring. There were no group differences in sleeping or eating patterns and symptoms of drug withdrawal were not observed. Complications to the mother were virtually non-existent.

The first follow-up after the neonatal period (Follow-up I), took place when the offspring were between the ages of 5.1 and 6.8 years of age.[10] It was impractical to include the entire original sample. Thus, using a technique to avoid bias 719 (36%) of the 2003 infants from Lesbos were designated for follow-up. The composition of the original sample, those re-examined and those finally analyzed are presented in the Table 1.

There were almost equal numbers of boys and girls in the sample of 415 children at Follow-up I. There were, however, significant differences between the PB and control group regarding the place of residence (PB had a greater number of urbanites), social class (PB had a higher social class), age distribution at examination (PB group was younger), and the incidence of neonatal jaundice (with the PB group having a much lower incidence as compared with those in the control group).[10] These differences made it necessary to use least square multiple regression analysis for the various outcome variables. In those analyses, the degree of jaundice and the group (PB vs control) were introduced separately, as compared with those in the control group.[10] The history and physical examination revealed that in general, the health of the offspring was good.

In physical growth the sex, the social class and place of residence were found to be important variables particularly for height. Beyond the effect of these factors, there appears to be an advantage conferred by PB which reached the level of statistical significance only for height. There were very few children who had neurological abnormalities and again, no group differences were noted. Children with moderate or marked neonatal jaundice had a higher rate of sensorineural hearing defects than those children with slight or no jaundice. In the three intelligence tests, there was a trend for the phenobarbital group to score higher than the control group but the difference was significant in only the Visuo-Motor Integration Test. The degree of jaundice was not a significant determinate of IQ.

Table 1. Analysis of the initial and follow-up populations of the Lesbos prenatal phenobarbital (PB) trial

MATERIAL	TOTAL	CONTROL	PB≥1.0gm	PB≤1.0gm
INITIAL POPULATION	2003	1040	822	141
DESIGNATED FOR FOLLOW-UP	719 (35.9%)	350 (33.6%)	325 (39.5%)	44 (31%)
EXAMINED AT FOLLOW-UP I*	490	206	249	35
INCLUDED IN THE ANALYSIS**	415	182	233	---
EXAMINED AT FOLLOW-UP II***	341	143	198	---

* Of the 229 (32%) children that were not examined only 28 could not be traced (9 in the PB group)

** The 75 children removed from analysis included: (a) PB <1.0 gm (35); (b) Preterm, low birth weight, twins (30); (c) incomplete examination (4); (d) conditions and syndromes affecting growth and development (6).

*** The 74 (18%) children not examined, included 12 not traced, 37 traced but not available at the time of examination and 25 refusals.

Follow-up during the period of adolescence[11] was particularly important because of the reports in the animal literature regarding the effects of phenobarbital on the reproductive system as well as the effect of the drug on behavior. Preliminary findings regarding growth and development, cognition, and behavior are summarized here. Complete reports in each area are being prepared. The 415 children used in the analyses of the first follow-up were designated for participation in the second follow-up approximately 10 years after the first follow-up. There was 18% attrition but the sample of 341 adolescents examined was not significantly different from the 74 children that were not examined on the basis of their first follow-up key variables (e.g., gender, place of residence, social class, and degree of jaundice).

The examination for growth and development included measurements of height and weight and pubertal staging using Tanner criteria[12,13] (genital and pubic hair stage for boys and breast and pubic hair stage for girls). Testicular volume was measured in boys using a Prader orchidometer. Age at menarche was obtained from the girls during the history and physical examination. In addition, serum hormone levels were drawn for testosterone (T) for boys and estradiol (E_2) for girls and androstenedione (delta 4-A) and cortisol (F) for both boys and girls. Three samples of blood were drawn at time 0, 20, and 40 minutes.

After controlling for age in the analyses of growth and development using analysis of covariance, the following group differences were noted: boys in the PB group were significantly taller than boys in the control group and boys in the PB group had a significantly lower mean testicular volume than boys in the control group. However, in Follow-Up I it was reported that social class and place of residence were related to some of the outcome variables. Thus, using multiple regression we also included these factors in the adolescent follow-up. For height percentile in boys, social class and place of residence were not related to height. Pubertal stage was positively related to height and the group differences remained, however. When we used age, social class and place of residence in a regression model to predict testicular volume, age was significantly related but the group differences did not hold up.

There were no significant group differences in genital or breast stage or pubic hair stage, once age was used as a control. There was, however, a trend for the phenobarbital group to be at a lower pubertal stage than the control group in both males and females. The majority of the adolescents were at Stage 4 or 5 at the time of the examination which may have influenced the findings. Age at menarche also was not different between the two groups. Thus, it appears that prenatal administration of phenobarbital may have some long-term effects on linear growth in boys. The finding regarding testicular size warrants further examination. At this point, we cannot determine if reproductive function has been influenced.

For hormones, all levels were within a normal clinical range for both boys and girls. Girls, had significantly higher hormone concentrations for delta 4-A and cortisol. Estradiol was assayed on girls only and T was assayed on boys only, thus, gender differences could not be examined in these two hormones. In our preliminary analyses, group differences were noted in only one hormone. Boys in the phenobarbital group had significantly lower cortisol levels than boys in the control group. Although this relationship is significant, even when age is controlled, we are currently exploring other factors which may influence this difference and the importance it may have.

Our preliminary analyses of cognitive function show that in virtually all of the subtests of the Wechsler Intelligence Scale for Children (WISC-R)[14] the PB group scored significantly higher than the control group. However, the picture seems more complex because cognitive development is influenced by multiple factors. In a regression model run separately by gender and which included age, social class and hormones as statistical predictors, age was not related to cognitive development while social class was. Of particular interest was that the subjects who were exposed to phenobarbital in utero differed from control subjects but this effect was only seen in the males. Boys in the PB group remained at significantly higher cognitive levels. There also were a number of hormone relations for cognitive development for boys and not for girls. Using this model, it appears that prenatal administration of PB may have differentiating effects on the brain depending on gender of the offspring.

In the adolescent follow-up the Child Behavior Checklist[15] was completed by the mother and the adolescent completed the Offer Self-Image Inventory for Adolescents[16] and the Youth Self-Report. Although these data have not been analyzed completely, it appears that the findings also are quite complex. For example, in the most strict sense, there are very few group differences on the psychological and behavioral measures. However, it appears that on preliminary analysis using multiple regression, when the groups (PB versus control) are examined separately, different hormone relations emerge. We are continuing to explore these relations to determine whether there are long-term effects of the prenatal administration of phenobarbital on the psychological and behavioral measures.

SUMMARY

Phenobarbital, when administered prenatally in a small dose to animals, produced profound, and permanent effects on reproductive function in the offspring. Preliminary analysis of a unique cohort of adolescents who were exposed to phenobarbital in utero, suggests that long-term effects are also evident in the human. The precise nature of these effects is currently being determined and will be reported separately. These effects may be qualitatively and quantitatively different from effects seen in animals because of species difference in the timing or neuroendocrine differentiation. Of greater importance, however, is the fact that biologic and pharmacologic effects can be seen in the human following exposure to xenobiotics perinatally. Implications for other pharmacologic agents await further investigation. The rat model appears to have validity for extrapolation to man.

REFERENCES

1. N. M. Gregg, Congenital cataract following German measles in the mother, Trans Ophthalmol Soc Aust. 3:35-41 (1941).
2. C. Gupta and H. Karavolas, Lowered ovarian conversion of ^{14}C-pregnenolone and other metabolites during phenobarbital block of PMS-induced ovulation in immature rats: Inhibition of 3α-hydroxysteroid dehydrogenation, Endocrinol. 92:117-124 (1973).
3. D. Trolle, Decrease of total serum bilirubin concentration in newborn infants after phenobarbital treatment. Lancet 2:705-710 (1968).
4. C. Gupta and S. J. Yaffe, Reproductive dysfunction in female offspring after prenatal exposure to phenobarbital: Critical period of action. Pediatr Res 15:1488-1491 (1981).

5. C. Gupta, B. R. Sonawane, S. J. Yaffe, B. H., Phenobarbital exposure in utero: Alterations in female reproductive functions in rats, Science 208:508-510 (1980).
6. B. R. Sonawane and S. J. Yaffe, Delayed effects of drug exposure during pregnancy: Reproductive function. Biol Res in Pregnancy 4(2):48-55 (1983).
7. C. Gupta, S. J. Yaffe, and B. H. Shapiro, Prenatal exposure to phenobarbital permanently decreases testosterone and causes reproductive dysfunction. Science 216:640-642 (1982).
8. C. Gupta, B. H. Shapiro, and S. J. Yaffe. Reproductive function in male rats following prenatal exposure to phenobarbital. Ped Pharmacol 1:55-62 (1980).
9. T. Valaes, S. Petmezaki, and S. A. Doxiadis SA, Effect on neonatal hyperbilirubinemia of phenobarbital during pregnancy or after birth: practical value of the treatment in a population with high risk of unexplained severe neonatal jaundice, in: "Bilirubin Metabolism in the Newborn." Birth Defects, Original Article Series, The National Foundation, March of Dimes 6:46 (1970).
10. T. Valaes, K. Kipouros, S. Petmezaki S, M. Solman, S. A. Doxiadis. Effectiveness and safety of prenatal phenobarbital for the prevention of neonatal jaundice. Pediatr Res 14:947 (1980).
11. The second follow-up of the Lesbos cohort was designed and organized by T. Valaes, S. J. Yaffe, G. Chrousos and E. Susman. It was carried out under NIH Contract No. NICHD-CRMC-84-35.
12. W. A. Marshall and J. M. Tanner, Variation in the pattern of pubertal change in boys. Arch Dis Child 45:13-23 (1970).
13. W. A. Marshall, J. M. Tanner, Variation in the pattern of pubertal change in girls. Arch Dis Child 44:291-301 (1969).
14. D. Wechsler, Wechsler Intelligence Scale for Children Revised. San Antonio, TX: The Psychological Corporation, 1974.
15. T. M. Achenbach, C. S. Edelbrock, Manual for the Child Behavior Checklist and the Revised Child Behavior Profile. Burlington, VT: University of Vermont, Department of Psychiatry, 1983.
16. D. Offer, E. Ostrov, and K. I. Howard, The Offer Self-Image Questionnaire for Adolescents: A Manual. Chicago: Michael Reese Hospital, 1977.
17. T. M. Achenbach and C. S. Edelbrock, Manual for the Youth Self-Report and Profile. Burlington, VT: University of Vermont, Department of Psychiatry, 1987.

CELL PLASTICITY DURING IN VITRO DIFFERENTIATION OF A HUMAN NEUROBLASTOMA CELL LINE

F. Clementi, C. Gotti, E. Sher, A. Zanini

Department of Medical Pharmacology, CNR Center of Cytopharmacology, University of Milano, via Vanvitelli 32, 20129 Milano, Italy

INTRODUCTION

The differentiation of nervous cells is a crucial process in the normal development of nervous tissue during ontogenesis, as well as for the regulation of cell plasticity during cell growth and regeneration.

Over the last years we have been interested in studying the process of differentiation in human neurons because this could be of great help in clarifying the changes occurring not only during ontogenesis but also in some pathologies, such as *senile dementia* and Alzheimer's disease.

The mechanisms of differentiation are difficult to study *in situ* at the cellular level, because of the complexity of the interactions between different cell types at different stages of maturation. Most cell-culture models of differentiation involve the use of primary cultures of either embryonic neurons or continuous cell lines derived from nervous tumors[1]. These cell lines retain part of the developmental program that occurs during certain phases of normal development and, furthermore, this program can be experimentally modulated. In fact, these cells undergo morphological and biochemical differentiation when subjected to various treatments, including exposure to agents that affect cell growth[2], increase intracellular concentrations of cAMP[3], or modify plasma-membrane properties[4]. Neuroblastoma cell lines consist of a homogeneous cell population, and thus provide a convenient model for studying the process of neuronal differentiation[1,5]. In the search for an appropriate cellular model, several human neuroblastoma cell line were screened for their biochemical properties; the IMR32 cell line, established in vitro by Tumilowicz[6] was selected because it can be induced to differentiate by the addition of two drugs (5-bromo-2'-deoxyuridine (BrdU) and N^6-O^2-dibutyryl cyclic adenosine 3':5' monophosphate (Bt_2cAMP) while it is not affected by the nerve growth factor[7,8].

Several biochemical and functional markers which are

characteristic of a nerve cell before and after the differentiation induced by the two different pharmacological agents were studied: in particular, morphology, neurotransmitter receptors, voltage gated ion channels and the processes of catecholamine synthesis, storage and secretion[5,14]. From the results it is possible to conclude that the process of neuronal differentiation is very complex. It is probably a multistep phenomenon that proceeds not in cascade but along parallel pathways, the sum of which produces the complex features of differentiation. Morphological and biochemical features can be independently activated at different steps of differentiation by different stimuli. This could be relevant for a possible *in situ* modulation of a differentiating neuron.

MORPHOLOGICAL ASPECTS OF CELL DIFFERENTIATION

The most striking feature of differentiated cells was the development of a large network of neurites connecting several cells together (Figs. 1, 3 and 4). The neurites were long, thin, regular structures with some swellings along their length and at the points of cell contact. At the ultrastructural level, the neurites were rich in microtubules and filaments, and contained numerous dense-core granules. The terminal boutons of these neurites were particularly numerous and were rich in dense-core granules, mitochondria, microtubules and intermediate filaments (Fig.1).
The results obtained by using the two differentiating agents, were very similar both in terms of the number of differentiated cells and of the number of neurites per cell.

These morphological data indicate that differentiation involves a large-scale reorganization of the cell structure with the appearance of new organelles (secretory granules) and the synthesis and redistribution of cytoskeletal elements. These modifications are the morphological counterpart of the functional and biochemical modifications reported below.

Na^+ AND K^+ CHANNELS

Voltage-operated channels, which are the basis of cell excitability, are a typical component of nerve cell membrane, and it could be of great interest to know how neurons

Fig.1. IMR32 cells treated with BrdU. The neurite network connecting the cells is dramatically increased. The cells often have more than one dendrite per cell, and the neurites are very long and branched. (a,b) The contacts between the cells and neurites are often characterized by a terminal expansion (c-f). At electron-microscopy, an increase in cytoskeletal elements and polysomes was observed, as well as a large increase in the dense-core vesicles of the cell body and, especially, along the neurites. (d,e,f) These organelles are particularly numerous in terminal boutons (f). a X 2400; b X 500; c X 18000; d X 10000; e X 10000; f X 35000 (from Gotti et al.[5]).

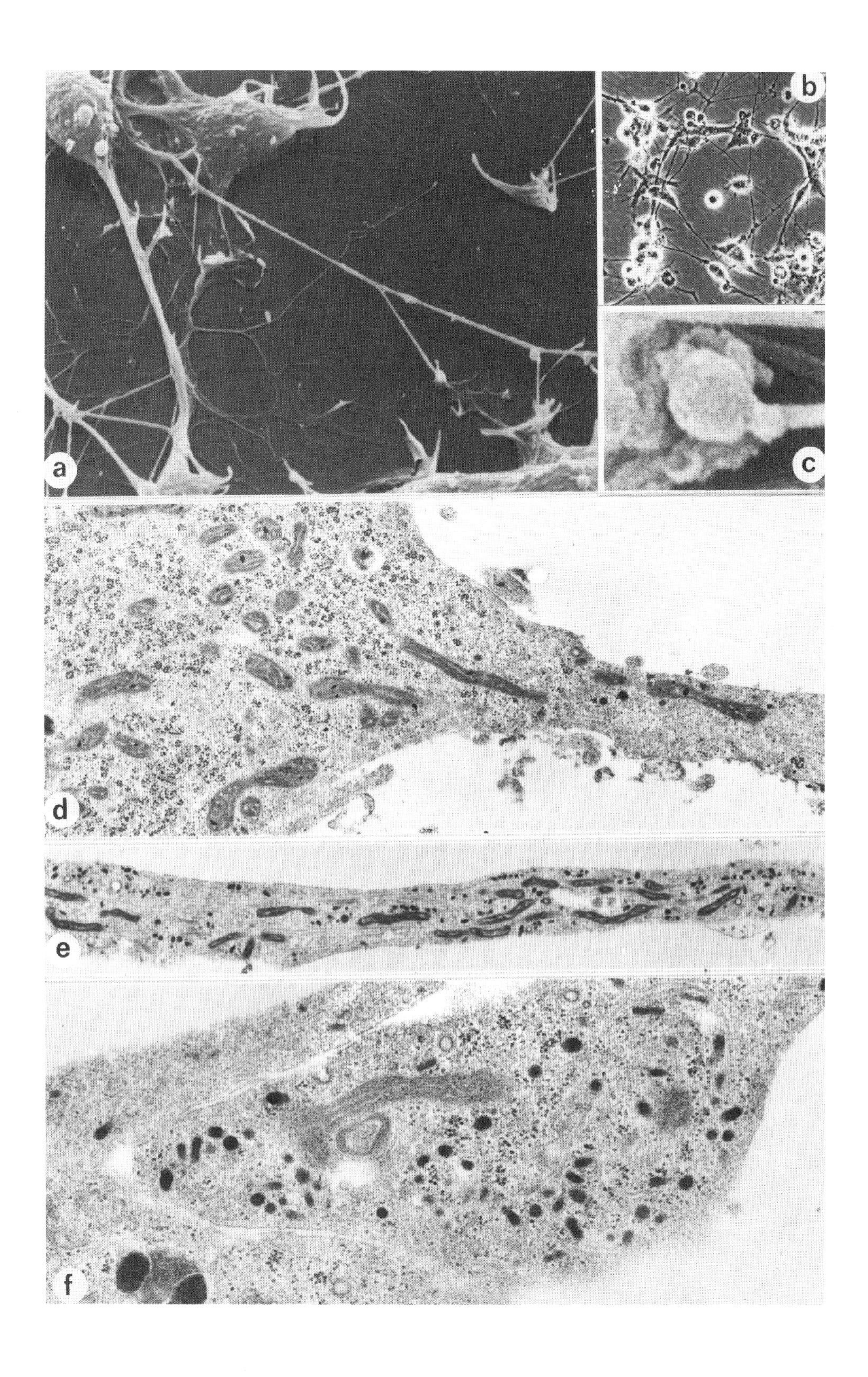
a
b
c
d
e
f

modulate their expression during differentiation.
By using electrophysiological techniques, Na^+, K^+,[5] and Ca^{2+} currents were studied. Furthermore for the Ca channels binding studies with ω-conotoxin, a toxin which only binds to a subtype of the Ca channels present in neurons[9,10] were also performed.

It was observed that Na^+ and K^+ currents were considerably greater after differentiation, and that the properties of voltage-gated channels were selectively affected by the drug used to induce differentiation. The most interesting finding concerned K^+ currents, which were inactivated in undifferentiated and BrdU-differentiated cells, but not in Bt_2cAMP-differentiated cells[5].

Ca^{2+} CHANNELS

More recently we have also investigated in some detail the Ca^{2+} currents and we found that IMR32 cells have both low-threshold (LVA) and high-threshold (HVA) Ca currents[9]. The time course and size of Ca currents changed remarkably according to the stage of morphological differentiation (Fig.2).
In the early days after exposure to differentiating agents, neurons with neurites longer than cell diameter showed a predominance of LVA (T-type) Ca currents (Fig.2a). Ca or Ba currents transiently activated between -50 mV and -20mV and displayed only a small sustained component positive to 0 mV when HVA Ca channels begin to activate. The size of the HVA currents increased with time during cell differentiation (Fig. 2 b,c) and reached, at 15 days, maximal amplitudes of 0.2-0.4 nA, comparable to the HVA current densities of other cultured neurons (0.04 mA cm^{-2})[11]. In parallel, a specific enrichment in the number of ^{125}I-ω-CgTx binding sites with respect to total cellular proteins during cell differentiation was found. The average amount of bound toxin was: 5.3, 10.1, 13.7 fmol/mg proteins on days 5, 10 and 15 of differentiation, respectively (from Carbone et al.[9]).

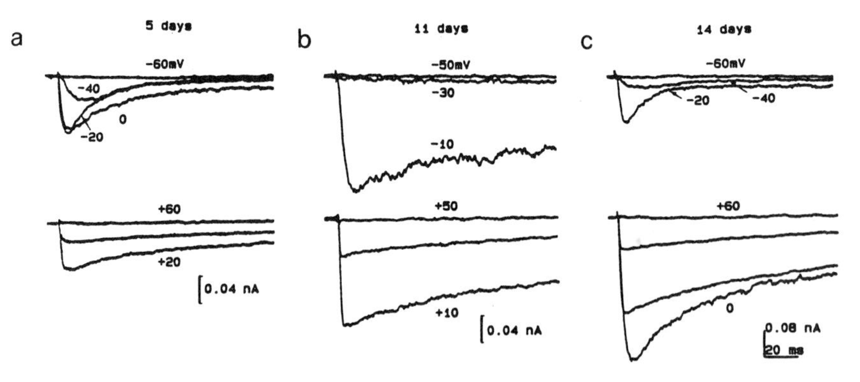

Fig.2. Families of Ba currents recorded from three human neuroblastoma IMR32 cells 5 (a), 11 (b) and 15 days (c) after exposure to differentiating agents. Membrane depolarization started from -60 or -50 mV with step increments of 20 mV. Holding potential was -90 mV. Note the different vertical scale in c.

During these studies we were able to demonstrate that ω-conotoxin binds and blocks selectively only one type of HVA channel which is clearly distinct from the classical L channel and seems to be present only in neuronal cells[10,11].

The expression of voltage-operated ion channels was selectively affected during cell differentiation. For example, in Bt_2cAMP-treated cells, a non-inactivating K channel prevailed, whereas in BrdU-treated cells a marked increase in ωCtx-labeled channels was found. It is conceivable that, in a complex situation such as *in situ* differentiation, the topological distribution of these channels is also affected by segregation, as in the case of the ωCtx channels to the presynaptic structures in adult differentiated neurons.

NEUROTRANSMITTER RECEPTORS

Neurotransmitter receptors are key moleules for the normal neuronal function and also these molecules are affected by the differentiation process. We concentrated on the modifications induced by differentiation on both muscarinic and nicotinic cholinergic receptors and on adenosine receptors.

Cholinergic receptors. The muscarinic receptors were studied by means of ^3H-scopolamine binding and the nicotinic receptors either by ^{125}I-αBgtx binding, or by electrophysiology[5].

In the Bt_2cAMP or BrdU differentiated cells there was a significant time dependent increase the number of ^{125}I-αBgtx binding sites. In fact the B_{max} values for ^{125}I-αBgtx binding were double than those of control cells, while the Kd values were not significantly different in treated cells.
Very few nicotinic cationic channels were detected in the nondifferentiated cells (an average of 1 channel per cell) but, the number of functional channel increased to 3±1 per cell in differentiated cells.

Measurements of channel conductance revealed a double population of channels. The following values were obtained: 18±3 and 30±4 pS for control cells; 25±3 and 50±6 pS for BrdU-treated cells, and 35±4 and 58±6 pS for Bt_2cAMP-treated cells. Furthermore, the channels of differentiated cells also had a shorter lifetime. These ACh-opened channels were of the nicotinic type, because the action of ACh was blocked by 1 mM hexamethonium, and not by αBgtx[5].

The number of muscarinic receptors revealed by ^3H-scopolamine binding, was not significantly affected by Bt_2cAMP treatment but, when the cells were treated with BrdU their number increased. The Kd values for ^3H-scopolamine binding were similar in both control and drug-treated cells (Table 1).

Adenosine receptors. In 5-BrdU differentiated IMR32 cells adenosine A_2-receptor-mediated stimulation of adenylate cyclase was markedly reduced. Under these conditions, the

Table 1. Modifications of neurotransmitter receptors complement in IMR32 cells after differentiation

	NICOTINIC RECEPTORS			MUSCARINIC RECEPTORS		ADENOSINE RECEPTORS
	^{125}I-αBgtx binding		No.of Na^+ channels/cell	^3H-scopolamine binding		°R-PIA stimulated adenylate cyclase activity
Treatment	B_{max}	Kd		B_{max}	Kd	
None	189±24	7.7 ±1.7	1	473± 73	3.32±0.77	225±25
BrdU (2.5 μM)	370±29*	8.11±1.4	3±1	1116±150*	4.80±1.5	50± 7
Bt_2cAMP (1 mM)	319±36*	7.35±4.6	2±1	423± 80*	3.85±1.7	80± 7

*P<0.01 (vs.controls)

° Adenylate cyclase activity is expressed as % stimulation of basal activity. No significant modification of basal activity was found up to 10^{-7}M R-PIA.

B_{max} values are expressed as fentomoles per milligram of membrane protein; Kd values are expressed in nanomoles. All values are the mean values ±SE of five to eight experiments performed on control and differentiated cells after 13 days of culture with or without the drugs.

stimulatory effects induced by R-PIA on cAMP production were reduced by about 75% with no apparent modifications of the relative potencies of the two adenosine agonists. Theophylline was able to eliminate the stimulatory effects induced by 100 μM A_2 agonists also in differentiated cells[12]. These results indicate that, in these cells, only A_2 subtypes of adenosine receptors are present and that, during differentiation, the cells lose part of these receptors. This event seems to be specific for adenosine receptors: other receptors (e.g., the nicotinic and muscarinic receptors) actually increased after drug-induced *in vitro* differentiation of this cell line.

The data here reported suggest that during differentiation cells can up- or down- regulate their membrane molecule responsible for the exchanges with the external milieu in a specific and discrete manner. In fact, during differentiation we have observed a down regulation of adenosine receptors, a no effect on the number of muscarinic receptors and an up regulation of functional nicotinic receptors. Furthermore the expression of the different receptors can be modulated in a different way according to the drugs used for cell differentiation.
It is not yet known how these *in vitro* modifications correlate with the process of differentiation occurring in physiological and pathological *in vivo* situations. However, these findings indicate that nerve cells can easily modify their surface receptors in response to external stimuli. This is a different phenomenon from that of the up or down regulation induced by receptor agonists or antagonists and provides nerve cells with further mechanisms of plasticity and adaptation to the external environment.

SYNTHESIS, STORAGE AND SECRETION OF NEUROTRANSMITTERS

One of the characteristic functions of neuronal cells is the synthesis and secretion of neurotransmitters.
In IMR32[5] cells we found that both differentiating agents, BrdU and Bt_2cAMP, induced a large increase in the Dopamine content and of tyrosine hydroxylase, the rate limiting step enzyme for its synthesis (Table 2).

Table 2. Dopamine content and Tyrosine hydroxylase activity in IMR32 cells

Treatment	Dopamine (pmol/mg protein)	Amount of L-dopa (pmol) formed/mg protein/10 min
None	8083±2167	8.5±0.5
BrdU (2.5 μM)	28724±5936*	122.8±4.5*
Bt_2cAMP (1 mM)	20161±1769*	68.5±2.7*

* P<0.01 (vs.control)
Dopamine content was determined by HPLC. All values represent the mean±SE of four to nine determinations.

In addition to the increase in neurotransmitter content it was also found that the cells were able to assemble the amine storage apparatus[13].

This process was followed biochemically, pharmacologically and morphologically. In non-differentiated cells it was found that Dopamine is not stored in secretory granules: the secretory granules are not visible by electron microscopy (Fig.1), Dopamine is not protected by monoaminooxydase activity, nor is released by reserpine (a drug able to stimulate release from secretory granules[14]). Furthermore integral membrane proteins typical of secretory granules, such as synaptophysin[15], are present only in small amounts and are not compartmentalized in any part of the cell (Fig.4).

After differentiation it was observed that Dopamine is stored in a compartment resistant to monoaminooxydase activity and that it is concentrated in discrete subcellular structures (Fig.3).

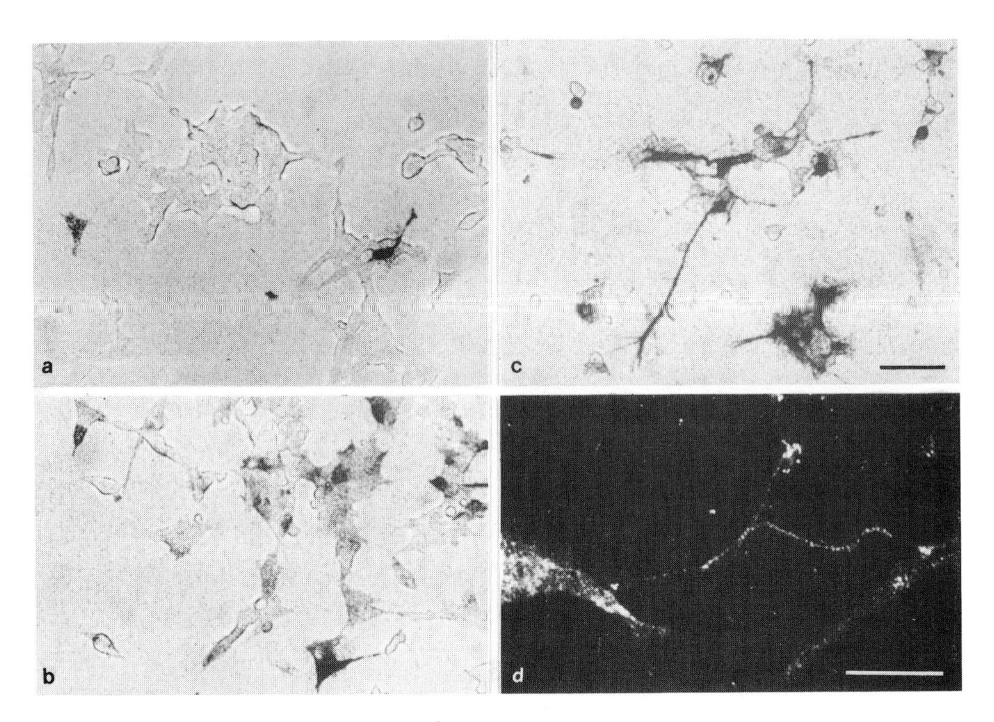

Fig.3. Autoradiography of ^{3}H-DA taken up from the medium and immunofluorescence with anti-DA antiserum. (a) Control, (b) Bt_2cAMP and (c) BrdU treated IMR32 cells were loaded with ^{3}H-DA and processed for autoradiography. Very few cells are labeled in control cultures (a), whereas both types of differentiated cultures (b,c) accumulate a large number of neurotransmitter. Note the particular compartmentalization of labeled structures, in the short neurites in Bt_2cAMP treated cells (b), and along the long processes of BrdU treated cells (c). In (d) the immunofluorescence pattern obtained with the anti-DA antiserum in BrdU treated cells is shown. A punctate labeling in the cytoplasm and along neurites is clearly noticeable. Bars: 50 μm (from Sher et al.[13]).

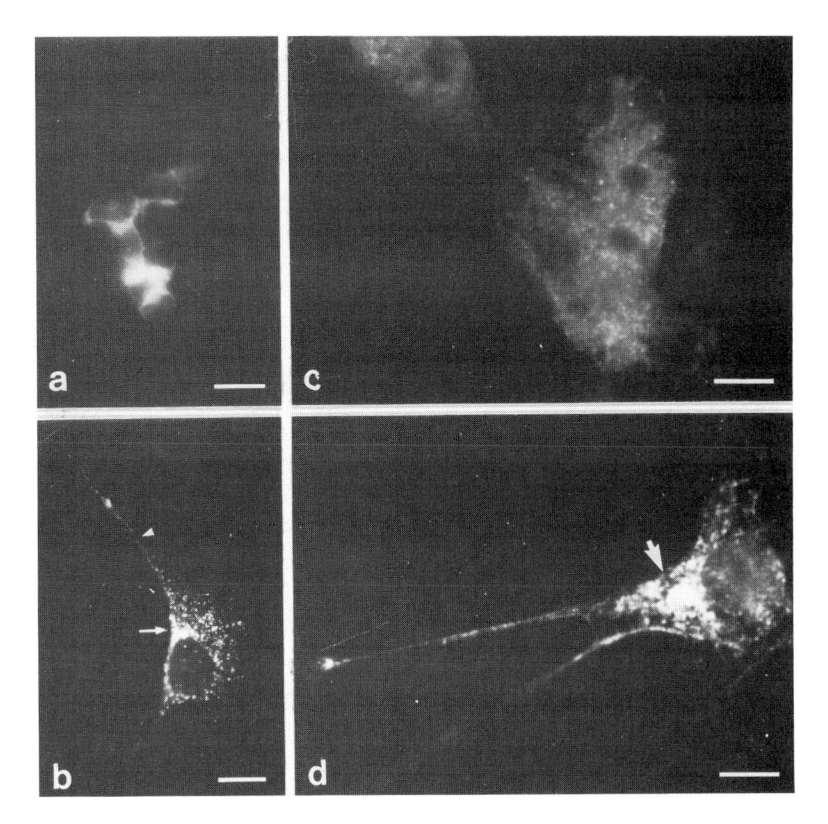

Fig.4. Immunolocalization of chromogranins and synaptophysin in control and BrdU differentiated cells. In control cells (a and c) a pale immunolabeling for both chromogranins (a) and synaptophysin (c) is present. In BrdU differentiated cells (b,d) typical punctate immunolabeling, with similar distribution of both chromogranins (b) and synaptophysin (d), is clearly evident in the Golgi area and in the neurites. Bars: 25 µm (from Sher et al.[13]).

There was a parallel appearance of numerous secretory organelles, revealed by electron microscopy (Fig.1), and of a large amount of synaptophysin, localized in discrete organelles, mainly in the Golgi region and in neurite and nerve endings (Fig.4).
Together with these events, chromogranins (which are proteins localized in secretory granules together with Dopamine[16]), appeared to have a localization similar to that of synaptophysin and Dopamine (Fig.4).
These data indicate that during differentiation, cells acquire the machinery for neurotransmitter storage, and that there is a complex reorganization of the cell structure.

Together with the assembly of the storage apparatus, the differentiated cells also acquire neurotransmitter releasing properties.

In fact, non differentiated IMR32 cells do not release ^{3}H-Dopamine, while BrdU-differentiated cells are able to release it in response to both ionomycin and αLTx (Table 3).

Table 3. Release of ^{3}H-DA from non differentiated (A) and differentiated (B,C) cells

Drugs	None	Bt_2cAMP	BrdU
	increase of release		
	A	B	C
α-Latrotoxin ($2x10^{-9}$M)	2.0±1	16±2	42±2
Ionomycin ($1x10^{-6}$M)	4.1±1.5	2±1	27±3
TPA ($0.1x10^{-6}$M)	3.0±1	1±0.5	6±2
Ionomycin+TPA	6.0±2	3±2	33±6
α-Latrotoxin + EGTA	1.5±0.7	10±2	31±5

The cells were loaded with ^{3}H-DA (0.8μM) and the amount of release determined after 8 min. Values are expressed as % increase in release over basal ±SE values, and represent the average of 5 to 8 different experiments.

TPA alone induced only a slight ^{3}H-Dopamine release which was additive to the ionomycin effect.
In contrast, Bt_2cAMP-differentiated cells were not able to release ^{3}H-Dopamine in response to either calcium ionophore ionomycin or TPA (Table 3). Only αLTx was able to induce a significant release of ^{3}H-Dopamine from such cells, although at a reduced level when compared with that of BrdU-differentiated cells. Since the level of spontaneous release was similar both in control and differentiated cells, we can exclude the possibility that this difference in the ability to release ^{3}H-Dopamine in response to secretagogues is due to a modification of the basal rate of release among different phenotypes. αLTx in Bt_2cAMP and BrdU-treated IMR32 cells, as in other systems[17], was able to induce neurotransmitter release even in the absence of extracellular calcium and at levels only slightly lower than those observed when extracellular calcium was present (Table 3).

These data indicate that differentiated cells acquire the ability to synthesize, store and release neurotransmitters. But the most fascinating finding was that related to Dopamine release. In fact, Bt_2cAMP-treated cells acquire only calcium independent release and not the more common calcium dependent release. This is interesting because it indicates that the activation of different secretion pathways is regulated in a different way from that controlling the complex pathway of neurotransmitters synthesis and storage. Furthermore, this cell model could be a very useful tool for the study of this uncommon Ca-independent neurotransmitter secretion.

CONCLUSIONS

We have shown that, in IMR32 cells, differentiation-inducing agents lead to substantial modification both in morphological, biochemical and physiological features; and that these modifications may differ according to the drug used to induce differentiation.

The mechanisms by which the two tested drugs induce differentiation are not completely known although they probably act differently. Bt_2cAMP mimics a dramatic increase in cAMP content, and this is known to induce differentiation in a large number of cells[1,3,7]. The effect of BrdU may depend on a block of DNA synthesis and on cell division[18], and on cAMP accumulation[8], but may also involve other modifications, such as the adhesion of cells to the surface of culture dishes, which may indirectly trigger the differentiation of neuroblasts[18]. All of these findings suggest that the process of differentiation in IMR32 neuroblastoma cells, in relation to their secretory properties, is probably a multistep phenomenon that proceeds not in cascade, but along parallel pathways, the coordination of which may produce the complex features of functional differentiation.

This *in vitro* model could be very far away from the process of differentiation that occurs *in vivo*, but it could be helpful for unraveling some of the steps and cellular processes that lie at the basis of this complex phenomenon.

REFERENCES

1. K.N. Prasad, Differentiation of neuroblastoma cells in culture, Biol. Rev. 50:129 (1975).
2. M. Gupta, M.D. Notter, S. Felten, D.M. Gash, Differentiation characteristics of human neuroblastoma cells in the presence of growth modulators and antimitotic drugs. Dev. Brain Res. 19:21 (1985).
3. M.T. Rupniak, G. Rein, J.F.B. Powell, T.A. Ryder, S. Carson, S. Povey, B.T. Hill, Characteristics of a new human neuroblastoma cell line which differentiated in response to cyclic adenosine 3' 5' monophosphate. Cancer Res. 44:2600 (1984).
4. V.Z. Littauer, M.Y. Giovanni, M.C. Glick, Differentiation of human neuroblastoma cells in culture. Biochem. Biophys. Res. Comm. 88:933 (1979).
5. C. Gotti, E. Sher, D. Cabrini, G. Bondiolotti, E. Wanke, F. Clementi, Cholinergic receptors, ion channels, neurotransmitter synthesis and neurite outgrowth are independently regulated during the in vitro differentiation of a human neuroblastoma cell line. Differentiation 34: 144 (1987).
6. J.J. Tumilowicz, W.W. Nichols, J.J. Cholan, A.E. Green, Definition of a continuous human cell line derived from neuroblastoma. Cancer Res. 30:2110 (1970).
7. S.W. De Laat, P.T. van der Saag, The plasma membrane as a regulatory site in growth and differentiation of neuroblastoma cells. Int. Rev. Cytol. 74:1 (1982).
8. K.N. Prasad, B. Mandel, S. Kumar, Human neuroblastoma cell culture: effect of 5-bromodeoxyuridine on morpho-

logical differentiation and levels of neural enzymes. Proc. Soc. Exp. Biol. Med. 44:38 (1973).

9. E. Carbone, E. Sher, F. Clementi, Ca currents in human neuroblastoma IMR32 cells: kinetics, permeability and pharmacology, Pflügers Arch. 416:170 (1990).
10. E. Sher, A. Pandiella, F. Clementi, ω-Conotoxin binding and effect on calcium channel function in a human neuroblastoma, FEBS Lett. 235:178 (1988).
11. E. Carbone, H.D. Lux, A low voltage activated, fully inactivating Ca channel in vertebrate sensory neurons, Nature 310:501 (1984).
12. M.P. Abbracchio, F. Cattabeni, F. Clementi, E. Sher, Adenosine receptor linked to adenylate cyclase activity in human neuroblastoma cells: modulation during cell differentiation. Neuroscience 30:819 (1989).
13. E. Sher, S. Denis Donini, A. Zanini, C. Bisiani, F. Clementi, Human neuroblastoma cell acquire regulated secretory properties and different sensitivity to Ca and α-latrotoxn after exposure to differentiating agents, J. Cell Biol. 108:2291 (1989).
14. B. Langhey, N. Kirschner, Effect of reserpine and tetrabenazine on catecholamine and ATP storage in cultured bovine adrenal medullary chromaffin cells. J. Neurochem. 49:563 (1987).
15. F. Navone, R. Jahn, G. di Gioia, H. Stukenbrok, P. Greengard, P. De Camilli, Protein P38: an integral membrane protein specific for small vesicles of neurons and neuroendocrine cells. J. Cell Biol. 103:2511 (1986).
16. P. Rosa, A. Hille, P.W. Lee, A. Zanini, P. De Camilli, W.B. Huttner, Secretogranin I and II: two tyrosine sulphated secretory proteins common to a variety of cell secreting peptides by the regulated pathway, J. Cell Biol. 101:1999 (1985).
17. J. Meldolesi, H. Scheer, L. Madeddu, E. Wanke, Mechanisms of action of α-latrotoxin: the presynaptic stimulatory toxin of the black widow spider venom, Trends Pharm. Sci. 7:151 (1986).
18. D. Shubert, F. Jacob, 5-Bromodeoxyuridine-induced differentiation of a neuroblastoma. Proc. Natl. Acad. Sci. USA 67:247 (1970).

LN-10, A BRAIN DERIVED cDNA CLONE: STUDIES RELATED TO CNS DEVELOPMENT

E.D. Kouvelas[1], I. Zarkadis[2], A. Athanasiadou[2], D. Thanos[3] and J. Papamatheakis[3]

Departments of [1]Physiology and [2]Biology, Medical School, University of Patras, Patras and [3]Institute of Molecular Biology and Biotechnology Heraklion, Greece

INTRODUCTION

The development and maturation of the nervous system are the results of highly specific interactions among developing neurons with their cellular and substrate enviroment. These interactions regulate cell proliferation and differentiation, neuroblast migration, axonal growth and guidance, target recognition and synapse formation. The development, over the past decade, of the methods of molecular biology, gives the opportunity for the study of these interactions in the molecular level. This kind of studies will give insight not only in the regulatory mechanisms of the establishment of form and pattern within the central nervous system (CNS) but also in very important properties of it, as it is plasticity and regeneration. Within this frame of studies it has been recently reported the isolation of cDNA clones from a λgt11 cDNA expression library made from NGF-treated PC12 cells[1]. The screening of the library was done with polyclonal rabbit and guinea pig anti-NILE (NGF Inducible Large External-Glycoprotein) antibodies the characteristics of which are previously described[2,3]. Two of these clones are modestly but reproducibly up-regulated by NGF in PC12 cells.

As we have previously described, a similar rabbit anti-NILE antibody stained the developing chick cerebellum and optic lobes in a specific age depending manner[4,5]. Therefore we thought that it would be of interest if, using the above probes, we could isolate neural specific cDNAs. These clones could be then used for studies related to CNS development. For such developmental studies chick brain provides certain advantages. It develops mostly during the in ovo period of life and is therefore a closed system, which offers also the opportunity for making experimental manipulations during embryogenesis. Furthermore, certain areas of chick brain have been well characterized developmentally and anatomically. The results which we report here describe the isolation and characterization of a cDNA clone (LN-10) from a λgt11 cDNA library made from 2 days old chick cerebellum.

ISOLATION AND CHARACTERIZATION OF LN-10 CLONE

In initial experiments genomic chick DNA was digested with the restriction enzymes HindIII and EcoRI and probed with the P12-13 cDNA clone which had been isolated from NGF-treated pheochromocytoma cells. The Southern blot analysis revealed cross-hybridization between the two heterologous DNAs (Fig.1).

An aliquot of the λgt11 cDNA library from poly $(A)^+$ RNA of 2 day old chich cerebellum was screened using as probe the heterologous P12-13 cDNA clone. In this experiment approximately 200000 recombinant phage-plaques were screened. Three positive clones were isolated with inserts of 0.5Kb, 1.8Kb and 2.0Kb long. The LN-10 clone, 2.0Kb long, was chosen for molecular characterization because it gave the strongest hybridization signal and hybridized to the DNA of the remaining clones. The nucleotide sequence of LN-10 clone contains 1941 nucleotides with a single open reading frame that starts with a methionine codon at nucleotide 43 and ends with a TAA stop codon at nucleotide 1629.

In order to investigate the number of genes coding for LN-10 molecule and the number of introns involved, we performed Southern blot analysis with chick genomic DNA and restriction enzymes for some of which. LN-10 DNA contains recognition sites (BamHI, SacI, RstI) and for some does not (EcoRI, BglII,

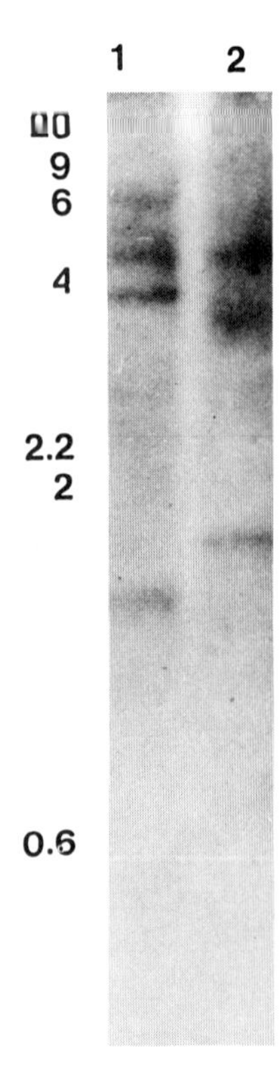

Fig. 1. Chick genomic DNA was digested with restriction enzymes EcoRI (lane 1) and HindIII (lane 2) and analyzed to 0.9% gel agarose. The hybridization with heterologous probe P12-13 was carried out at 60°C.

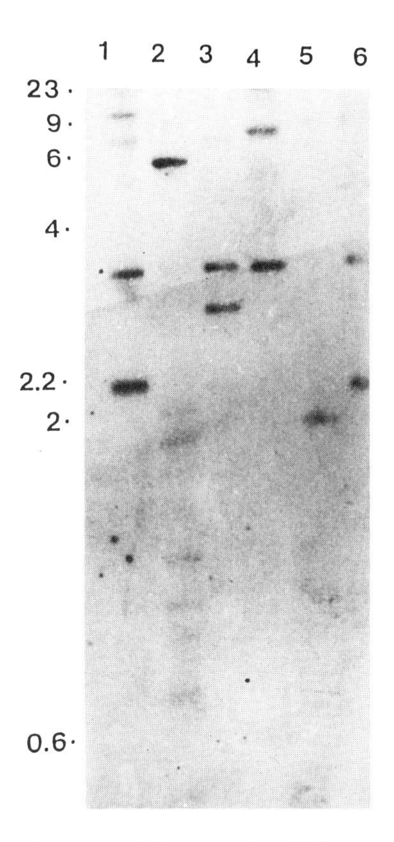

Fig. 2. Chick genomic DNA digested with enzymes 1.SacI, 2. PstI, 3. HindIII, 4. EcoRI, 5. BglII and 6. BamHI, and analyzed to 0.9% agarose gel. The hybridization with LN-10 probe was carried out at 65oC.

HindIII). One hybridization band with BglII and three bands with BamHI was found, whereas two bands were detected in the case of BamHI and EcoRI enzymes (Fig.2). These findings are compatible with the explanation that possibly LN-10 clone is encoded by a single gene, in which the LN-10 sequence is interrupted by at least one intervening sequence, containing restriction sites for EcoRI and HindIII enzymes.

NORTHERN BLOT ANALYSIS PROVIDES EVIDENCE FOR NEURAL TISSUE SPECIFICITY OF LN-10 CLONE

Total RNA from spleen, blood, intestine, optic lobes, hemispheres and liver was analyzed. All tissues were isolated from two days old chick. No trancript was identified from non-neural tissues, whereas three bands from optic lobes and four from hemispheres were detected. The detected transcripts from optic lobes are 2.0, 4.2 and 5.2Kb long and from hemispheres are 2.0, 4.2, 5.2 and 6.8Kb long (Fig. 3).

If it is true that one single gene is related to LN-10 clone then probably the different mRNAs detected, are derived from alternative splicing of a single transcript.

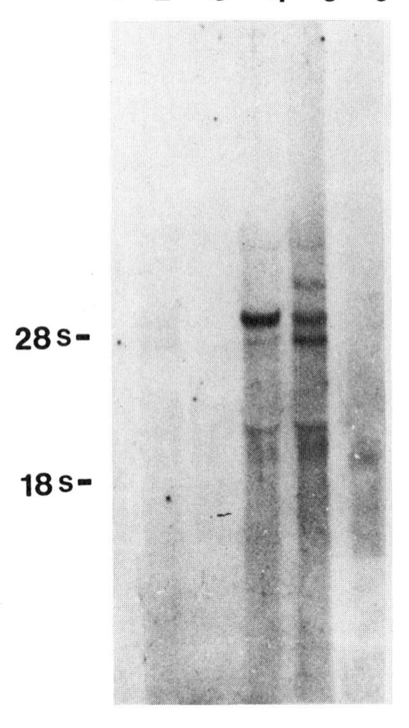

Fig. 3. Total RNA from tissues of two days old chick 1. spleen, 2. blood, 3. intestine, 4. optic lobes, 5. hemispheres, 6. liver analyzed to 1.3% formaldehyde-agarose gel. The hybridization was carried out with LN-10 cDNA probe at 65oC. Markers: 18S and 28S ribosomal RNA. Exposure: 36hrs.

NORTHERN BLOT AND IN SITU HYBRIDIZATION STUDIES IN CHICK CEREBELLUM PROVIDE EVIDENCE FOR DEVELOPMENTAL STAGE SPECIFICITY OF LN-10

The expression of LN-10 gene during development of chick cerebellum was studied with Northern blot analysis experiments. The LN-10 probe recognizes two transcripts of 1.5 and 2.0Kb long in chick cerebellar tissue total RNA as well poly $(A)^+$ RNA (Fig. 4).

Total RNAs from chick cerebellum of different ages were analyzed with Northern blots. (Fig. 5) At embryonic day 14 both messages of 1.5 and 2.0Kb were expressed. The expresion of these messages shows a gradual increase from embryonic day 14 until post-hatching day 2 and from this day a decrease follows. At post-hatching day 17 both messages are almost not expresed. These results together with previous developmental studies[6-8] show that the two detected messages in chick cerebellum are expressed during the period when the molecular and the internal granular layers show a rapid development, the number of parallel fibers rises and a significant number of synaptic connections can be detected between parallel fibers and the dendritic spines of Purkinje cells.

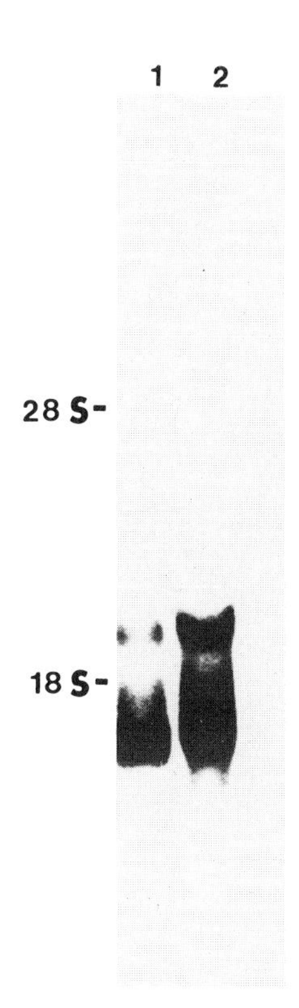

Fig. 4. Total RNA (1) and poly $(A)^+$ RNA (2) from two days old chick cerebellum analyzed to 1.2% formaldehyde-agarose gel and hybridized with LN-10 cDNA probe. Exposure: 12hrs.

Assuming that these results suggest that LN-10 related messages play a role in these processes we proceeded to in situ hybridization experiments. Sections of 2 days old chick cerebellum hybridized in situ with the antisense strand of LN-10 mRNA, while as control was used the sense strand. The LN-10 mRNA expression was detected mainly in the external granular layer and in the peripheral part of the internal granular layer and it declines in the deeper parts of the internal granular layer. LN-10 mRNA was also detected in cells localized in the molecular layer and which we assume are migrating granule cells (Fig. 6). Thus the in situ hybridization studies further support the hypothesis that mRNAs which are detected with LN-10 probe may encode for protein(s) which are related with processes associated with the migration of granule cells. In these processes are included the neural-glial cell recognition, elongation of parallel fibers, formation of parallel fiber bundles and establishment of synaptic connections. More in situ hybridization experiments on tissue slices of chick cerebellum from different developmental stages are now in progress in our laboratory.

THE LN-10 DEDUCED AMINO ACID SEQUENCE HAS HOMOLOGIES WITH NEURAL SPECIFIC MOLECULES CONTAINING EGF-LIKE REPEATS

The deduced protein sequence from LN-10 clone

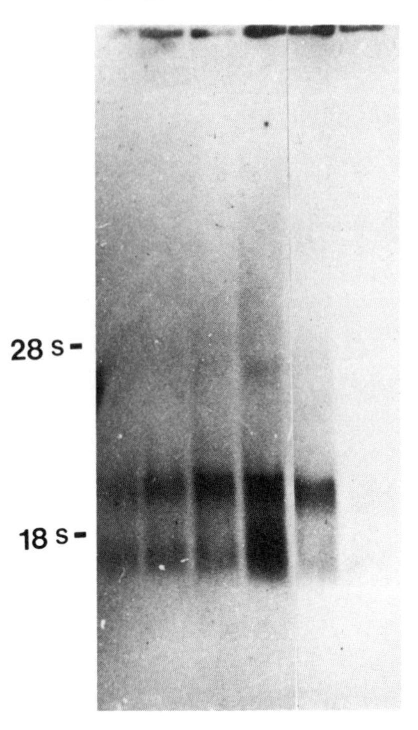

Fig. 5. Total RNA from chick cerebellum of different ages, E14 (1), E18 (2), E20 (3), P2 (4), P8 (5), P17 (6), analyzed to 1% formaldehyde-agarose gel and hybridized with LN-10 cDNA probe.

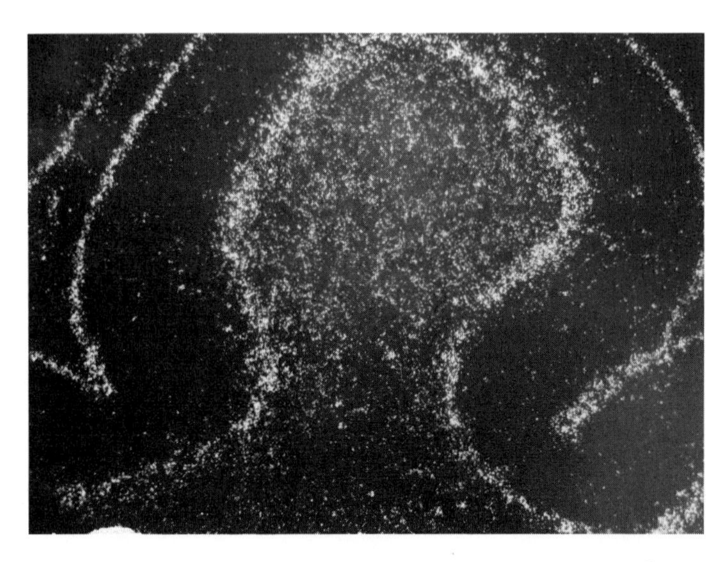

Fig. 6. RNA-RNA in situ hybridization. Sections from two days old chick cerebellum hybridized with ^{35}S labelled antisense LN-10 RNA probe at 55oC. Exposure: 6 Days.

contains 540 amino acid residues and the molecular weight was calculated to be 60800 daltons. The amino acid sequence was compared with other proteins using the wordsearch program, an algorithm similar to the algorithm of Wilbur and Lipman[9]. No significant homology was found with cell adhesion molecules of the inmmunoglobulin super family, a member of which is NILE glycoprotein[10]. However an homology of about 40% was found with neural specific molecules of different species, N-Cadherin (chick)[11], Notch (Fruit fly)[12], Delta (Fruit fly)[13]. It is of interest that both Notch and Delta have very important developmental properties and contain in their molecule EGF-like repeats. Another interesting homology from developmental and functional point of view is the one detected with chick villin. Villin is a major cytoskeletal protein in microvilli from brush-border cells of intestine and kindney. The LN-10 derived protein shows a 37% similarity with the carboxyl-terminal domain of this protein, which includes the villin "headpiece" a fragment involved in bundling of actin filaments[14].

CONCLUSIONS AND FUTURE DIRECTIONS

In the present article we report the isolation of a neural specific cDNA clone (LN-10) from a λgt11 cDNA library made from two days old chick cerebellum. The results which we present here suggest that LN-10 may have important functions in the development of the nervous system. In vitro translation experiments which are in progress in our laboratory will give to us the possibility to produce antibodies against the LN-10 related protein. These antibodies will be used for experiments which hopefully will give some clues for the function of this protein. Furthermore the study of genomic clones which we have isolated will give insight in the regulatory mechanisms which govern the expression of these molecules.

ACKNOWLEDGEMENTS

Supported from a grant of the Secretariat of Research and Technology of Greece (E.D.K.).

REFERENCES

1. P. Sajovic, D.J. Ennulat, M.L. Shelanski and L.A. Greene, Isolation of NILE glycoprotein-related cDNA probes, J. Neurochem 49:756 (1987).

2. S.R.J. Salton, C. Richter-Landsberg, L.A. Greene and M.L. Shelanski, Nerve growth factor inducible large external (NILE) glycoprotein: Studies of a central and peripheral neuronal marker, J. Neurosci, 3: 441 (1983).

3. S.R.J. Salton, M.L. Shelanski and L.A. Greene, Biochemical properties of the nerve growth factor-inducible large external (NILE) glycoprotein, J. Neurosci. 3: 2420 (1983)

4. A. Batistatou, L.A. Greene, M.L. Shelanski and E.D. Kouvelas, Development of NILE GP in chick brain, Neuroscience, 22:S230 (1987).

5. A. Batistatou and E.D. Kouvelas, Development of NILE glycoprotein in chick brain in: "Molecular aspects of development and aging of the nervous system " J.M. Lauder ed. Plenum Press, New York (1990).

6. J. Hanaway, Formation and differentiation of the external granular layer of the chick cerebellum, J. Comp. Neurol. 131:1 (1967).

7. E. Mugnaini, Ultrastructural studies on the cerebellar histogenesis. II. Maturation of nerve cell populations and establishment of synaptic connections in the cerebellar cortex of the chick, in: "Neurobiology of cerebellar evolution and development", R. Llinas, ed. American Medical Association, Chicago (1969).

8. R.F. Foelix and R. Oppenheim, The development of synapses in the cerebellar cortex of the chick embryo, J. Neurocyt. 3:277 (1974).

9. N.J. Wilbur and D.J. Lipman, Rapid similarity searches of nucleic,acid and protein data banks, Proc. Natl. Acad. Sci. (USA) 80:726 (1983).

10. J.T. Prince, N. Milona and W.B. Stallcup, Characterization of a partial cDNA clone for the NILE glycoprotein and identification of the encoded polypeptide domain. J. Neurosci, 9:1825 (1989).

11. K. Hatta, A. Nose, A. Nagafuchi, and M. Takeichi, Cloning and expression of cDNA encoding a neural Calcium-dependent cell adhesion molecule: its identity in the Cadherin gene family. J. Cell Biol. 106:873 (1988).

12. K.A. Wharton, K.M. Johansen, T. Xu and S. Artavanis-Tsakonas, Nucleotide sequence from the neurogenic locus Notch implies a gene product that shares homology with proteins containing EGF-like repeats , Cell 40:567 (1985).

13. H. Vassin, K.A. Bremer, E. Knust and J.A. Champos-Ortega, The neurogenic gene Delta of Drosophila melanogaster is expressed in neurogenic territories and encodes a putative transmembrane protein with EFG-like repeats, EMBO J. 6:3431 (1987).

14. W.L. Bazari, P. Matsudaira, M. Waller, T. Smeal, R. Jakes and Y. Ahmed, Villin sequence and peptide map identify six homologous domains Proc. Natl. Acad. Sci. (USA) 85:4986 (1988).

SPINAL CORD SLICES WITH ATTACHED DORSAL ROOT GANGLIA: A CULTURE MODEL FOR THE STUDY OF PATHOGENICITY OF ENCEPHALITIC VIRUSES

A. Shahar, S.Lustig, Y.Akov, Y.David, P.Schneider, and R.Levin

Israel Institute for Biological Research, Ness-Ziona, Israel 70450

SUMMARY

We describe here a culture system for long-term growth of organotypic slices of spinal cord, with attached dorsal root ganglia (DRG) derived from 13-14 day mouse fetuses. This is a unique in vitro tool in which both central and peripheral nervous tissue grow and differentiate in culture to become heavily myelinated. During cultivation the slices and the ganglia become flattened so as to allow microscopical and immunocytochemical staining. When both central and peripheral myelin had been formed (usually around the third week of cultivation), cultures were infected with $5x10^6$ PFU of West Nile Virus (WNV). Progeny virions appeared first in about 10% of the neurons and were subsequently observed between lamellae in the central myelin sheath of several axons. Such viral arrangement in relation to myelin membranes, might provide a novel concept for a possible mechanism underlying slow viral infection.

Key Words: Slice cultures, West Nile Virus, Myelin lamellae.

INTRODUCTION

West-Nile Virus (WNV) is an arbovirus which belongs to the Flaviviridae family and has neurotropic potential. (Goldblum 1959; Weiner et al. 1970; Monath 1986). This virus which is widely disseminated by mosquitoes and wild birds (Goldblum et al. 1954; Hayes et al. 1982), is capable of epidemic spread. In man WNV confers immunity and causes commonly a sub-clinical infection rather than a severe encephalitis, the latter manifested at a higher rate if contracted at old age (Marburg et al. 1956; Chamberlain 1980). Demyelination is not a typical feature of human Flaviviral infection (Monath 1986). Viral infection of susceptible adult mice by intracerebral and intraperitoneal injection, causes acute encephalitis and death of animals after 7 days.

Plasticity and Regeneration of the Nervous System
Edited by P.S. Timiras *et al.*, Plenum Press, New York, 1991

Different encephalitic viruses induce distinct cytopathic effects and cause different pathogenicity when added to organotypic spinal cord/dorsal root ganglia (DRG) cultures (Shahar et al. 1986; 1990). It seemed interesting to study the cytopathic changes induced by WNV in these slice cultures. They preserve basic central (CNS) and peripheral (PNS) architecture, but have no blood brain barrier and therefore are not under the influence of any humoral or immunological factors.

In the present study we describe an in vitro model of combined spinal cord/DRG cultures and its advantages in the research of neurotoxicity. In addition, we assess the light microscopic and ultrastructural cytotoxical alterations, associated with WNV pathogenicity in these cultures.

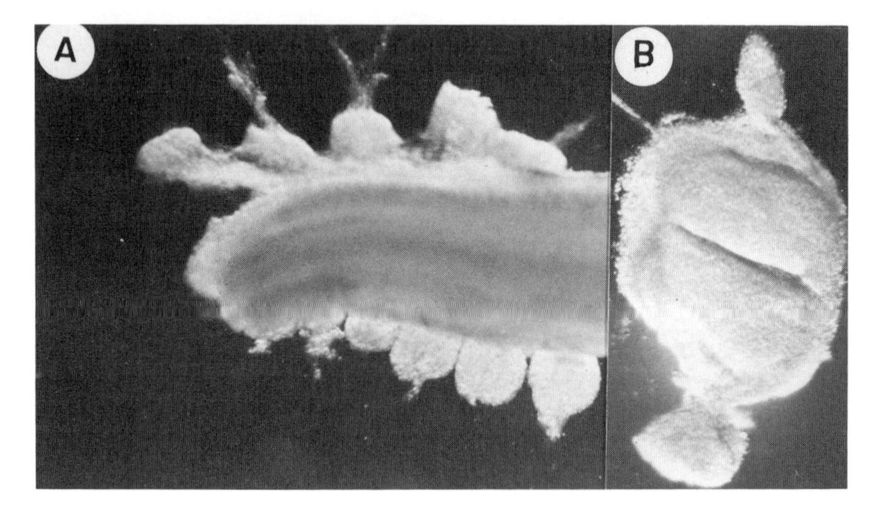

Fig. 1 . A, Part of a dissected spinal cord and B, a single slice with attached dorsal root ganglia (DRG). A = x66; B = x105.

MATERIALS AND METHODS

When dissected with meninges, the entire spinal cord from 13-14d mouse fetus came out with most of the DRG attached (fig 1A). Uniform slices 400 micron thick, obtained with a McIllwain tissue sectioner, usually contained, beside the spinal cord, also 1 or 2 attached DRG (fig 1B). Single slices were explanted on 12x24 mm coverslips and were grown as organotypic cultures by means of the roller tube technique as previously described (Gahwiler 1981; Shahar et al. 1986).

Cultures were selected for experiments after being cultivated for about three weeks. By this time both CNS and PNS culture components have already accomplished most of their growth and differentiation pattern and the cultures exhibited heavy central and peripheral myelination.

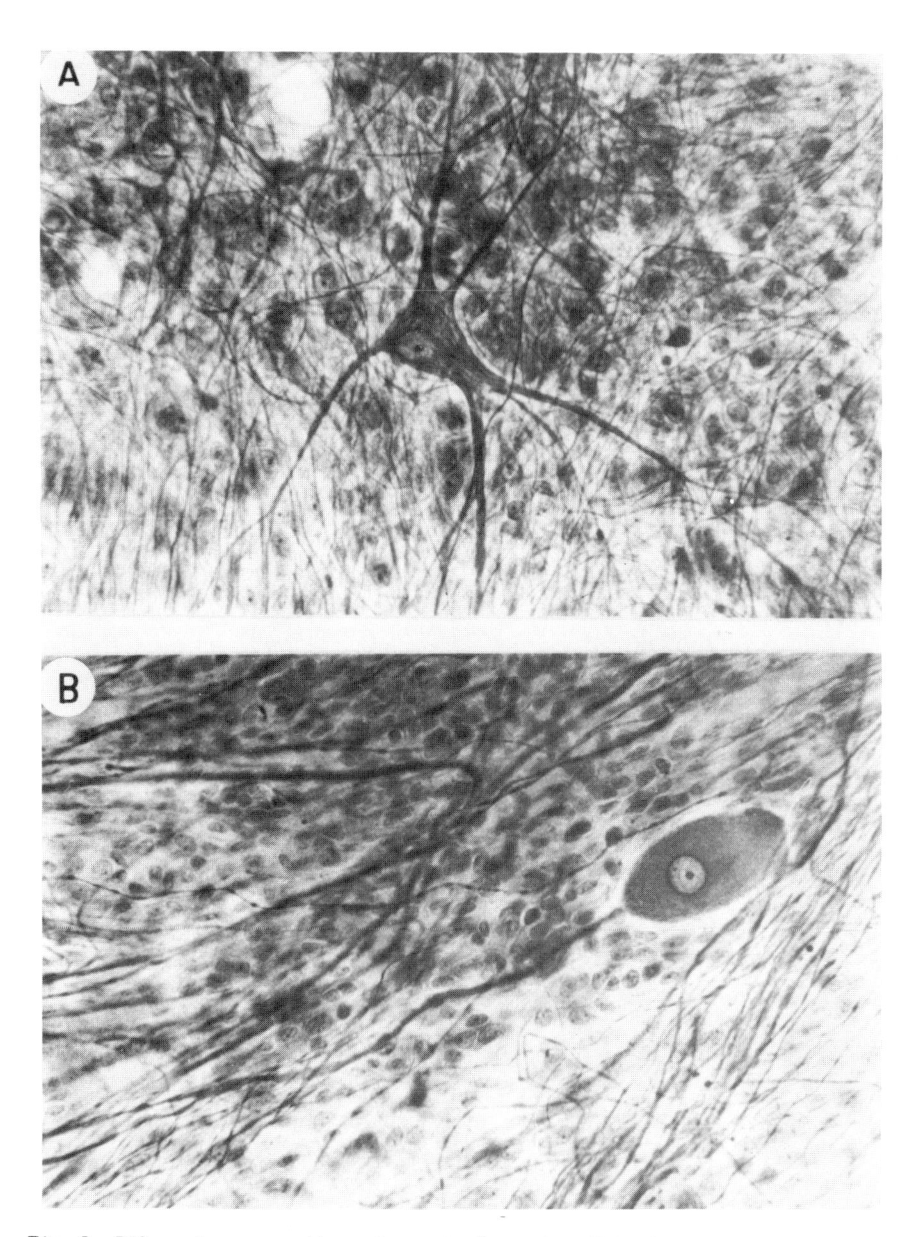

Fig 2. Silver impregnation of a spinal cord multipolar neuron in A and a pseudounipolar neuron from the DRG in B. 24 days in vitro (div) x400

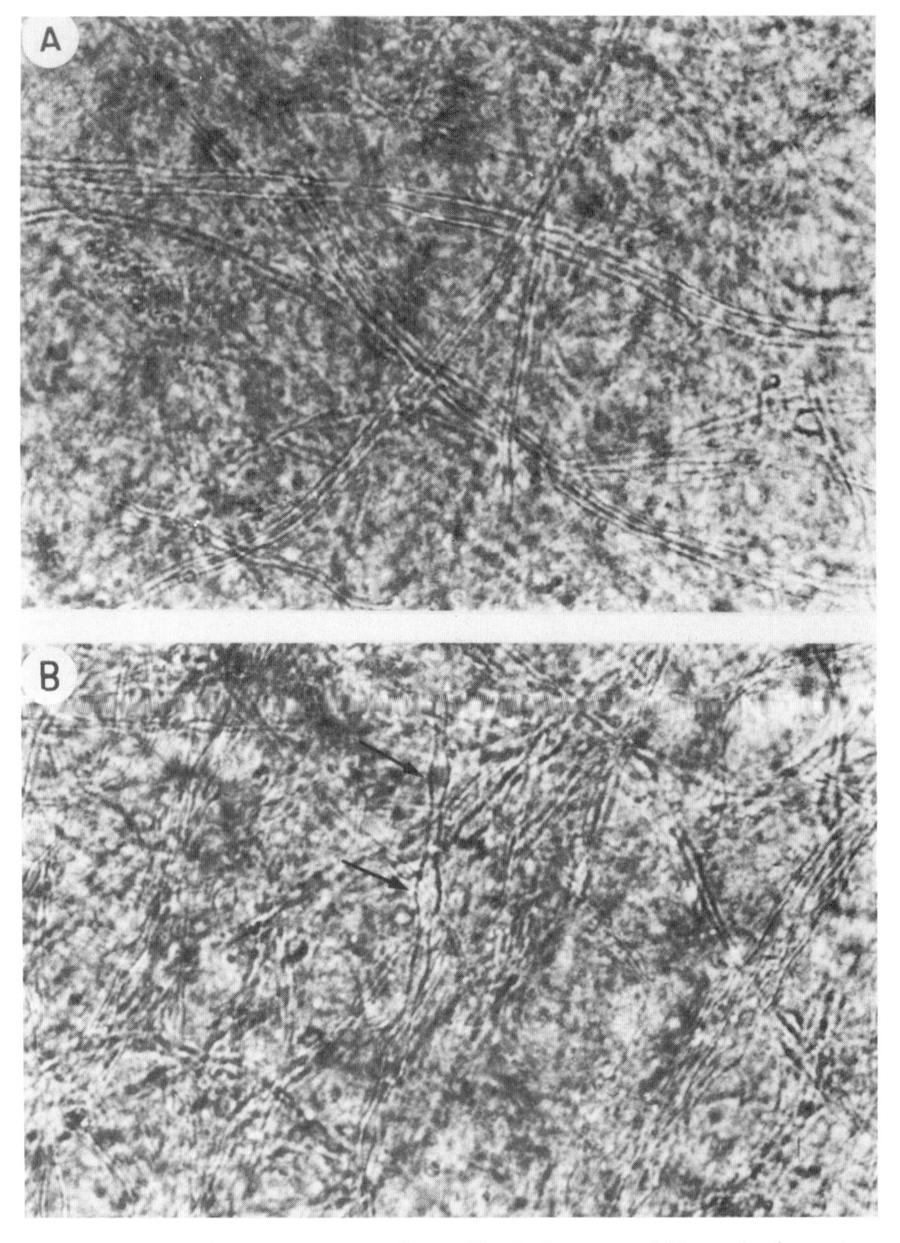

Fig.3 . Bright-field microscopy of myelinated axons of the spinal cord (central myelin):

A. Control culture

B. Swelling of myelin (arrows) 4 days after viral infection. 27 (div) x400

Cultures were infected by adding 0,1 ml of WNV ($5x10^6$ PFU) into their nutrient medium. The infected cultures were placed in a roller drum at 37¬C for 3h. At the end of the absorbtion period the medium was replaced with fresh culture medium.

At designated time points, triplicates of cultures were processed for light microscopical examination and for electron microscopy. Simultaneously, samples of the culture medium were harvested for virus plaque assay in vero cell cultures. (Ben-Nathan et al. 1989).

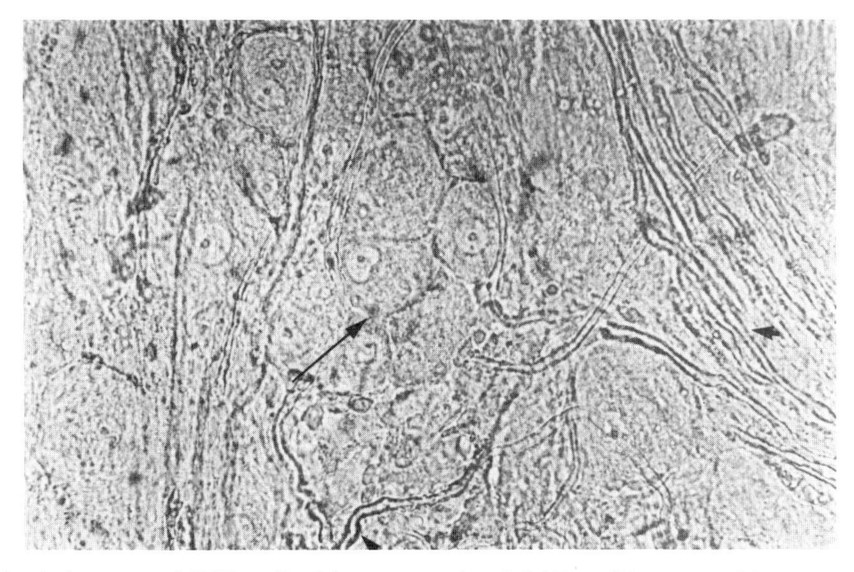

Fig.4. A group of DGR cells (long arrow) exhibiting fibers myelinated with rippled peripheral myelin (short arrows), 24 (div) x400.

RESULTS AND DISCUSSION

By the end of the first week, there was an intensive fiber outgrowth and cell migration in the cultures causing the slices to flatten; this facilitated visualization of perikarya, glia cells and nerve fibers (Fig 2). During the second week in culture synaptic interconnections were established and spinal cord axons became progressively myelinated with central myelin (fig 3A). The large neurons of the DRG, recognized by their round form and central euchromatic nuclei, appeared usually in groups. They became myelinated, with thick and rippled peripheral myelin, toward the third week in culture (fig 4).

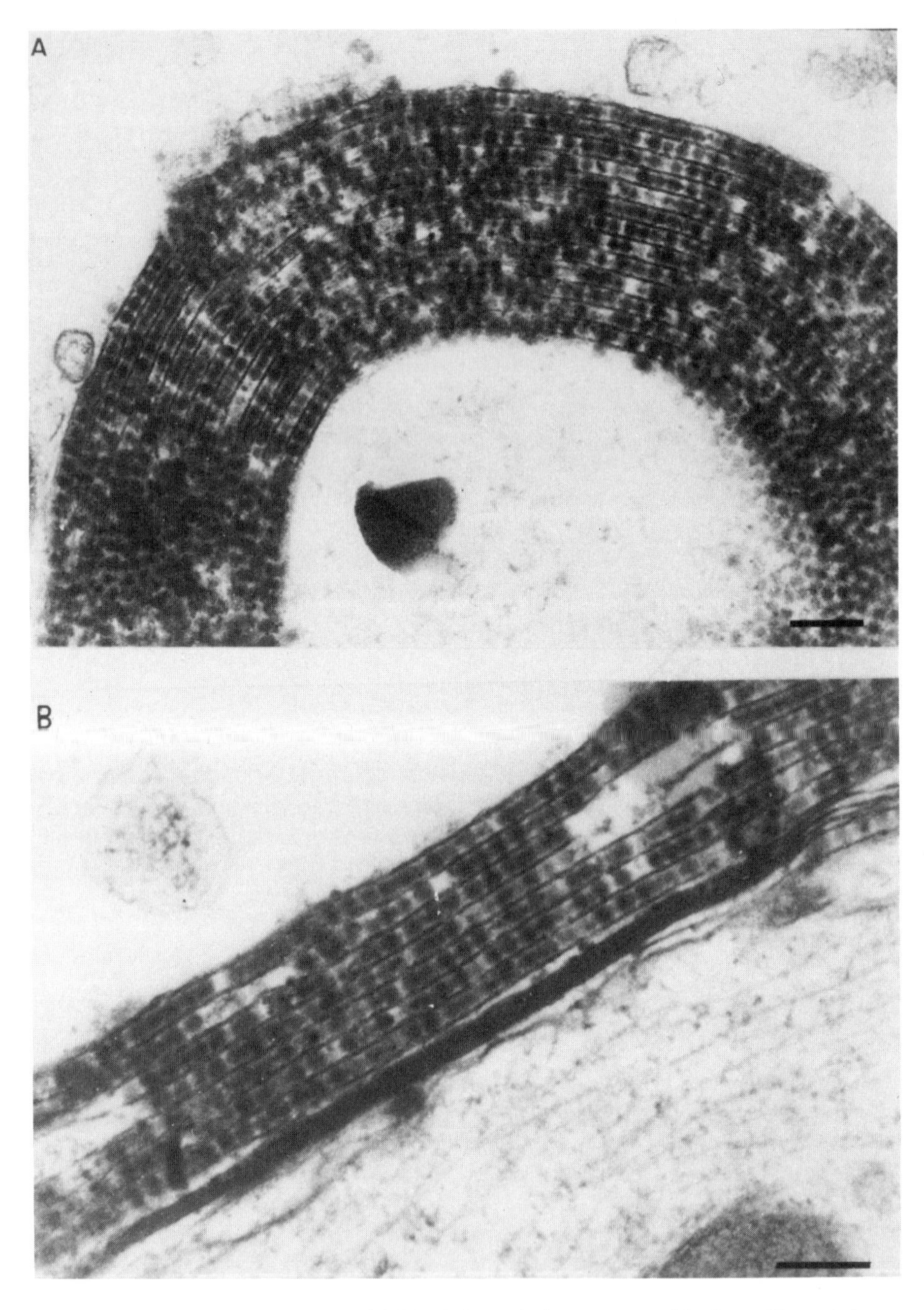

Fig.5 . Electron microscopy of WN virions between myelin lamellae 4 days after viral infection :

A. Virions aligned among all lamellae composing the myelin sheath

B. Arrangement of virions only between peripheral lamellae of the myelin sheath. Bars = 0.2 μ

This type of slice culture has several advantages: In the first place it allows a combined growth and differentiation of the two components: the CNS component which is the spinal cord and the DRG which represents the PNS. In addition, although the organotypic organization of the culture is preserved, there is an easy and direct access to individual neurons. Immunocytochemical and other kinds of staining are possible due to flattening of the slices. Finally, central and peripheral myelinated axons can be visualized and distinguished by bright-field microscopy, and events such as demyelination or reversible remyelination can be followed over a considerable period of time.

Light microscopy of cultures infected with WNV showed, during the first three days in vitro, only minor cytopathic alterations which consisted in granulation of a few cells and swellings along some of the central myelinated axons (Fig. 3B). However, the myelin sheath of most central and peripheral axons appeared undamaged. Electron microscopy, made during the first three days after viral infection, did not reveal WNV in any of the myelin forming cells but only in neurons. Furthermore, WN virions were found in the cytoplasm and axons of only about 10% of the neurons (Shahar et al. 1990). This fact may indicate that special type(s) of neuron(s) support WNV replication, presumably nerve cells with a definite type of neurotransmitter. If so, neurotransmitter receptors may simultaneously serve also as receptors for WNV. A similar explanation has been suggested for the adhesion and uptake mechanism of Borna and rabies viruses by neuronal elements (Gosztonyi and Ludwig 1984).

Nevertheless, a number of undamaged neurons, which did not contain WN virions, as well as several myelinated central and peripheral axons, were observed in cultures even 7 days after viral infection.

The most striking electron microscopical observation was made on day 4 after WNV infection, which coincided with the highest rate of viral replication (Shahar et al. 1990). In several central myelinated axons, virions were arranged in long rows between myelin lamellae (Fig 5). Virions were aligned either between all lamellae composing the myelin sheath or only between the outermost lamellae. This depends probably on the myelinating state in which the axons were at the time of viral infection.

The peculiar arrangement of virions along the interperiod lines is indicative of their exterior location. It is possible that following their replication in neurons, WN virions adhere also to oligodendrocytes. Such arrangement of WN virions between myelin lamellae, was observed only in the central myelin but not in the peripheral myelin sheath, which is surrounded by a basement membrane.

The fact that the alignment of virions between myelin lamellae is observed only in a number of central axons might indicate the possible existence of certain oligodendroglia to which the WN virions are preferably attracted.

Similar alignment of virions between adjacent myelin lamellae, has been reported by Mazlo and Tariska 1980 in a study on replication of Polyoma virus in oligodendroglia cells of a human brain, in a case of Progressive Multifocal Leucoencephalopathy.

Unlike WNV, other encephalitic viruses such as Sindbis and Theiler's virus, induced severe cytotoxicity and demyelination already within 48h after infection of cultures.

Theiler's virus has been shown to replicate specifically in the myelin forming cells (Frankel et al. 1986; Shahar et al. 1986).

Finally this entrapment of WN virions between myelin lamellae from the extracellular space, without necessarily entering or infecting the oligodendrocyte, might provide a novel concept for a possible mechanism underlying slow viral infection.

REFERENCES

Ben-Nathan D., Lustig S., and Feverstein G., 1989, The influence of cold or isolation stress on neuroinvasivness and virulance of an attenuated variant of West Nile virus. Arch. Virol., 109:1-10.

Chamberlain R.W., 1980, Epidemiology of arthopod-borne togaviruses: the role of arthopods as hosts and vectors and of vertebrate hosts in natural transmission cycles, in: The togaviruses biology, Structure replication. Schlesinger R.W.(ed), Academic Press New York, pp 157-227.

Frankel G., Friedmann A., Amir a., David Y., and Shahar A., 1986, Theiler's virus replication in isolated Schwann cell cultures. J. Neurosci. Res., 15: 127-136.

Gahwiler B.H., 1981, Organotypic monolayer cultures of nervous tissue. J. Neurosci. Methods., 4: 329-342.

Goldblum N., 1959, West Nile fever in the Middle East. Proc. 6th Int. Congress Tropical Med. Malaria., 8: 112-125.

Goldblum N., Strek V.V., Paderski B., 1954, West Nile fever: The clinical features of the desease and the isolation of West Nile virus from the blood of nine human cases. Am.J. Hyg., 59: 89-103.

Gosztonyi G., and Ludwig H., 1984, Neurotransmitter receptors and viral neurotropism. Neuropsychiatr. Clin., 3: 107-114.

Hayes C.G., Bagar S., Ahmed T., Chowdhry M.A., Reisen W.K., 1982, West Nile virus in Pakistan, I. Sero epidemiological studies in Punjab province. Trans.R.Trop.Med.Hyg., 76: 431-435.

Marburg K., Goldblum N., Sterk V.V., Jasinka-Klinberg W., Klinberg M.A., 1956, The natural history of West Nile fever, I. Clinical observations during an epidemic in Israel. Am.J.Hyg., 64:259-269.

Mazlo M., and Tariska I., 1980, Morphological demonstration of the first phase of Polyoma virus replication in oligodendroglia cells of human brain in Progressive Multifocal Leucoencephalopathy (PML). Acta Neuropathol. (Berlin)., 49: 133-143.

Monath T.P., 1986, Pathobiology of the Flaviviruses, In: The Togaviridae, Schlesinger S., & Schlesinger M.J., (eds) Plenum New York pp375-440.

Shahar A., Frankel G., David Y., and Friedmann A., 1986, In vitro cytotoxicity and demyelination induced by Theiler viruses in cultures of spinal cord slices. J.neurosci.Res., 16: 671-681.

Shahar A., Lustig S., Akov Y., David Y., Schneide P., and Levin R., 1990 A, Different pathogenicity of encephalitic Togaviruses in organotypic cultures of spinal cord slices. J.Neurosci.Res., 25: 345-352.

Shahar A., Lustig S., Akov Y., David Y., Schneider P.,

Friedmann A., and Levin R., 1990 B, West Nile virions aligned along myelin lamellae in organotypic spinal cord cultures. J.Neurosci.Res., 26:

Weiner L.P., Cole G.A., Nathanson N., 1970, Experimental encephalitis following peripheral inoculation of West Nile virus in mice of different ages. J.Hyg., 68: 435-446.

HUMAN FETAL BRAIN CULTURES: A MODEL TO STUDY NEURAL PROLIFERATION, DIFFERENTIATION AND IMMUNOCOMPETENCE

Silvia Torelli, Valeria Sogos, Maria Grazia Ennas, Costantino Marcello*, Domenico Cocchia** and Fulvia Gremo

Dept. Cytomorphol., *Inst. Obst. & Gynecol., Cagliari, **Inst. Anat. 2nd Med. School, Rome, Italy

INTRODUCTION

In vitro cultures provide a very useful tool to investigate problems related to the development of the Central Nervous System (CNS). Since the beginning of this century (Harrison 1907, Levi 1929) the properties of neuronal and glial cells have often been studied both in explants and dissociated cultures. However, most of these studies utilize avian or non-human mammalian sources hence the results obtained do not necessarily apply to the human species.

In the last few years, a limited number of laboratories have studied some aspects of the morphological, biochemical and electrophysiological properties of *in vitro* human fetal CNS (Kennedy et al., 1980; Kato et al., 1985; Kim et al., 1988). However, most of the data have been obtained from spinal cord and the range of gestational weeks has been restricted to 8-12. We have developed human fetal brain cultures from 8 up to 20 weeks of gestation (Gremo et al., 1986; Ennas et al., 1990; Torelli et. al., 1990a)

In this paper, we discuss the usefulness of this type of cultures to approach the problem of neuronal proliferation, differentiation and immunocompetence during the development of the human CNS.

MATERIALS AND METHODS

Preparation of Tissue Cultures

Cultures have been prepared as previously described (Gremo et al.,1986; Ennas et al.,1990).Human brains, obtained from either spontaneous or medically induced abortions, were freshly dissected and, when possible, separated into brain areas. Tissues were incubated with 0.2% sterile trypsin and dissociated into single cells by flushing through a Pasteur pipette. The cell suspension was diluted to 2×10^6 cells/ml in DMEM (Dulbecco's Modified Eagle Medium)

Plasticity and Regeneration of the Nervous System
Edited by P.S. Timiras *et al.*, Plenum Press, New York, 1991

plus 10% FCS (fetal calf serum). Cells were plated on plastic culture dishes pretreated with polylysine (PL). PL-dishes were prepared according to Pettman et. al. (1979).

Immunocytochemistry

Glass coverslips (12 mm in diameter) were dropped into the plastic dish before PL treatment. At different times, coverslips were removed from dishes and routinely fixed at -20° C for 4 minutes and kept at -20° C until staining. For the staining, samples were washed with Tris-HCl buffer (0.05 M,. pH 7.4) plus Triton X-100 (0.4%). The excess of liquid was dried out and samples were covered with 20 µl of antibody solution (2 µg/ml final concentration in buffer). After incubation at room temperature for 30 minutes, samples were washed three times with buffer. Then, they were incubated at room temperature for 30 minutes with 20 µl of fluorescent anti-mouse antibodies. For double staining, after the two-step incubation with the monoclonal plus fluorescent anti-mouse antibodies, a second two-step incubation with polyclonal plus rhodamine-labeled anti-rabbit antibodies (1:20) was performed. After three washes, samples were dried and mounted. In order to test the effects of fixation on antigen preservation, some samples were incubated unfixed with either the first or with the first and the second antibody and then fixed with methanol as above. In both cases preincubation with goat serum was used in order to block non specific labeling.

Autoradiography

Cells grown on PL pretreated coverslips were incubated with different doses (0.01-2 µCi/ml) of ^{3}H-thymidine (25 Ci/mmole, New England Nuclear) for 1-6 days and fixed in cold methanol (4°C for 4 minutes) at different times. The coverslips were mounted on glass slides, stained with monoclonal or polyclonal antibodies for neurofilaments (NF), and glial fibrillar acidic protein (GFA-P) and then processed for autoradiography.
Slides were dipped in Ilford K emulsion, dried and stored in the dark at 4° C for 15-20 days. The slides were then developed in Kodak Dektol developer for 5 minutes at 20° C and fixed in Kodak Unifix fixer for 5 minutes at 20° C. After that, slides were washed and dried, and a second coverslip was placed over emulsion covered cells using a drop of glycerol-PBS and sealed with nail polish.

Scanning Electron Microscopy (SEM)

Cultured cells were prefixed in a solution containing 0.1% glutaraldehyde in 0.1 M cacodylated buffer plus 0.1% sucrose for 15 minutes at 37° C and for additional 45 minutes at room temperature. Samples were then post-fixed with 2% glutaraldehyde in cacodylate buffer for 60 minutes at room temperature. After 3 washes in buffer and 3 more washes in distilled water, samples were dehydrated with acetone and critical point dried using carbondioxide. Specimens were then briefly coated (1.5 minute) with gold palladium in a 5100 "cool" Polaron sputtering apparatus and observed with an ISI SS40 Scanning Electron Microscope operating at 20 Kv.

Transmission Electron Microscope Immunocytochemical Staining (TEM)

Cultured cells were fixed in a solution containing 4% paraformaldheyde in 0.1 M phosphate buffer (pH 7.4) for sixty minutes. After several washes with buffer, cells were processed for immunoperoxidase reaction using the peroxidase-antiperoxidase (PAP) method (Sternberger et al., 1970). Dilution of the reagents in PBS and reaction times were as follows: monoclonal antibody (1:20); preimmune goat serum (1:50); anti-mouse serum (1:50) for 1 h; PAP complex (1:100) for 1 h; 3,3'-diamino-benzidine-4HCl in 0.05 M Tris-HCl buffer (ph 7.4) (0.4 mg/ml), containing 0.05% H_2O_2 for 15 minutes. At the end of each step, the cells were washed 5-6 times in PBS over a period of 30 minutes. The cells were postfixed in 1.5% OsO_4 in 0.1 M phosphate buffer for 30 minutes, dehydrated in ethanol and propylene oxide and embedded in Epon 812. Sections were prepared using a LKB ultratome III, counterstained with lead cytrate, and observed with a Philips EM 300.

RESULTS

Growth and Morphological Features of Brain Cultures

Dissociated cells from 8 to 20 week-old embryo brains attached within a few hours to the pretreated surface. Within 24-30 hours, there were small processes extending from the cells and these increased in length and thickness with time in culture. Fig.1a is a representative picture of human brain cells prepared from 9 week-old embryonic tissue and cultured for 7 days. Neuronal cells form aggregates which generally grow on the top of flat, non-neuronal cells. Seldom, a few neurons develop isolated on the substratum (arrow). After 20 days in culture (fig.1b), a well developed network of processes, some of which showed varicosities (arrow) could be observed. In fig.1c, neurons with processes extending through a thick layer of cells and fibers are showed. As described in materials and methods, cells were submitted to cell passages through trypsinization, which, however, did not affect neuronal maturation and morphological differentiation.

Immunocytochemistry of Human Cultures

Results about immunocytochemical characterization of brain cultures are summarized in Table I. As shown by positivity for both neurofilaments (NF) and neuron-specific-enolase (NSE), neurons grew in all the cultures regardless the age of the embryo. Their percentage increased up to the 4th week in culture, then it started decreasing. However, neurons could be observed up to 3 months *in vitro* (Table II). Positivity for neurofilament subunit 200K preceeded positivity for NSE and appeared already after 24 hours in culture, in correlation with the development of processes described before. This observation is in agreement with our previous results on chick embryo retina cultures in which positivity for neurofilaments could be seen in a few hours in culture (Torelli et al., 1989).

Early positivity for vimentin (VT) could also be observed.

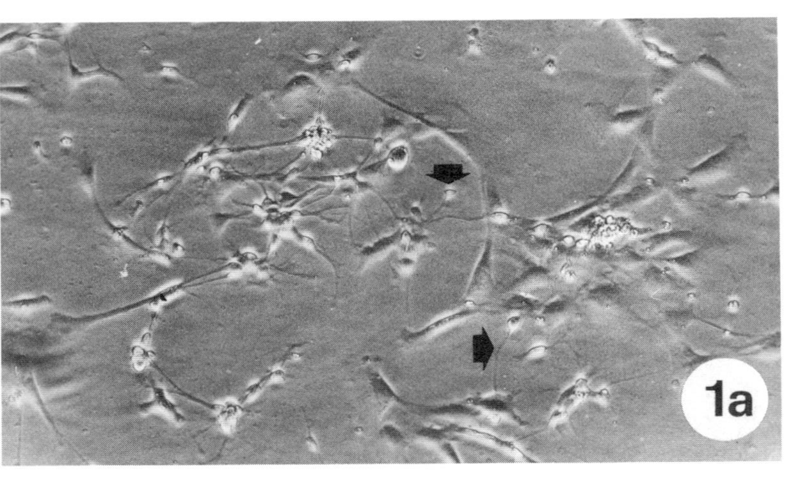

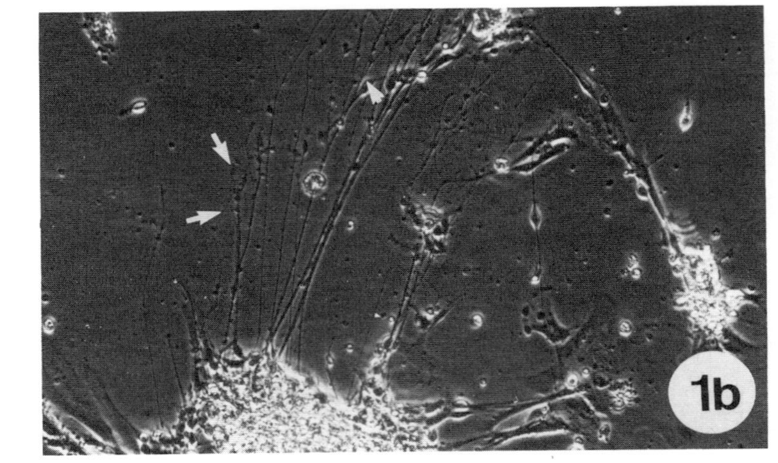

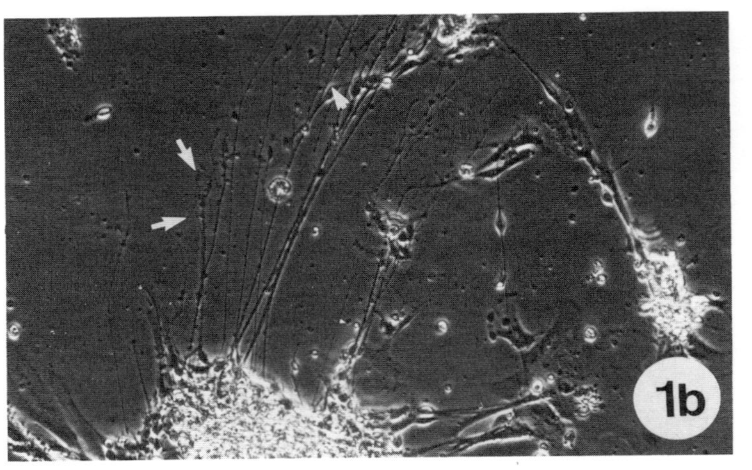

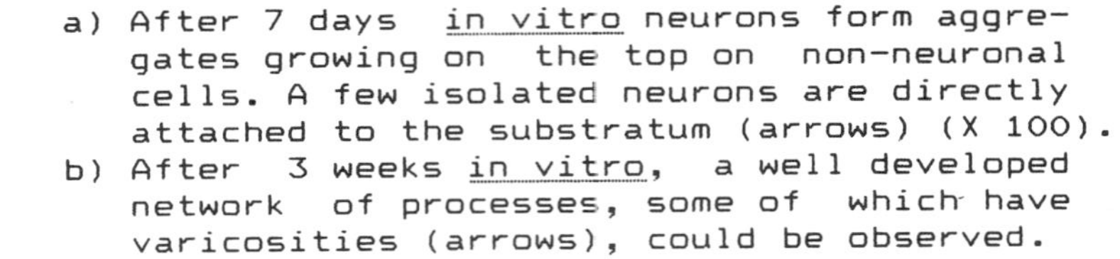

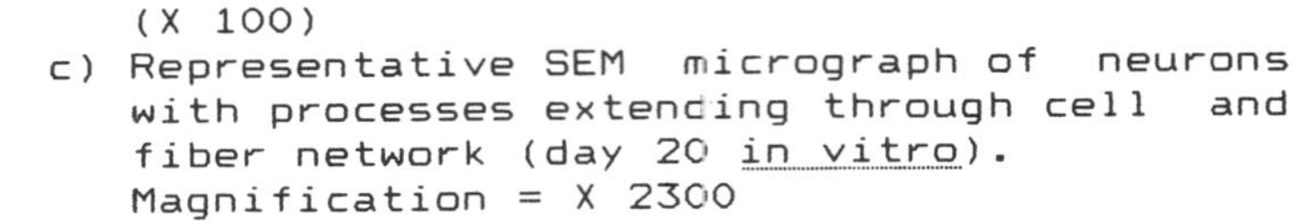

Fig.1 Morphological features of human fetal brain cultures (9 week old embryo)

a) After 7 days <u>in vitro</u> neurons form aggregates growing on the top on non-neuronal cells. A few isolated neurons are directly attached to the substratum (arrows) (X 100).

b) After 3 weeks <u>in vitro</u>, a well developed network of processes, some of which have varicosities (arrows), could be observed. (X 100)

c) Representative SEM micrograph of neurons with processes extending through cell and fiber network (day 20 <u>in vitro</u>). Magnification = X 2300

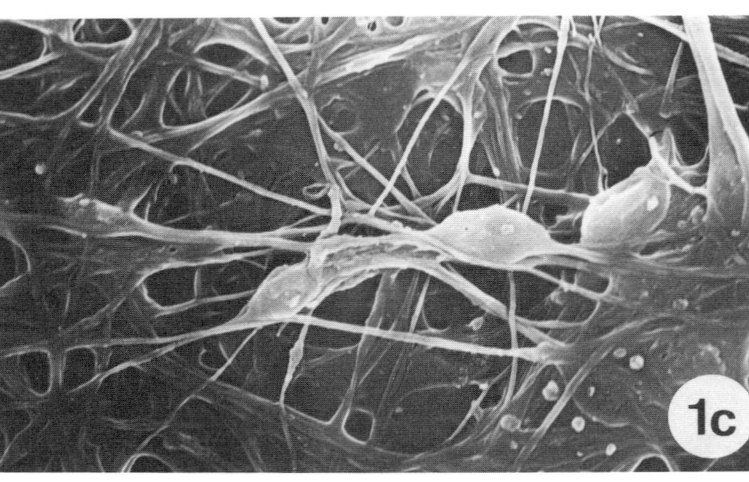

TABLE I

Immunocytochemical characterization of human fetal brain cultures

Embryo Age (Weeks)	n	NF	NSE	VT	GFA	FN	Fc	VIII
7 - 8	2	+	+	+	+	–	–	–
10 - 11	2	+	+	+	+	–	–	–
12	8	+	+	+	+	–	–	–
13 - 15	5	+	+	+	+	–	–	–
16 - 17	2	+	+	+	+	–	–	–
18 - 20	4	+	+	+	+	–	–	–

Immunocytochemical staining of cultures was performed as described in materials and methods at day 15 *in vitro* n = number of cultures. For other abbreviations see the text.

VT antibodies labeled neuron-like as well as non-neuronal cells within a few hours in culture. Interestingly, some cells bearing very long and thin processes which stretched from one to the other flat cell were positive for VT. Later, these or similar cells became positive for glial-fibrillar acidic protein (GFA-P) also. Indeed, double staining for NF and VT or GFA-P and VT (not shown) demonstrated that some of VT^{+} cells were either neurons or astrocytes. With time in culture, the percentage of VT^{+} cells decreased, presumably as a consequence of *in vitro* cell maturation. Negativity for factor VIII (VIII) excluded contamination of our cultures from entothelial cells. Also fibronectin (FN) was virtually absent in our cultures, when cells were stained unfixed and Triton X-100 was omitted from the washes. This result means that no fibroblasts were present in the cultures. When cells were fixed with methanol and permeabilized with Triton, positivity for FN was detected in the cytoplasm of non-neuronal cells. Double staining for GFA-P indicated that they were astrocytes. After 7-8 weeks in culture, FN was detectable outside the cells, presumably as a consequence of secretion from astrocytes, as suggested by Tomaselli et al. (1986). Galactocerebroside (GC) negativity excluded the presence of differentiated oligodendrocytes. This result is in agreement with the observations of Elder and Major (1988), who also noticed the absence of oligodendrocytes in fetal human brain cultures.

TABLE II

Neurofilament positivity in human brain cultures

Embryo Age (Weeks)	n	Days in culture 0 - 5	15	30	60	90
10 - 12	4	+++	+++	+++	+++	+
18 - 20	3	++	+++	+++	++	+

+ = weakly positive ++ = positive +++ = very positive. n = number of cultures.

Timing in GFA-P expression

As shown in Table III, some GFA-P positive staining was detected after 10-15 days in culture in younger embryos. In older embryos, the expression of GFA-P was more precocious. GFA-P+ astrocytes increased with time in culture and cell passages, so that after 2.5 months over 90% of cells were GFA-P+ astrocytes. GFA-P positivity was mantained up to 6 months, after which GFA-P was not detectable anymore. Our data are in agreement with Kato et al. (1985), who also observed a delay in non-neuronal cell morphological differentiation in 8-9 week-old embryo dissociated human spinal cord cultures. On the other hand, Paetau (1988) reported the absence of GFA-P positivity after repeated subcultures of human embryo glial cells. However, the latter study was performed on two 16-week-old fetuses only and no polylysine was used as a substratum, which may account for these different results. Interestingly, differences in GFA-P distribution was observed through time in culture. In the first 10-14 days in culture or a few days after cell passages, GFA-P positivity was restricted to a few either round-shaped cells or elongated, bipolar cells (fig. 2a). Alternatively, some diffuse staining could be detected in the cytoplasm of a few cells (fig. 2b), but the protein was not organized in filaments. On the contrary, a full, well recognizable positivity of intermediate filaments for GFA-P was detectable only between the second and the third week in culture (fig.2c)

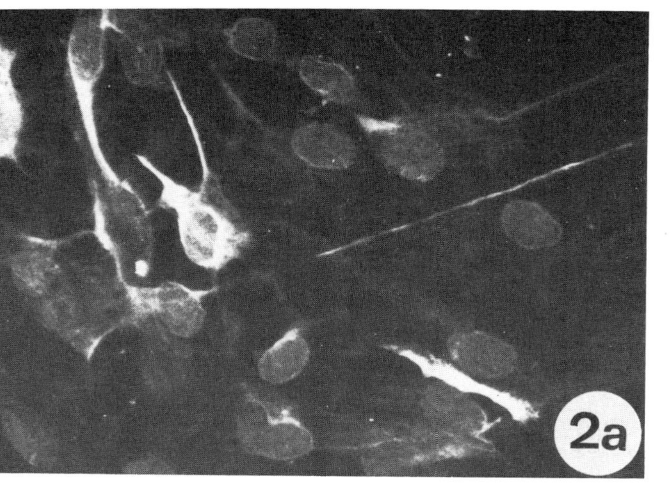

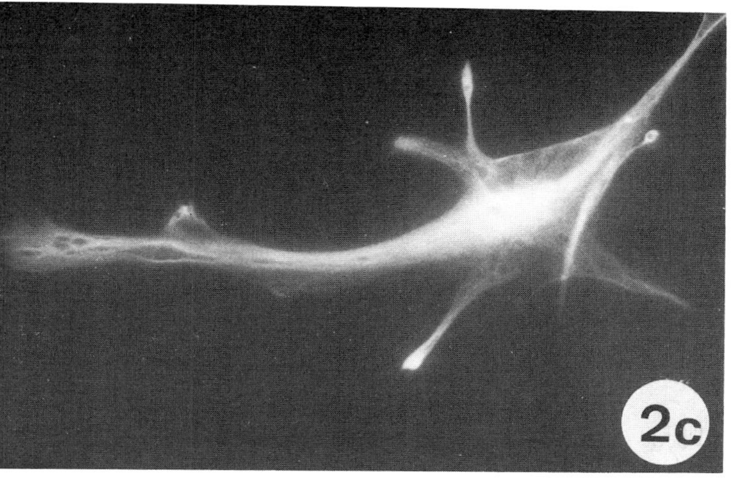

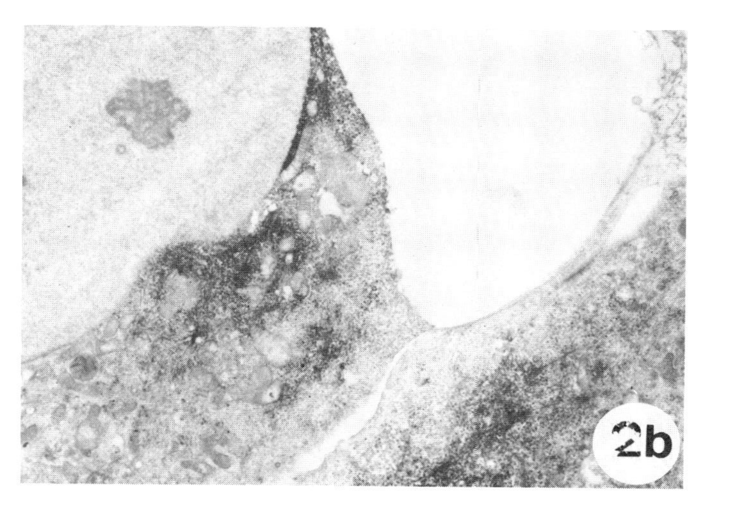

Fig.2 GFA-P staining of human brain cultures.

a) After a few days *in vitro*, GFA-P is present only in some round-shaped or bipolar cells (X 400).
b) TEM micrograph showing that GFA-P (arrow) is detectable in the cytoplasm, but it is not organized in filaments (X 4800).
c) A typical GFA-P positive astrocyte (day 21 *in vitro*) (X 500).

TABLE III

Timing in GFA-P expression in human fetal brain cultures

Embryo Age (Weeks)	n	Days in culture 3-5	10	15	30	120
7-8	2	-	-	+	++	+++
9-10	1	±	±	+	++	+++
11-12	9	n.d.	±	+	++	+++
13-15	5	n.d.	+	+	++	+++
18-20	4	±	+	+	++	+++

± = partially positive ; + = positive ; ++ = very positive; +++ = intensively positive ; n = number of cultures ; n.d. = not determined

Presence of dividing neuroblasts in fetal human brain cultures

In the past, a common assumption was that neurons did not divide in cultures prepared from fetal and neonatal CNS tissues, but only differentiated under proper conditions (Shubert et al., 1973). More recently, the presence of proliferating brain cells has been shown in dissociated cultures of chick embryo (Sensenbrenner et al. 1980; Fedoroff et al., 1982) and fetal rat brain (Schrier and Shapiro, 1974; Kriegstein and Dichter, 1984; Asou et al., 1985). We consequently addressed the question whether or not human fetal cultures contained dividing neuroblasts. Results obtained with double staining with anti-proliferating cell antibody (PC) and neurofilaments demonstrated that in 12-13 and 18 week-old brain cells PC^+/NF^+ cells were present after 6-10 days in culture. As shown in fig. 3, the dividing cells were mainly in the aggregates. In order to confirm these data, we performed a group of experiments with incorporation of ^{3}H-thymidine (Sogos et al., 1989). As explained in materials and methods, cells were stained for NF or GFA-P before autoradiography. Labeling of neuronal nuclei was observed when cells were incubated with 1-2μCi/ml ^{3}H-thymidine for 5 days (fig. 4). Lower doses or less time of exposure to the nucleoside were ineffective. On the contrary, non-neuronal cells were always heavily stained. Since some neurons incorporated the labeled nucleoside, it presumably means they were ready to undergo mitosis. Even if the percentage of these labeled neurons was small (<2%) and much lower than the percentage of dividing astrocytes, these results demonstrated that some cells of neuronal lineage were retained in the culture and they were capable of synthesizing DNA. These data are in agreement with Sensenbrenner et al. (1980) and with Kriegstein and Dichter (1984) who also observed ^{3}H-thymidine incorporation in neuronal precursors in vitro.

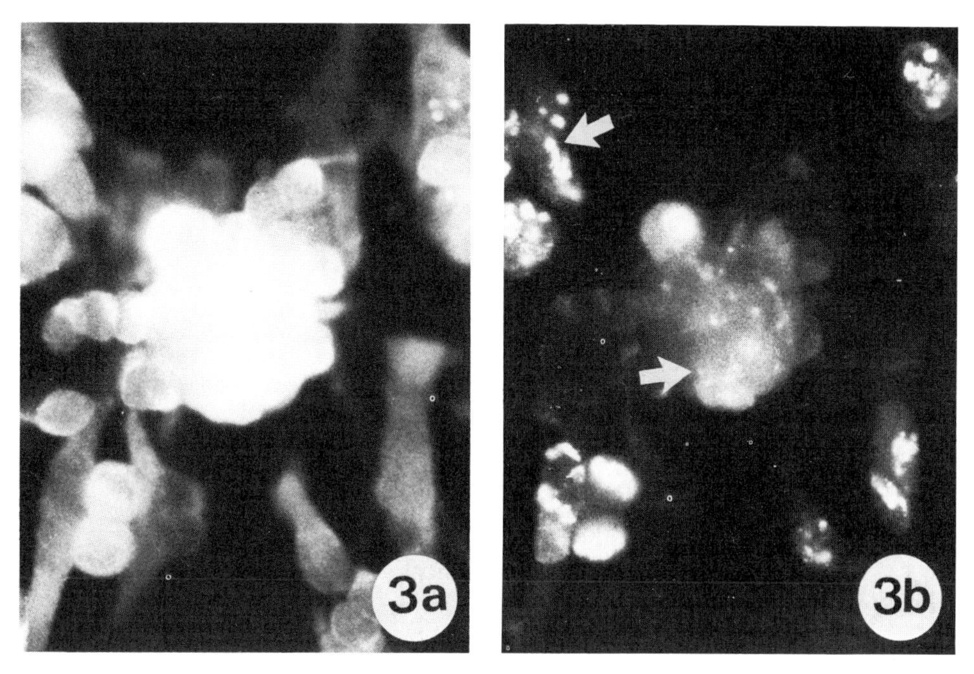

Fig.3 Proliferating neuroblasts in human fetal brain cultures

a) NF antibodies labeled neuronal aggregates
b) PC antibodies labeled neuronal cells at the bottom of aggregates (arrows)
(X 400)

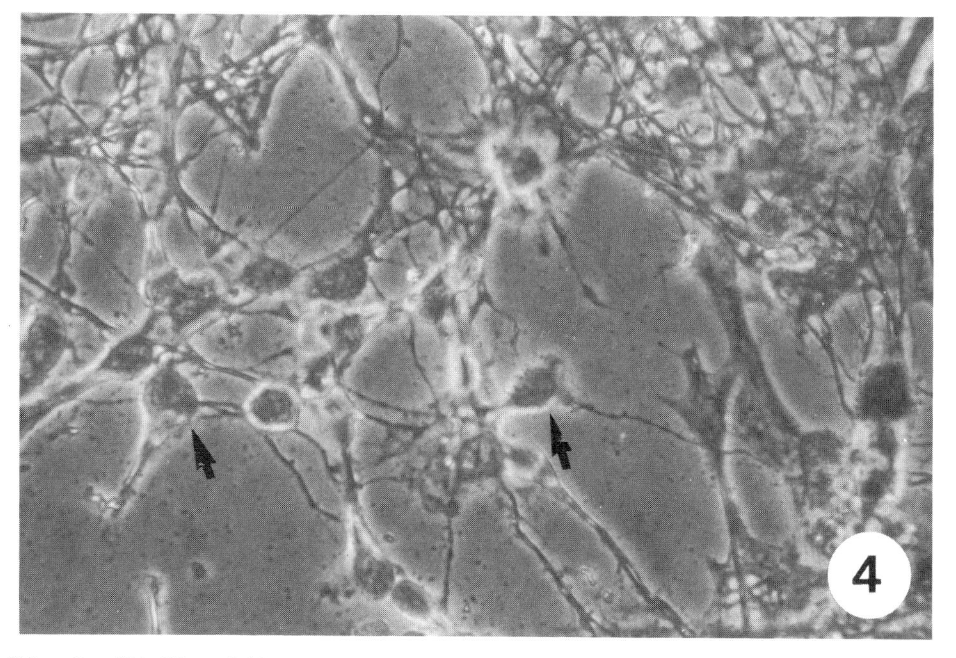

Fig.4 ^{3}H-Thymidine incorporation in human cultured neurons. Some neurons (identified through NF positivity) show autoradiographic granules in the nuclei (arrows). (X 400)

Immunocompetence of cells of the Central Nervous System

The concept that psychological states can affect the outcome of human desease is an old one. However, only recently data on interactions among nervous system and immune system have been accumulated (for an overview see Dunn, 1988). In particular, one of the most exciting hypothesis which has been advanced is about the possible role played by astroglia as immunoregulatory and as antigen-presenting cells. Indeed, some authors demonstrated that astrocytes could present the antigen to T-lymphocytes (Fontana et al., 1984) and the ability of antigen presentation could be enhanced by interferon-gamma (Fierz et al., 1985;Takiguchi and Frelinger, 1988). Moreover, it has been shown that interleukin-I and II were present in the supernatants of cultured astrocytes (Fontana et al., 1982) as well as interleukin III-like factor after stimulation by lipopolysaccharide (Frei et al., 1985). All these data have been collected with animal models and much less is known about the human species. Cultured human glial cells can express antigens of the major histocompatibility complex class II either spontaneously (Marrosu et al., 1990) or under exogenous stimulation (Pulver et al., 1986, Mauerhoff et al., 1988).

We have previously shown that several antigens typical of macrophages and of B and T lymphocytes were present in human brain cells in vitro (Ennas et al., 1990). Whereas only a few of macrophage markers were detectable on human astrocytes, CD21 and CD24, markers of B lymphocytes, could be observed in both astrocytes and neurons in a wide variety of stages of development (Ennas et al., submitted). Moreover, CD4, the cellular receptor for the virus of acquired immunodeficiency syndrome, HIV, was detectable in both astrocytes and neurons in culture. Interestingly, CD4 was present first in neurons, after 7-15 days in culture (Torelli et al., 1990b), whereas $CD4^+$ astrocytes could be detected after 3 weeks in vitro only. Moreover, in neurons CD4 was detectable both on surface and in the cytoplasm.

DISCUSSION

Our data clearly show that it is possible to use human brain fetal cultures as a model to approach several aspects related to CNS development. In particular, they show the possibility to keep long term cultures of both human neurons and astrocytes, without loosing their properties. Indeed, neurons can still be detected after 3 months in culture, which might account either for a continous proliferation of neuroblasts, not affected by cell passages, and/or by continous differentiation of precursor cells. As already mentioned,several authors generated human fetal CNS cultures from spinal cord. The use of the brain instead of the spinal cord could be of some advantage, since we were able to obtain neuron-enriched cultures from fetuses much older then 9-11 weeks. Kato et al. (1985) reported the absence of neurons in cultures after 11 weeks of age. Also Kennedy et al. (1980) reported that cultures prepared with embryonic tissue from the 15th to 21th week contained astrocytes, oligodendrocytes, fibroblasts and macrophages, but none of the cells were positive for tetanus toxin binding. Using the same age of

embryonic human spinal cord, Dickson et al. (1984) found that less than 2% of the cells in culture bound an antibody for neurofilament protein. Kim et al. (1988) grew human neurons up to 6 months. However, they did not report the percentage of neurons present in their cultures, which were explants and not dissociated cultures. Thus, our method allows to obtain long-term, neuronal enriched cultures also from older embryos. This difference might be due to the use of polylysine as a substratum and/or to the source of tissue (brain versus spinal cord).
Our cultures also provide the first evidence that, beside astrocytes, also human neurons can divide in culture. Therefore, it might be possible to develop methods to encourage such a proliferation and ultimately to establish cell lines of the human CNS. In our opinion, it would be a more suitable model then neuroblastoma, or the human cortical line recently established from a patient with megalencephaly (Ronnet et al., 1990).
Moreover, our data demonstrate that human fetal brain cultures can provide useful informations of the possible role of both astrocytes and neurons in the immunological event of the nervous system. As already discussed, the role played by astrocytes as macrophages (Kusaka et al., 1986) or as antigen presenting cells (Fontana et al., 1984) has been demonstrated in rodents. In man, several authors beside us have shown that astrocytes can express the antigen of major histocompatibility complex-class II, HLA-DR (Pulver et al., 1986; Mauerhoff et al., 1988) which might mean that also human astrocytes can present the antigen. However, the expression of HLA-DR could also play a different role. Since in the peripheral immune system lymphocyte CD4 interacts with MHC-class II of antigen presenting cells, it can be postulated that when astrocytes are induced to express HLA-DR, its interaction with neuronal CD4 can provide a signal of activation. Thus, our cultures provide a tool to understand some of the interactions among neurons and glia during the development of the human CNS as well as during pathological events in the adult.

ACKNOWLEDGEMENTS

This work was supported by MPI (40%-60%) and Regione Autonoma Sardegna Grants to F.G. We thank Dr.A.Riva for SEM observations.

REFERENCES

Asou, H. Iwasaki, N., Hirano, S., and Dahl, D., 1985, Mitotic neuroblasts in dissociated cell cultures from embryonic rat cerebral hemispheres express neurofilament protein, Brain Res., 332:355

Dickson, J. G., Flanigan, T.P., and Walsh, F.S, 1984, Antigen expression in human neuronal primary cell cultures and human X mouse neuronal cell hybrids. In "Research Progress in Motor Neuron Disease", F.C. Rose, ed., Pitman Press, Bath

Dunn, A. J., 1988, Nervous system-immune system interactions: an overview, J.Recept. Res., 8: 589

Elder, G. A., and Major, E. O., 1988, Early appearance of

type II astrocytes in developing human fetal brain, Dev. Brain Res., 42: 146

Ennas, M. G., Torelli, S., Sogos, V., Marcello, C., Riva, A.,and Gremo, F., 1990, Immunocompetent-like cells in human fetal brain cultures, In "Trends in Neuroimmunology" M.G. Marrosu, C. Cianchetti, B. Tavolato Eds, Plenum Press, New York, p. 87

Fedoroff, S., Krukofft, T., and Fisher, K., 1982, The development of chick spinal cord in time culture: III. Neuronal precursor cells in culture. In vitro, 18: 183

Fierz, W., Endler, B., Reske, K., Wekerle, H., and Fontana, A., 1985, Astrocytes as antigen-presenting cells. 1. Induction of Ia antigen expression on astrocytes by T cells via immune interferon and its effect on antigen presentation, J. Immunol., 134: 3785

Fontana, A., Kristensen, F., Dubs, R., Gemsa, D., and Weber, E., 1982, Production of prostaglandin E and interleukin 1 like factor by cultured astrocytes and C6 glioma cells, J Immunol.,129: 2413

Fontana, A., Fierz, W., and Wekerle, H., 1984, Astrocytes present myelin basic protein to encephalitogenic T-cell lines, Nature, 307: 273

Frei, K., Boomer, S., Schwerdel, C.,and Fontana, A., 1985, Astrocytes of the brain synthesize interleukine 3-like factor, J. Immunol., 135:4044

Gremo, F., Torelli, S., Sogos, V., Riva, A., Marcello, C., and Lecca, U.,1987, Morphological and immunocyto-chemical characterization of human fetal brain cultures, Soc. Neurosci. Abstract, Vol. 13, part II, p. 1119

Harrison, R. G., 1907, Observations on the living develop-ing nerve fiber, Proc. Soc. Exp. Biol. Med. 4: 140

Kato, A. C., Touzeau, G., Bertrand, D., and Bader, C.R., 1985, Human spinal cord neurons in dissociated monolayer cultures: morphological, biochemical and electrophysiological properties, J. Neurosci, 5: 2750

Kennedy, P. G. E., Lisak, R. P., and Raff, M. C., 1980, Cell type-specific markers for human glial and neuronal cells in culture. Lab. Invest., 43: 342

Kim, S. U., Osborne, D. N., Kim, M. W., Spigelman, I.,Puil, E., Shin, D. H., and Eisen, A., 1988, Long-term culture of human fetal spinal cord neurons: morphol-ogical, immunocytochemical and electrophysiological characteristics, J. Neurosci, 25: 659

Kriegstein, A., and Dichter, M. A., 1984, Neuron generation in dissociated cell cultures from fetal rat cerebral cortex, Brain Res., 255: 184

Kusaka, H., Hirano, A., Borstein, M. B., Moore, G.R.W., and Raine, C.S., 1986, Trasformation of cells in organotypic cultures of mouse spinal cord, J. Neurol. Sci., 72: 77

Levi, G., 1929, Il contributo portato dal metodo della coltivazione in vitro alla conoscenza della strut-tura del tessuto nervoso, Mon. Zool .Ital. XL: 302

Marrosu, M. G., Ennas, M. G., Torelli, S., Sogos V., Puligheddu, P., Lecca, U., and Gremo, F., 1990, Spontaneous expression of Ia antigen (HLA-DR), in cultured cells of human fetal brain at different

stages of development , In "Trends in Neuro-immunology", M. G. Marrosu, C. Cianchetti, B. Tavolato Eds, Plenum Press, New York, p. 103
Mauerhoff, T., Pujol-Borrel, R., Mirakian, R., and Bottazzo, G. F., 1988, Differential expression and regulation of major histocompatibility complex (MHC) products in neural and glial cells of the human fetal brain, J. Neuroimmunol., 18: 271
Paetau, A., 1988, Glial fibrillary acidic protein, vimentin and fibronectin in primary cultures of human glioma and fetal brain, Acta Neuropathol., 75: 448
Pettman, B., Louis, J. C., and Sensenbrenner, M., 1979, Morphological and biochemical maturation of neurons cultured in the absence of glial cells, Nature, 281: 378
Pulver, M., Carrel, S., Mach, J. P., and De Tribolet, N., 1987, Cultured human fetal astrocytes can be induced by interferon-gamma to express HLA-DR, J. Neuroimmunol., 14: 123
Ronnet, G. V., Hester, L. D., Nye, J. S., Connors, K., and Snyder, S. H., 1990, Human cortical neuronal cell line: establishment from a patient with unilateral megalencephaly, Science, 248: 603
Schrier, P., and Shapiro, D., 1974, Effects of fluorodeoxy-uridine on growth and choline acetyltransferase activity in fetal rat brain cells in surface culture, J. Neurobiol., 5: 151
Sensenbrenner, M., Wittemporp, E., Bakakat, I., and Rechenmann, R., 1980, Autoradiographic study of proliferating brain cells in culture, Develop. Biol. 75: 268
Shubert, D., Harris, A. J., Heinemann, S., Kidokoro, Y., Patrick, J., and Steinbach, J. H., 1973, Differentiation and interaction of clonal cell lines of nerve and muscle, In "Tissue Culture of the Nervous System", G. Sato ed., Plenum Press, New York, p. 55
Sogos, V.,Gremo, F., Ennas, M.G., and Torelli, S., 1989, Presence of proliferating neuroblasts in human fetal brain cultures, Bas. Appl. Histochem., 33: 96
Sternberger, L. A., Hardy, P. H. Jr., Cuculis, J. J., and Meyer, H. G., 1970, The unlabeled antibody enzyme method of immunocytochemistry. Preparation and properties of soluble antigen-antibody complex (horseradish peroxidase-antihorseradish peroxidase) and its use in identification of spirochetes, J. Histochem. Cytochem., 18:315
Takiguchi, M., and Frelinger, J. A., 1986, Induction of antigen presentation ability in purified cultures of astroglia by interferon-gamma, J. Mol. Cell. Immunol., 2: 269
Tomaselli, K.J.,Reichardt, L.F., and Bixby, J.L., 1986, Distinct molecular interactions mediate neuronal processes outgrowth on non-neuronal cell surface and extracellular matrix, J. Cell Biol., 1303: 2659
Torelli, S., Sogos, V., Marzilli, M. A., D'Atri, M., and Gremo, F., 1989, Developmental expression of intermediate filament proteins in the chick embryo retina: in vivo and in vitro comparison, Exptl. Biol., 48 187
Torelli, S., Dell'Era, P., Ennas, M. G., Sogos, V., Gremo,

F., Ragnotti, G., and Presta, M., 1990a, Basic fibroblast growth factor in neuronal cultures of human fetal brain, J. Neurosci. Res., 27:000
Torelli,S., Ennas, M.G., Sogos, V., Pilia, E., and Gremo, F., 1990b, *In vivo* and *in vitro* expression of CD4 in fetal human brain, Proc. 20th Meet. Soc. Neurosci., St. Louis, 10/28-11/2

ORIGIN OF MICROGLIA AND THEIR REGULATION BY ASTROGLIA

S. Fedoroff, and C. Hao

Department of Anatomy
University of Saskatchewan, Saskatoon, Saskatchewan
Canada

INTRODUCTION

Astrocytes and oligodendrocytes develop from neuroectodermal epithelial tissue. Astroglia develop along with the neurons throughout embryogenesis (Fedoroff, 1986; Sturrock, 1986); whereas oligodendroglia precursor cells, in mouse and rat, are "dormant" during embryogenesis but early postnatally they proliferate and form mature oligodendrocytes (Wood and Bunge, 1984). However, the development of the third type of glia, the microglia, is still a controversial topic. Early evidence, based on morphology, suggested their development from mesodermal tissue outside the central nervous system. It was thought that the precursor cells originated from the pia mater or tela choroidea and then migrated into the CNS (del Rio-Hortega, 1932; Polak et al., 1982). Another suggestion, based on autoradiography using tritiated thymidine, was that microglia originate from neuroepithelium through glioblasts, as do astroglia and oligodendroglia (Kitamura, 1973; Kitamura et al., 1984).

The morphology of microglia is variable; they can appear as ramified cells with a number of thin processes, as bipolar, spindle-shaped cells or as ameboid or even phagocytic cells (Polak et al., 1982). Because microglia can resemble body macrophages in some respects, they have been included as members of the Mononuclear Phagocytic System (MPS) (Van Furth et al., 1972), implying that the microglia of the CNS originate from monocytes of the hematopoietic system (Ling 1981; Streit et al., 1988) in the same way as all other macrophages of the MPS, but so far there is no direct evidence for this.

Interest in microglia has recently been revived because of the finding that astrocytes and particularly microglia can express on their membranes major histocompatability complex (MHC) antigens, class II, thus qualifying them as antigen presenting cells (Fontana et al., 1987). This, together with the findings that microglia express complement (Mac-1) and immunoglobulin (Fc) receptors and some macrophage-specific surface antigens and that glia cells in the CNS communicate with each other using polypeptide growth factors (cytokines), the same factors used by hematopoietic cells, is leading to an unravelling and identification of the existence of a neuroimmunological network encompassing direct interaction between glia and cells of the immune system (Perry et al., 1985; Guilian, 1987; Hao et al., 1990; Fedoroff, 1990; Fierz and Fontana, 1986; Graeber and Streit, 1990).

The recent observation in our laboratory that macrophage-like cells (microglia) form in astroglia cultures subjected to nutritional deprivation, led to systematic investigation of the origin of macrophage-like cells and their function in culture (Hao et al., 1990a,b,c). The present paper briefly reviews our findings.

FORMATION OF MACROPHAGE-LIKE CELLS (MICROGLIA) IN ASTROGLIA CULTURES

We initiate astroglia cultures from newborn C3H/HeJ mouse neopallium. After 8 to 10 days in culture a monolayer of astroglia forms. In such cultures, better than 95% of the cells are GFAP-positive and less than 1% of the cells have macrophage-specific complement receptor Mac-1. When such cultures are subjected to nutritional deprivation, i.e., when culture medium is no longer changed, after a few days pleomorphic, phase-dark cells appear on the surface of the astroglia. In some areas where the phase-dark cells are found, the astroglia tend to retract from the plastic substratum, thus providing free space for the newly formed cells. When the phase-dark cells adhere to the plastic surface they assume typical ameboid cell morphology and develop a number of vacuoles in their cytoplasm. Time-lapse cinematography shows them to be highly motile. After 10-12 days of nutritional deprivation, the whole culture is populated with these ameboid, macrophage-like cells. At the same time the astroglia retract from the substratum and form clumps of living as well as dead cells which can easily be washed from the cultures. Thus, in nutritionally deprived conditions, in a period of 8-10 days the monolayer culture of astroglia completely transforms into a culture of macrophage-like cells (Fig. 1) (Hao et al., 1990a).

We have observed such a transformation of astroglia in cultures initiated from neopallia of embryos of 9 days of gestation up to birth, or from newborn animals, and conclude that the formation of macrophage-like cells from astrocyte cultures is not related to any specific developmental stage at which cultures are initiated (Hao et al., 1990a).

CHARACTERIZATION OF THE MACROPHAGE-LIKE CELLS

Under scanning and transmission electron microscopes the macrophage-like cells have typical macrophage morphology including irregular nuclei, lysozomes and lamellipodia.The cells are highly phagocytic and ingest latex particles (10μm in size) which in electron micrographs can be seen in secondary lysozomes or surrounded by a membrane.

Astroglia cultures grown under normal conditions have in their medium no or very little detectable lysozyme, a macrophage-specific secretory product, after culture for more than 18-20 days; whereas comparative cultures under nutritional deprivation, which contain macrophage-like cells, have at least 15 times more lysozyme in the medium (Hao et al., 1990a). The lysozyme increases with duration of culturing and the number of cells present.

The macrophage-like cells have some similarlities to spleen and peritoneal macrophages: they do not express GFAP but do express vimentin; they express Mac-1, Mac-3, F4-80, and Ia antigens; they take up Dil-ac-LDL and contain non-specific esterase. They differ from macrophages derived from spleen and peritoneal exudate in that they do not attach as firmly to the plastic substratum (Hao, et al., 1990a).

ORIGIN OF MACROPHAGE-LIKE CELLS

Since the prevailing opinion is that microglia of the CNS form from hematopoietic cells, i.e., the monocytes-macrophages, we systematically investigated for the presence of hematopoietic precursor cells in our cultures. As stated previously, the 10 day monolayer astroglia cultures contained less than 1% of Mac-1 positive cells (macrophages). To eliminate all macrophages present in the cultures before subjecting them to nutritional deprivation, we treated the astroglia cultures twice with antibodies to Mac-1 and guinea pig complement. To prevent the possible escape of some macrophages from the cytotoxic antibody-complement reaction by hiding underneath the astroglia, the cells in suspension were treated first with the antibody and complement, then planted and grown for 10 days under nutritional deprivation. Regardless whether the cultures were treated with the antibody-complement or not, in all cultures macrophage-like cells formed and the lysozyme content in treated and non-treated cultures was approximately the same. This indicated that the few macrophages present in the astroglia cultures were not the source of the macrophage-like cells. Therefore, in our hands, the precursor cells were neither morphologically nor antigenically distinguishable from astroglia (Hao et al., 1990a).

In search of hematopoietic stem cells we used the monoclonal antibodies E3 81-2 (anti stem cell antigen 2) and E13 161-7 (anti stem cell antigen 1) which decorate primitive hematopoietic stem cells (Spangrude et al., 1988). No positively reacting cells were detected in the astroglia cultures.

To further determine whether the astroglia cultures contained hematopoietic precursor cells we used the spleen colony assay method. The cells from astroglia cultures were injected into the tail veins of irradiated (925 rads) C_3Hf/Bi x C57 Bl/6 F_1 hybrid mice. We found that injected cultured astroglia as well as disaggregated neopallial cells did form some spleen colonies. In these experiments we used F1 hybrids as hosts and C_3H/HeJ cells for injection because they had marker chromosomes. Marker chromosomes have no C-chromatin. Cells from male C_3H/HeJ mice, have two marker chromosomes but cells from F1 hybrids have only one marker chromosome. We found that the cells of spleen colonies formed in the irradiated animals had only one marker chromosome, and therefore were of host origin. We concluded that the astroglia cultures did not contain any monocyte-macrophage precursor cells (Hao et al., 1990a).

GM-CFC, (granulocyte-monocyte precursor cells) and monocytes in methyl cellulose cultures in the presence of colony-stimulating factor 1 (CSF-1) and interleukin 3 (IL-3), proliferate and form colonies. We did such experiments using disaggregated cells from our astroglia monolayer cultures. During 16 days of culturing no proliferation of cells was observed, indicating that no macrophage precursor cells were present in astroglia cultures. From all these experiments we concluded that our astroglia cultures do not contain hematopoietic precursor cells.

If the macrophage-like cells in astroglia cultures do not form from hematopoietic cells, the alternative would be neuroectodermal origin. We dissected the neural tube from C_3H/HeJ mouse embryos of Theiler stage 13 (8.5 days old), a stage at which there is no vascularization in the CNS and no myeloid precursor cells are present in the embryo or in the yolk sac (Pera and Feldman, 1977). The first hematopoietic progenitor cells in the mouse embryo are detected in the dorsal aorta at day 10 (Smith and Glomski, 1982). We used neural tissue from 8.5 day embryos; therefore it is reasonable to assume that no

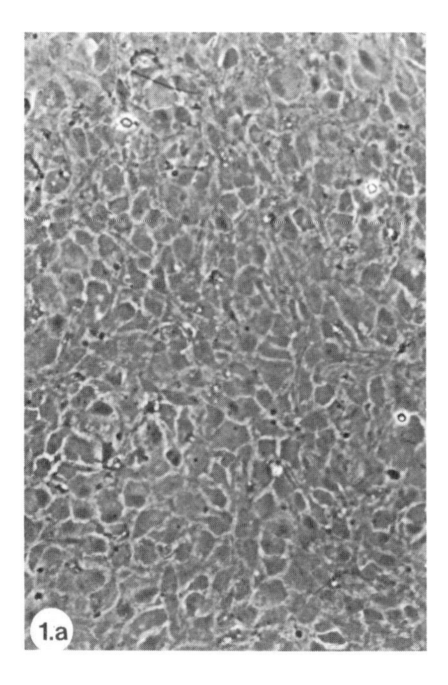

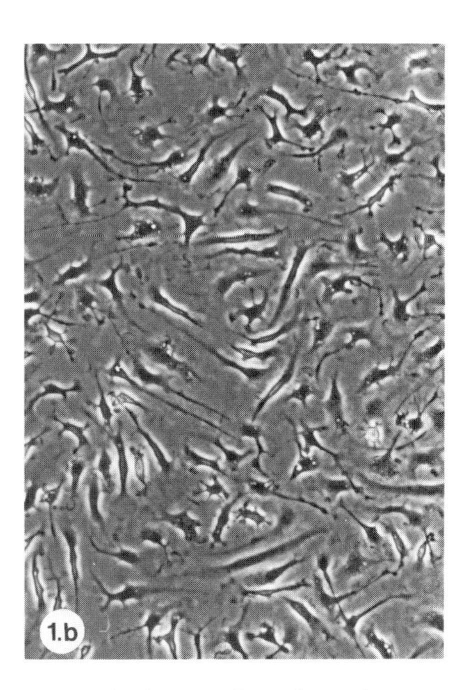

Fig. 1 a) Astroglia culture after 10 days in vitro, medium changed every 2 to 3 days.
b) Similar culture as in a), subjected for 10 days to nutritional deprivation (no medium change). Culture consists of pure populations of macrophage-like cells (microglia).

hematopoietic precursor cells had yet invaded the CNS. To avoid contamination by mesodermal cells we treated the dissected neuroepithelium briefly with trypsin solution (0.25%) to dislodge any mesodermal cells still attached. The neuroepithelium was cut into small fragments, and grown in culture for 20 days with regular feeding every 2 or 3 days. After 20 days the medium was not changed again, and after 10-14 days in such nutritionally deprived culture conditions, many ameboid, macrophage-like cells appeared. The supernatant medium contained lysozyme, indicating the presence of macrophage-like cells. These experiments suggest that neuroectodermal cells give rise to precursors for the macrophage-like cells. Whether these precursor cells belong to a separate lineage or are part of the astroglia-oligodendroglia lineage remains to be determined (Hao et al., 1990a).

We also studied the development of microglia in the cerebral hemispheres of mice. We used the histochemical method for nucleoside diphosphatase (NDPase) to demonstrate the microglia in tissue sections (Novikoff and Goldfischer, 1961; Fujimoto et al., 1989). In newborn animals the hemispheres contained only the rare microglia cell. At 5 days postnatally the number of microglia increased, but the main proliferation of microglia occurred 10 days postnatally. At that time the population of microglia in adult mice is reached. These findings indicate that the origin of microglia in vivo is developmentally regulated and that it occurs late in development. There is some indication that microglia may have a specific topographic arrangement within the CNS rather than random distribution (I. Ahmed, C. Hao and S. Fedoroff, unpublished).

PRODUCTION OF COLONY-STIMULATING FACTOR-1 (CSF-1) BY ASTROGLIA IN CULTURES

For survival and function macrophages require the cell-specific cytokine, CSF-1, and lineage-specific cytokines, IL-3 and GM-CSF. We investigated whether the macrophage-like cells in astroglia cultures require macrophage-specific cytokines. We tested the supernatant medium from astroglia cultures for the presence of CSF-1, IL-3 and GM-CSF using two cell lines: 5/10.14, a macrophage cell line that responds to CSF-1, IL-3 and GM-CSF, and DA1.K, a lymphoblastoma cell line that responds to Il-3, GM-CSF, G-CSF, erythropoietin and LIF (Branch and Guilbert, 1987; Branch et al., 1987; L.J. Guilbert, unpublished). The supernatant medium from the astroglia cultures stimulated proliferation of cells of the 5/10.14 cell line but not those of the DA1.K cell line, suggesting that the supernatant medium did not contain IL-3 or GM-CSF but most likely contained CSF-1. That CSF-1 is the macrophage growth factor produced by astroglia in cultures was confirmed by specific radio-receptor assay (RRA). Using RRA it was determined that astroglia in cultures produced between 150 and 300 CSF-1 units/ml, with the rate of production related to the number of cells present in the cultures. Additional confirmation was obtained at mRNA level. Total RNA was prepared from astroglia cultures and size-separated on agarose gel by electrophoresis. RNA blots were hybridized with mouse CSF-1 cDNA probe (Ladner et al., 1988). Astrocyte RNA, similarly to that of mouse L-929 fibroblasts (positive controls) contained 4kb CSF-1 mRNA (Hao et al., 1990b). Similar observations were made by Théry et al., 1990).

RESPONSE OF MACROPHAGE-LIKE CELLS AND ASTROGLIA TO CSF-1

We determined whether astroglia and microglia express CSF-1 receptor mRNA. CSF-1 receptor is a single subunit of glycoprotein (Mr 165,000), identical or closely related to the c-fms proto-oncogene protein product (Morgan and Stanley, 1984; Sherr et al., 1988). Northern blot analysis of RNA was carried out using murine c-fms cDNA probe. It showed that macrophage-like cells and 5/10.14 S1 cells (controls) contained detectable c-fms mRNA, but astroglia did not.

Astroglia and macrophage-like cells were also tested for the presence of receptors by a ^{125}I-labeled CSF-1 binding assay. As expected, no specific ^{125}I-labeled CSF-1 binding was found on the astroglia. In contrast, specific binding of ^{125}I-labeled CSF-1 was detected on macrophage-like cells, indicating that these cells have translated c-fms to functional cell surface receptor.

Addition of CSF-1 to astroglia and macrophage-like cell cultures resulted in CSF-1 stimulation of proliferation of macrophage-like cells but not astroglia. The growth response

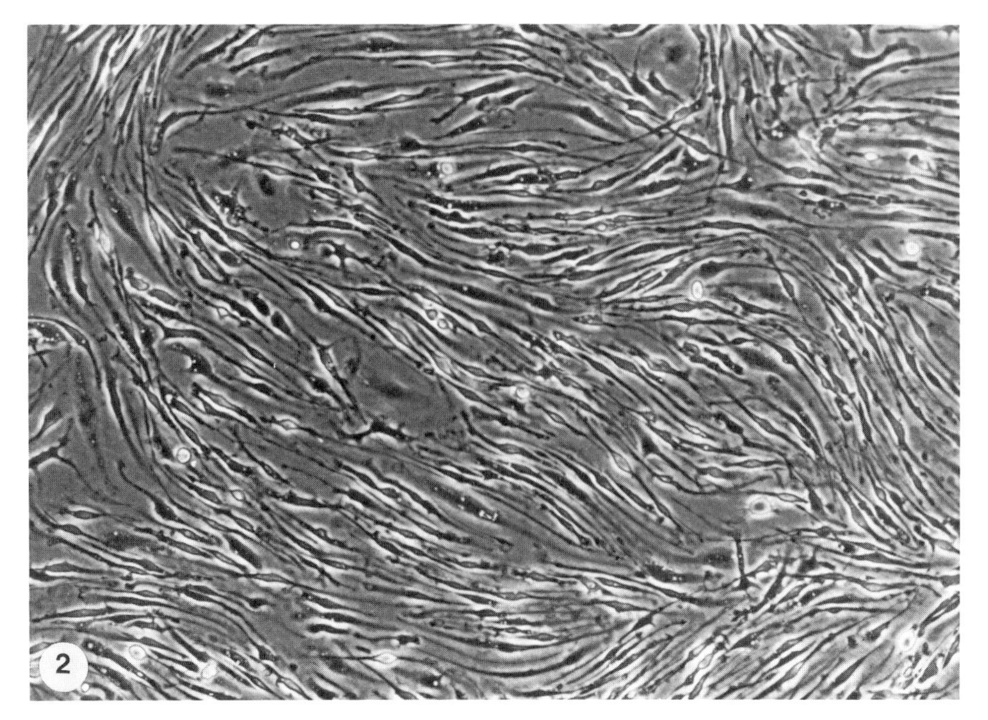

Fig. 2 Microglia culture, similar to the culture in 1b but tested with CSF-1 for 24 hours.

of macrophage-like cells was accompanied by a morphological change, i.e., in the presence of CSF-1 the macrophage-like cells became long, bipolar, and crescent-shaped (Fig. 2).

The above results indicate that c-fms proto-oncogene is expressed in the macrophage-like cells at the mRNA and functional cell surface receptor levels but is absent in astroglia (Hao et al., 1990b). Sawada et al. (1990) also found that the addition of CSF-1 to cultures of mouse microglia resulted in cell proliferation and change in cell morphology as well as increase in lysosome enzyme activity and superoxide anion formation. Using antibody to c fms protein product, they demonstrated the presence of CSF-1 receptor on the surface of microglia. They proposed that the CSF-1 effect may be mediated by activation of protein kinase C, because the protein kinase inhibitors, such as staurosporine or H-7, inhibited the effects of CSF-1.

Neither IL-3 nor GM-CSF stimulated the proliferation of macrophage-like cells. Tumor Necrosis Factor (TNF) inhibited the growth-promoting effect of CSF-1 on macrophage-like cells, quite different from its action on mouse bone marrow-derived and thioglycolate-elicited peritoneal exudate macrophages. TNF is synergistic to the CSF-1 in stimulating proliferation of mouse bone marrow-derived and thioglycolate-elicited peritoneal exudate macrophages (Branch et al., 1989; Shieh et al., 1989).

This indicates that microglia react differently to TNF than do bone marrow macrophages. In cultures, microglia are supported for survival and function by CSF-1 secreted by astroglia; the presence of TNF inhibits the function of CSF-1.

CONCLUSION

From the experiments described in this paper, we concluded that the macrophage-like cells are most likely microglia. The cultures initiated from the neural tube did not contain choroid plexus, meninges or blood vessels; the only cells present were neuroectodermal epithelial cells and the macrophage-like cells must have developed from them. Therefore the macrophage-like cells must correspond to neurectodermally derived microglia.

The macrophage-like cells differ from bone marrow macrophages in that they do not adhere to the substratum as firmly; moreover, TNF inhibits the proliferation-inducing effect of CSF-1 rather than having a synergistic effect on the action of CSF-1 as it does on bone marrow and peritoneal exudate-derived macrophages (Branch et al., 1989; Sawada et al., 1990).

The significant observation is that macrophage-like cells (microglia) in astroglia cultures depend for their survival on CSF-1 secreted by astroglia. We therefore conclude that a paracrine relationship exists between astroglia production of CSF-1 and the response of macrophage-like cells (microglia) to the cytokine in cultures (Hao et al., 1990c).

We hypothesize that astrocytes may regulate the production, differentiation, survival and function of neuroectoderm-derived microglia by secreting CSF-1.

Acknowledgments

Members of this laboratory are members of the Canadian Centre of Excellence in Neural Regeneration and Functional Recovery. This work was supported by the Medical Research Council of Canada Grant MT 4235. We would like to thank Osman Kademoglu for the Photography, Shirley West for typing and M.E. Fedoroff for reading the manuscript.

REFERENCES

Branch, D.R., Turner, A.R. and Guilbert, L.J., 1989, Synergistic stimulation of macrophage proliferation by the monokines tumor necrosis factor-alpha and colony-stimulating factor 1, Blood, 73:307-311.

Branch, D.R., and Guilbert, L.J., 1987, Practical in vitro assay systems for the measurement of haematopoietic growth factors, J. Tissue Culture Methods, 10:101-108.

Branch, D.R., Turc, J.M. and Guilbert, L.J., 1987, Identification of an erythropoietin-sensitive cell line, Blood, 69:1782-1785.

del Rio-Hortega, P., 1932, Microglia, in: "Cytology and Cellular Pathology of the Nervous System," Vol. 2, W. Hoeber, ed., Penfield, New York, pp. 483-534.

Fedoroff, S., 1990, Astrocyte reaction in injury and regeneration, in: "Advances in Neural Regeneration Research," F.J. Seil, ed., Wiley-Liss, New York, pp. 161-170.

Fedoroff, S, 1986, Prenatal ontogenesis of astrocytes, in: "Astrocytes," Vol. 1, S. Fedoroff, and A. Vernadakis, eds., Academic Press Inc., Orlando, pp. 35-74.

Fierz, W., and Fontana, A., 1986, The role of astrocytes in the interaction between the immune and nervous system, in: "Astrocytes," Vol. 3, S. Fedoroff, and A. Vernadakis, eds., Academic Press Inc., Orlando, pp. 203-229.

Fontana, A., Frei, K., Bodmer, S. and Hofer, E., 1987, Immune-mediated encephalitis: On the role of antigen-presenting cells in brain tissue, Immunol. Rev., 100:185-201.

Fujimoto, E., Miki, A. and Mizoguti, H., 1989, Histochemical study of the differentiation of microglial cells in the developing human cerebral hemispheres, J. Anat., 166:253-264.

Giulian, D., 1987, Ameboid microglia as effectors of inflammation in the central nervous system, J. Neurosci. Res., 18:155-171.

Graeber, M.B. and Streit, W.J., 1990, Microglia: Immune network in the CNS, Brain Path., 1:2-5.

Hao, C., Richardson, A., and Fedoroff, S., 1990a, Macrophage-like cells originate from neuroepithelium in culture: Characterization and properties of the macrophge-like cells., Int. J. Dev. Neurosci. (in press).

Hao, C. Guilbert, L.J. and Fedoroff, S., 1990b, Production of CSF-1 by mouse astrocytes in vitro, J. Neurosci. Res., (in press).

Hao, C., Guilbert, L.J. and Fedoroff, S., 1990c, Paracrine relationship between astroglia and microglia in cultures., Soc. Neurosci. Abstr. 16:995.

Kitamura, T., Miyake, T. and Fujita S., 1984, Genesis of resting microglia in the grey matter of mouse hippocampus, J. Comp. Neurol. 226:421-433.

Kitamura, T., 1973, The origin of brain macrophages - some considerations on the microglia theory of del Rio-Hortega, Acta. Path. Japan 23:11-26.

Ladner, M.B., Martin, G.A., Noble, J.A., Wittman, V.P., Warren, M.K., McGrogan, M. and Stanley, E.R., 1988, cDNA cloning and expression of murine macrophage colony-stimulating factor from L929 cells, Proc. Natl. Acad. Sci., U.S.A., 85:6706-6710.

Ling, E.A., 1981, The origin and nature of microglia, in: "Advances in Cellular Neurobiology," Vol. 2, S. Fedoroff, and L. Hertz, Academic Press, New York, pp. 33-82.

Morgan, C.J. and Stanley, E.R., 1984, Chemical crosslinking of the mononuclear phagocyte specific growth factor CSF-1 to its receptor at the cell surface, Biochem. Biophys. Res. Commun., 119:35-41.

Novikoff, A.B. and Goldfischer, S., 1961, Nucleosidediphosphatase activity in the Golgi apparatus and its usefulness for cytological studies, Proc. Nat. Acad. Sci., U.S.A., 47:802-810.

Perah, G. and Feldman, M., 1977. In vitro activation of the in vivo colony-forming units of the mouse yolk sac, J. Cell Physiol. 1:193-200.

Perry, V.H., Hume, D.A., and Gordon, S., 1985, Immunohistochemical localization of macrophages and microglia in the adult and developing mouse brain, Neurosci., 15:313-326.

Polak, M., D'Amelio F., Johnson J.E.J. and Haymaker, W., 1982, Microglial cell origins and reactions, in: "Histology and Histopathology of the Nervous System," Vol. 1, W. Haymaker and R.D. Adams, eds., Charles C. Thomas, Springfield, Illinois, pp. 481-559.

Sawada, M., Suzumura A., Yamamoto, H. and Marunouchi, T., 1990, Activation and proliferation of the isolated microglia by colony stimulating factor-1 and possible involvement of protein kinase C., Brain Res., 509:119-124.

Sherr, C.J., Rettenmier, C.W., Sacca, R., Roussel, M.F., Loon, A.T. and Stanley, E.R., 1985, The c-fms proto-oncogene product is related to the receptor for the mononuclear phagocyte growth factor, CSF-, Cell, 41:665-676.

Shieh, J.H., Peterson, R.H.F., Waren, D.J. and Moore, M.A.S., 1989, Modulation of colony-stimulating factor-1 receptors on macrophages by tumor necrosis factor, J. Immunol., 143:2534-2539.

Smith, P.A. and Glomski, C.A., 1982, "Hemogenic endothelium" of the embryonic aorta: does it exist? Dev. Comp. Immunol., 6:359-368.

Spangrude, G.J., Aihara, Y., Weissman, I.L. and Klein, J., 1988. The stem cell antigens SCA-1 and SCA-2 subdivide thymic and peripheral T lymphocytes into unique subsets, J. Immunol., 141:3697-3707.

Streit, W.J., Graeber, M.B. and Kreutzberg, G.W., 1988, Functional plasticity of microglia: A review, Glia, 1:301-307.

Sturrock, R.R., 1986, Postnatal ontogenesis of astrocytes, in: "Astrocytes," Vol. 1, S. Fedoroff and A. Vernadakis eds., Academic Press Inc., Orlando, pp. 75-103.

Théry, C. Hétier, E. Evrard, C. and Mallat, M., 1990, Expression of macrophage colony-stimulating factor gene in the mouse brain during development, J. Neurosci. Res., 26:129-133.

Van Furth, R., Cohn, Z.A., Hirsch, J.G., Humphry, J.H., Spector, W.G. and Langervoort, H.L., 1972. The mononuclear phagocyte system: a new classification of macrophages, monocytes and their precursor, Bulletin WHO, 46:845-852.

Wood, P. and Bunge, R.P., 1984, The biology of the oligodendrocyte, in: "Oligodendroglia," W.T. Norton, W.T. Plenum Press, New York, pp. 1-46.

NEURONAL-ASTROCYTIC INTERACTIONS IN BRAIN DEVELOPMENT, BRAIN FUNCTION AND BRAIN DISEASE

Leif Hertz

Departments of Pharmacology and Anaesthesia
University of Saskatchewan
Saskatoon, Sask. Canada S7N OWO

INTRODUCTION

The purpose of this review is to discuss neuronal-astrocytic interactions which appear to be of major importance in development and function of the brain as well as in brain disease. Initially, the role of these interactions during development will be reviewed followed by a description of their role in normal brain function; finally, the possible importance of a breakdown of these interactions during disease processes will be discussed. However, before starting the description of these interactions, some of the current knowledge of astrocytic functions will be briefly reviewed.

ASTROCYTIC FUNCTIONS

Ramon y Cajal wrote in 1909 that even now (1909) the functions of glial cells are unknown, and - what is worse - will remain so for a long time to come because no methods exist to study these functions. This visionary statement held true for exactly 50 years. Then, Holger Hyden (1959) was the first to develop preparations - microdissected samples of glial cells as well as of individual neurons - which could be used to investigate functions of living glial cells and neurons. Since then, the development has been rapid: preparations of neurons and different types of glial cells obtained by gradient centrifugation were developed by different groups, most notably by Norton and co-workers, and Booher and Sensenbrenner were the first to report a simple technique for preparation of highly enriched cultures of either neurons or astrocytes, based upon the developmental age at which the cultures were prepared.

Some physiological functions of astrocytes seem today to be quite generally accepted, maybe first and foremost 1) involvement in potassium homeostasis at the cellular level, 2) accumulation of glutamate and some other amino acid transmitters, 3) expression of receptors for a multitude of neurotransmitters, and 4) the recently established role in calcium signalling. It may be indicative of today's respectability of research on astrocytes that it lasted decades before their role in potassium homeostasis and transmitter function was accepted, whereas their contribution to calcium signalling in brain already is becoming well established after less than 5 years. Possibly, still more astrocytic functions may be unravelled during the coming decades, but at the same time research on astrocytes seems to be entering a new phase in which more emphasis can be placed on the role of these cells in the function of the intact central nervous system, mainly through interactions with neurons, but to a minor extent also through interactions with other cells in the central nervous system (CNS).

Functional interactions between astrocytes and neurons obviously require that the two cell types can "cross-talk". The talk from neurons to astrocytes occurs to a large extent in the

Plasticity and Regeneration of the Nervous System
Edited by P.S. Timiras *et al.*, Plenum Press, New York, 1991

same manner as that between neurons and other neurons or neurons and effector cells (e.g. smooth muscle), i.e., by release of neurotransmitters, often from "varicosities" rather than from genuine synapses (Beaudet and Descarries, 1984). This mechanism of impulse transmission is much more diffuse than that of classical synaptic transmission (Dismukes, 1977). In addition, each monoaminergic neuron, e.g., from locus coeruleus, affects target cells over large areas. The extracellular potassium concentration, which partly is determined by neuronal release and partly by astrocytic redistribution, constitutes another language in which neurons can communicate with astrocytes. The question of talk in the opposite direction, i.e., from astrocytes to neurons, is less established and more controversial. Again there is the possibility of cross-talk by aid of the extracellular potassium concentration, since even minor deviations from the physiological level of the potassium concentration in the extracellular fluid exert pronounced effects on neuronal excitability (Sykova, 1983; Walz and Hertz, 1983; Walz, 1989). Genuine transmitter release, i.e., a depolarization-induced release of a transmitter, requiring the presence of calcium, does not appear to occur from astrocytes. However, as will be described later in more detail, these cells do exhibit a potassium induced, non-calcium-dependent release of glutamate, which can be expected to exert an action on neuronal glutamate receptors on account of the close apposition between neurons and astrocytes (Schousboe et al., 1988). Also, due to the high potassium conductance of astrocytes, potassium ions move rapidly across astrocytic membranes in both directions as determined by the electrochemical potential. This, together with active transport of potassium (Walz and Hertz, 1983; Hertz, 1990a) and a wave of alterations in free intracellular calcium concentration evoked by local stimulation and spreading through adjacent astrocytes (Cornell-Bell et al., 1990) may provide astrocytes with the capability of altering neuronal excitability not only locally but also over longer distances. Moreover, neurons are metabolically handicapped cells (or metabolically highly specialized cells) since they exhibit little or no activity of the two enzymes glutamine synthetase (Norenberg and Martinez-Hernandez, 1979) and pyruvate carboxylase (Yu et al., 1983; Shank et al., 1985). The first of these catalyzes synthesis of glutamine from glutamate, and the second formation of oxaloacetate, a constituent of the tricarboxylic acid (TCA) cycle (or Krebs' cycle) from glucose. As will be discussed in more detail below, these enzymes appear to be essential for de novo synthesis of precursors for glutamate in neurons, a process that is essential because of the astrocytic uptake of glutamate and GABA. Astrocytes, in contrast, have relatively high activities of both enzymes and the inevitable conclusion is that in order to maintain normal glutamatergic and GABAergic transmitter function, glutamine and/or alpha-ketoglutarate must travel from astrocytes, where they are formed, to neurons, where they are utilized. Metabolic cycling is well established within individual cell types (e.g. the TCA cycle inside mitochondria and the malate-aspartate shunt connecting mitochondrial and cytosolic energy metabolism across the mitochondrial membrane), and the existence of a GABA-glutamate-glutamine cycle between a GABA producing and a GABA degrading metabolic compartment in the brain (which now are known to represent neurons and astrocytes, respectively) has also been recognized for twenty years (Van den Berg and Garfinkel, 1971). A glutamate-glutamine cycle between glutamatergic neurons and astrocytes was suggested somewhat later (Benjamin and Quastel, 1975) and is based on less direct experimental evidence. However, there is no doubt that glutamine is an effective precursor for release of both transmitter glutamate and transmitter GABA. During the last decade evidence has accumulated, suggesting a much modified and much less rigid glutamate-glutamine cycle in glutamatergic neurons (Fig. 1). Delineation of such a cycle is relatively easy because chemically different species are involved at different steps of the cycle. A much more difficult question is whether a similar neuronal-astrocytic interaction exists in potassium homeostasis at the cellular level, but some information is slowly accumulating about the function of such a cycle. These neuronal-astrocytic interactions will be described in more detail later in the review.

Compounds which in adult life serve a transmitter or signalling function are known to play a role as morphogens, i.e., differentiating agents in early development (Buznikoff and Shmukler, 1981, Lauder, 1988). This is also the case in brain and neuronal-astrocytic interactions have been found to be important in this regard. Within the last couple of years indirect evidence has also begun to appear that aberrations in neuronal-astrocytic interactions might play a major role in neurological disease, maybe first and foremost in neurodegenerative diseases (e.g., Alzheimer's Disease) and cerebral dysfunction during chronic exposure to certain toxins (e.g., hyperamonemia in hepatic encephalopathy). These developmental interactions will be discussed in the following section and the suggested impairment of neuronal-astrocytic interactions during disease processes towards the end of the review.

NEURONAL-ASTROCYTIC INTERACTIONS DURING DEVELOPMENT

A picture started to develop a couple of years ago that the monoamine transmitters dopamine, serotonin and noradrenaline as well as some polypeptide transmitters act as morphogens (promoting morphological differentiation) and functiogens (promoting functional differentiation) for astrocytes (Brenneman et al., 1984; Hatten, 1985; Hansson and Ronnback, 1988a, Hertz, 1990b; Meier et al., 1991), whereas glutamate, GABA and elevated concentrations of potassium ions, i.e., compounds which to a large extent are regulated by astrocytic uptake, are morphogenic and functiogenetic for neurons (Balazs et al., 1988, 1990; Mattson, 1988; Meier et al., 1991), but maybe not for astrocytes. However, this concept appears partly to be an oversimplification since increased extracellular K^+ concentrations do affect astrocytic proliferation (Reichelt et al., 1989) and differentiation (Peng et al., 1991), and since serotonin acts as a morphogenic compound by affecting both neurons and astrocytes, maybe in a sophisticated concert (see below).

The most detailed investigation in this area appears to be the recent elegant work by P. Whitaker-Azmitia and E. Azmitia (Whitaker-Azmitia and Azmitia, 1989; Azmitia et al., 1990). This work has shown that a factor released from astrocytes in response to serotonergic stimulation serves as a morphogen for neurons and that S-100 (the first astrocyte specific protein ever demonstrated (Hyden and McEwen 1966)) is 1) released from astrocytes during exposure to serotonin and 2) can replace the morphogenic factor released from the astrocytes. This obviously is in complete agreement with the concept that monoaminergic developmental cues are exerted on astrocytes and also with previous work by Lauder (1987) showing that serotonin is a morphogen for both neurons and astrocytes. The most recent work by the Azmitias has suggested that a $5\text{-}HT_{1A}$ receptor on astrocytes has to be activated together with a $5\text{-}HT_2$ receptor on neurons to evoke full morphogenic effect (P. Whitaker-Azmitia and E. Azmitia, personal communication). Moreover, evidence has been presented (Hertz 1990b; Meier et al. 1991) that a profound morphogenic and functiogenic effect of dibutyryl cAMP (dBcAMP) on cultured astrocytes may represent a substitute for activation of receptors operating via an intracellular increase in cyclic AMP (e.g. beta-adrenergic and $5HT_1$ receptors). A related morphological effect can also be seen in the presence of phorbolester, activating the phosphoinositol-protein kinase C-calcium system (Mobley et al., 1986). The functional effects of dBcAMP have been recently reviewed (Hertz, 1990b; Meier et al., 1991). Among the most pronounced alterations in astrocytes in primary cultures after exposure to dBcAMP for one week are a doubling of carbonic anhydrase activity and of Na^+, K^+-ATPase activity, a tripling of monoamine oxidase B activity, an alteration of a co-transport mechanism for GABA with one sodium ion to cotransport with two sodium ions (enabling a more concentrative uptake) and, maybe most dramatically, the appearance of a voltage sensitive L channel for calcium, indicated by calcium channel activity in astrocytes (MacVicar 1984, Barres et al. 1989), a high affinity binding of nitrendipine to astrocytes (Hertz, 1989a), a potassium stimulated uptake of calcium which is potently inhibited by nimodipine (Hertz et al., 1989a), and the ability of an elevated concentration of potassium to stimulate glycogenolysis (K.V. Subbarao and L. Hertz, unpublished results). These alterations might be essential for astrocytic function in the adult CNS.

NEURONAL-ASTROCYTIC INTERACTIONS IN BRAIN FUNCTION

1. Transmitter Function

Glutamatergic neurotransmission. Glutamate seems to be the major excitatory transmitter (Fonnum, 1984); it is released in large amounts by a potassium stimulated, calcium-dependent mechanism from glutamatergic neurons both in vivo, in brain slices and in primary cultures, e.g. cultured cerebellar granule cells (Hertz and Schousboe, 1986, 1987; Schousboe et al., 1988, Peng et al. 1991). Although there is a quite intense reuptake of glutamate into these cells, the uptake rate into astrocytes is even higher, reflecting a preferential uptake of glutamate into astrocytes in brain. This uptake has been demonstrated in glial cells obtained by gradient centrifugation, in glial cells in vivo and in cultured astrocytes (for a review see Hertz and Schousboe, 1986; Schousboe et al., 1988) and it drains glutamate from neurons. An impairment of the astrocytic uptake might well be involved in increases in the extracellular glutamate concentration leading to excitotoxicity.

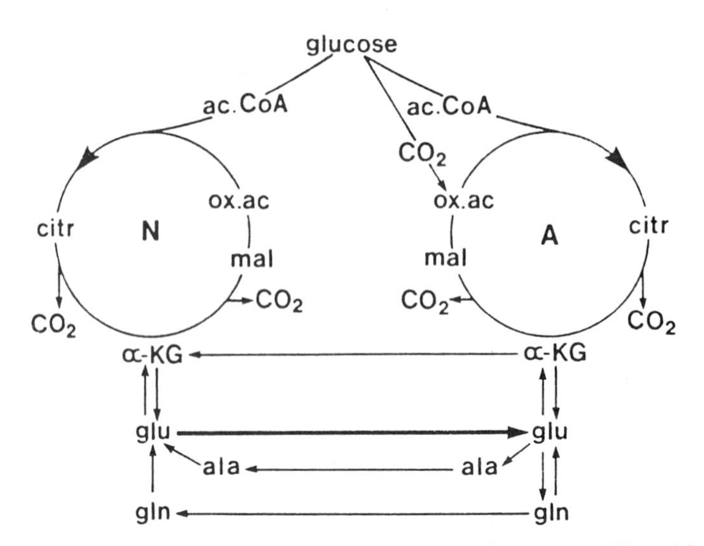

Fig. 1. Glutamate (glu) is released from glutamatergic neurons (N) and intensely acumulated into astrocytes (A) (heavy line). Synthesis of glutamine (gln), an astrocyte specific process, release of gln from astrocytes, its re-accumulation into neurons and hydrolysis to form glu contributes to replenish neuronal glu. However, astrocytic glu is also metabolized to CO2 and its amino group is partly transaminated into other amion acids, e.g. alanine (ala). Unless there is a net import of gln into the CNS, also other glu precursors must therefore exist in glutamatergic neurons. Only astrocytes have pyruvate carboxylase activity, allowing a net synthesis of oxaloacetate (ox.ac) and other TCA cycle constituents, e.g. alpha-ketoglutarate (α-KG) from glucose. α-KG and glu are released from astrocytes, accumulated into neurons and used to form the carbon skeleton of glu. Incorporation of amino nitrogen from ala (and/or other amino acids) supplies the amino group. From Hertz, 1989b.

Based upon the neuronal release of glutamate and the astrocytic uptake it has been suggested that glutamate released from neurons is quantitatively converted to glutamine, catalyzed by the glutamine synthetase, released to the interstitial fluid and reaccumulated into glutamatergic neurons, where it is hydrolyzed, catalyzed by phosphate activated glutaminase (PAG) to form glutamate (Fig. 1). This concept is supported by the findings that glutamine synthetase is an astrocyte specific enzyme (Norenberg and Martinez-Hernandez 1979), and that PAG is enriched in, although not restricted to glutamatergic neurons (Patel et al., 1982; Hogstad et al., 1988). However, experiments using astrocytes in primary cultures have shown that accumulated glutamate is only partly converted in astrocytes to glutamine, whereas the major part (at least under our experimental conditions) is utilized as a metabolic substrate (Yu et al., 1982). In addition, Yudkoff and coworkers (1986) have demonstrated that the amino group of glutamate is transaminated to several other amino acids in astrocytes. Thus, the amount of glutamine which is synthesized in astrocytes is too small to replace the amount of glutamate which has been accumulated into the astrocytes. Moreover, glutamine is degraded to glutamate in astrocytes and there is no preferential uptake of glutamine in neurons (Schousboe et al., 1979, 1988; Hertz and Schousboe, 1986, 1988). The inevitable conclusion of these findings is that there must either be a net import of glutamine into the central nervous system (which is in disagreement with most experimental findings), or other compounds than glutamine must also be utilized as precursors for transmitter glutamate in neurons. This applies both to the carbon skeleton and to the amino group. Possible precursors for the carbon skeleton include TCA cycle constituents like alpha-ketoglutarate. These compounds can be synthesized from glucose in astrocytes (Fig.1) due to the presence of pyruvate carboxylase activity. This enzyme is the major anaplerotic enzyme in brain (Shank and Aprison 1988) and catalyzes the formation of the TCA cycle constituent oxaloacetate. In contrast, pyruvate carboxylase activity is absent in neurons (both glutamatergic and GABAergic), indicating that these cells cannot perform a net synthesis of TCA cycle constituents from glucose (Yu et al., 1983; Shank et al., 1985; Kaufman and Driscoll, 1990). Thus, it is likely that a TCA cycle constituent, formed in astrocytes, supplements glutamine as a precursor for the carbon skeleton of transmitter glutamate in glutamatergic neurons. Shank and Campbell (1984) have found that both malate and alpha-ketoglutarate are accumulated into synaptosomes and Peng et al. (1990) have recently demonstrated that alpha-ketoglutarate, in the presence of an amino group donor (alanine) is, indeed, an efficient precursor for synthesis of transmitter glutamate in cerebellar granule cells. Different reactions (glutamine hydrolysis and transfer of an amino group, respectively) are obviously involved in the utilization of glutamine and alpha-ketoglutarate as glutamate precursors in cerebellar neurons. These differences may be functionally important, and are discussed elsewhere (Palaiologos et al., 1989; Hertz, 1989a; Peng et al., 1990).

GABA-ergic neurotransmission: GABA is also accumulated into astrocytes but a larger fraction of released GABA is reaccumulated into neurons (Hertz and Schousboe, 1987; Schousboe et al., 1988). Moreover, GABA oxidation in astrocytes occurs much more slowly than glutamate oxidation (Yu and Hertz, 1983). Glutamine is an effective precursor for releasable GABA, but GABA-ergic neurons may *not* be able to utilize alpha-ketoglutarate as a precursor for GABA (Kihara and Kubo, 1989).

Monoaminergic neurotransmission: Several aspects of the interaction between neurons and astrocytes in glutamatergic transmission are affected by monoaminergic transmitters (Table 1). Thus, glutamate uptake into astrocytes is stimulated by $alpha_1$-adrenergic agonists (Hansson and Rönnbäck, 1988b). Subsequent oxidative metabolism of glutamate is stimulated both by $alpha_1$ and $alpha_2$ agonists and the effect of noradrenaline is inhibited by the corresponding antagonists (Subbarao and Hertz, 1991). The same subtype antagonists also inhibit deoxyglucose utilization in the rat brain in vivo (Savaki et al., 1982). Serotonin as well as some trace amines (phenylethylamine), on the other hand, decrease glutamate release from both astrocytes and neurons (Fig. 2). In mature neurons (e.g. 8-day-old cerebellar granule cells) the non-stimulated glutamate release is unaffected, whereas the potassium-induced calcium sensitive release is abolished or reduced. In younger neurons, where there is no potassium induced increase in glutamate release, the release is reduced by 75% both in the presence and absence of excess potassium but there appears to be a rebound increase in release after removal of serotonin or phenylethylamine. In mature astrocytes (18-day-old) there is little effect on the release in the absence of excess potassium but the normal increase in release by an elevated potassium concentration is abolished (Hertz et al., 1989b); in younger astrocytes there is also a reduction of the release in the absence of excess potassium. At both ages there appears to be an increased glutamate release after the removal of the monoamine .

Table 1. Subtype of adrenergic receptor stimulating energy metabolism and transmitter amino acid uptake in primary cultures of astrocytes and [^{14}C]deoxyglucose utilization in vivo.

	$alpha_1$	$alpha_2$	beta
glycogenolysis a)		+	+
glycolysis b)	+		
TCA cycle activity b)	+	+	
glutamate uptake c)	+		
GABA uptake c)			+
glucose utilization d)			

a) From Subbarao and Hertz, 1990; b) From Subbarao and Hertz, 1991; c) From Hansson and Ronnback, 1988b; d) From Savaki et al., 1982. In a) and b) subtype determination was determined by both agonist and antagonist studies; in c) by agonist studies and in d) by antagonist studies.

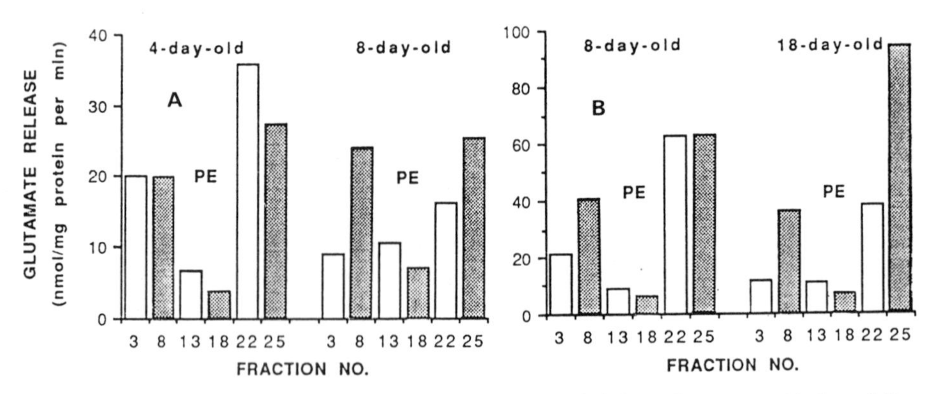

Fig. 2. Release of glutamate (nmol/min per mg protein) from immature (4-day-old) and mature (8-day-old) cerebellar granule cells (A) and from 8-day-old and 18-day-old cerebellar astrocytes (B). Each graph (fractions 3-25) represents one experiment in which a culture was superfused with a saline medium (containing glucose, glutamine and calcium) for 26 min and 1 min fractions were collected and analysed. During the period corresponding to fractions 1-5, 11-15 and 21-26 the medium contained a normal potassium concentration (open bars) and during the remaining periods a depolarizing potassium concentration (hatched bars). The trace monoamine phenylethylamine (PE) was present while fraction, 11-20 were collected. Representative fractons only are indicated. Note potassium-induced increase in mature neurons and in astrocytes as well as inhibition in the presence of PE. From Hertz et al., 1989b.

Experiments using 1-^{14}C labelled glutamate have demonstrated that TCA cycle activity is enhanced at the step of the alpha ketoglutarate dehydrogenase complex (Fig. 1). This obviously means that oxidative metabolism of glucose is also enhanced and such an effect has been directly demonstrated (Hertz, 1989b) . In addition, both glycolysis and glycogenolysis occur at a higher rate in the presence of adrenergic agonists (Subbarao and Hertz, 1990, 1991). Experiments using either antagonists or agonists have shown that there are differences with respect to the subtypes of adrenergic receptors which are involved, since glycogenolysis is enhanced by activation of either a beta receptor or an alpha$_2$ receptor, whereas the glycolytic rate is stimulated only by alpha$_1$ receptor activation (Table 1).

It is not known whether GABA oxidation in astrocytes it is affected by adrenergic agonists. However, as shown in Table 1, GABA uptake is enhanced by stimulation of ß-adrenergic receptors (Hansson and Ronnback, 1988b).

Astrocytes are not the only target for monoaminergic agonists. It is well established that brain microvessels have receptors for noradrenaline and other neurotransmitters (Joo, 1983, Kalaria et al., 1989a) and that stimulation of locus coeruleus (the nucleus from which noradrenergic fibers reach the entire cerebral cortex) affects vascular permeability and contractility (Hartman et al., 1979). Moreover, in spite of the fact that neuronal receptors for beta-adrenergic receptors may have a very low density (Stone and Ariano, 1989), noradrenergic synaptic profiles are found, in addition to varicosities, in the cerebral cortex (Aoki et al., 1987).

2. Potassium and calcium homeostasis

Potassium: It has now been overwhelmingly well demonstrated that the potassium concentration in the extracellular space of the central nervous system increases measurably (from 3 to about 4 mM) during physiological stimulation, to a larger extent (from 3 up to a "ceiling level" of 12 mM) during seizures or direct simulation of afferent pathways, and to exceedingly high values (up to 80-100 mM) during anoxia and spreading depression (Sykova, 1983; Walz and Hertz, 1983; Walz, 1989). It is also well established that the potassium concentration in the extracellular space after stimulation, seizures, spreading depression or short-lasting anoxia rapidly returns to its resting level of 3 mM. Potential mechanisms for potassium removal from the extracellular clefts would be that potassium either leaves the brain by accumulation into blood vessels or cerebrospinal fluid, or that potassium is redistributed within the brain tissue itself. Such a redistribution might either occur by diffusion within the extracellular space, or it might involve adjacent cells, be it neurons or astrocytes. Of these possibilities diffusion through the extracellular space or removal into blood vessels or cerebrospinal fluid can be ruled out as the major mechanism(s) for potassium removal from the mammalian brain in vivo, since these processes are not fast enough to explain the actually measured rates of potassium removal (for references, see Walz and Hertz, 1983). Thus, one or more kinds of cellular mechanisms must participate in the redistribution of potassium.

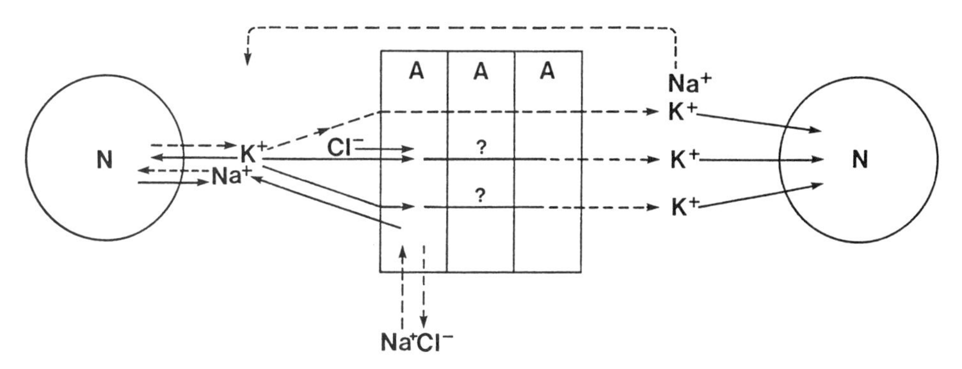

Fig. 3. Interactions between neurons (N) and astrocytes (A) in potassium (K^+) homeostasis at the cellular level. Active transport processes are shown as fully drawn lines and passive processes as stippled lines. An excited neuron loses K^+ and gains sodium (Na^+). Inward active transport of K^+ and outward active transport of Na^+ in neurons partly restore prestimulated ion levels. However, in addition K^+ passes *through* electrically coupled astrocytes by a passive, current-carried transport (the 'spatial buffer') from a region with elevated extracellular K^+ to a different extracellular location. The electrical circuit is closed by an extracellular transport of Na^+ back to the origin. Moreover, active transport of K^+ together with chloride (Cl^-) or in exchange with Na^+ occurs *into* astrocytes. It is unknown whether these K^+ ions can be further transported through astrocytes to different locations and whether they are subsequently released to the extracellular space or, maybe, transported directly from astrocytes to neurons. Na^+ enters astrocytes passively in spite of a low membrane permeability, due to a high electrochemical driving force, and Cl^- leaves passively. From Hertz, 1989b.

Already in the mid-1960s, it was suggested that glial cells might be involved in potassium homeostasis at the cellular level of the central nervous system, and that this mechanism may be of major importance for brain function (Hertz, 1965; Orkand et al., 1966; Kuffler, 1967). Two different hypothesis were proposed, i.e., 1) that extracellular potassium is accumulated into glial cells by an active uptake process (Hertz, 1965), or 2) that an increased local concentration of extracellular potassium is redistributed through glial cells by a current-carried transport mechanism ("spatial buffering") (Orkand et al., 1966). The former of these hypotheses was suggested on the basis of findings in isolated rat or cat glial cells and the latter on the basis of biophysical membrane characteristics (selective potassium permeability and electrical coupling) of glial cells in intact non-mammalian nervous systems. Both processes are outlined in Fig. 3. Potassium transport through glial cells by a current carried mechanism (stippled line through all A's) removes a local excess of extracellular potassium, resulting from excitation of neurons, and redistributes this potassium to other regions. The process is passive, and the circuit is closed by return of extracellular sodium (the predominant extracellular ion) back to the origin or by a transport of extracellular chloride in the opposite direction (Dietzel et al., 1989). This redistribution of potassium is not directly dependent upon energy metabolism, although it does depend on membrane polarization which, in turn, is a result of energy metabolism. It does not remove potassium from the extracellular space to glial cells, since virtually all extracellular potassium after the process is completed, remains extracellular potassium. However, it does remove potassium ions from *the* neuron(s) which have suffered a potassium loss during excitation. By redistributing potassium ions and thus influencing neuronal membrane potential, it may be of importance in regulation of neuronal excitability.

Active uptake into glial cells (fully draws lines into the first A) is conceptionally quite different. It is energy dependent and does transfer potassium ions from neurons to glial cells. Two different mechanism exist: a ouabain inhibitable uptake in exchange with sodium and a furosemide-sensitive co-transport mechanism (Walz and Hertz, 1983, 1984; Tas et al., 1987, Walz, 1989). The former reaches its maximum around an extracellular potassium concentration of ≈12mM and the latter at even higher extracellular potassium concentrations. The neuronal uptake is, in contrast , not increased by an elevation in the extracellular potassium concentration above its normal level (L. Hertz, unpublished experiments). Therefore, after excitation neuronally released potassium ions will primarily be accumulated into astrocytes. Such a neuronal-glial transfer of potassium has been directly demonstrated in non-cultured invertebrate preparations (Schlue and Wuttke, 1983; Coles, 1989). Any potassium ions taken up into glial cells will eventually have to be reaccumulated into neurons in order to prevent neuronal depletion of potassium. Thus, the entire process is, in principle, similar to that which appears to exist for glutamate turnover. There is, however, one major difference, i.e., that glutamate can be converted to other compounds (Fig.1), whereas potassium ions remain potassium ions. Thus, if potassium is first accumulated by an active uptake into astrocytes and subsequently reaccumulated into neurons, probably by a second energy-requiring uptake, the whole process is costly energetically. This may be a reason that the concept of an active potassium uptake into astrocytes (Hertz, 1965) was met with disbelief, and even hostility, by some leading Danish physiologists. However, there is *now* no doubt that potassium ions in the intact brain are both actively accumulated into astrocytes and redistributed by the spatial buffer mechanism through astrocytes (Grafe and Ballanyi, 1987; Walz, 1989; Dietzel et al. 1989). What remains undetermined is the relative importance of each of the two processes. This probably differs under different conditions and will eventually have to be determined under in vivo conditions since potassium uptake appears to be considerably decreased in brain slices (Hertz and Franck, 1978).

Calcium: Calcium plays an important role in the regulation of cellular functions in the CNS. Free intracellular calcium is involved in the control of a variety of physiological processes, including membrane permeability (e.g. calcium activated potassium channels), activation of second messengers and of proteinkinases (e.g., proteinkinase C), release of neurotransmitters (Janis et al., 1987; Miller, 1987; Turner and Goldin, 1988), and breakdown of glycogen (K.V. Subbarao and L. Hertz, unpublished experiments) as well as in pathological reactions, eg., following ischemia, when the extracellular calcium concentration in brain falls from about 1.0 to about 0.1 mM (Harris and Symon, 1984). Calcium enters cells via either voltage-sensitive (of the L, T, or N type) or receptor-operated calcium channels. Of the available organic calcium-channel blockers, the 1,4-dihydropyridines interact specifically and potently with voltage-sensitive calcium channels of the L type (Janis et al., 1987; Miller, 1987).

Lazarewicz et al. (1977) and Albrecht et al. (1988) have reported a potassium-induced stimulation of uptake of ^{45}Ca in a human glioma cell line and/or astrocytes obtained by gradient centrifugation, and both a potassium-induced uptake of ^{45}Ca and a potassium-induced increase in free intracellular calcium concentration have been observed in mouse astrocytes in primary cultures (Hertz et al., 1989a, 1991; Code et al., 1991). An almost maximum effect was seen already at 10-15 mM potassium, i.e., the "ceiling level" of the interstitial potassium concentration during seizures. This observation is in contrast to potassium-stimulated ^{45}Ca uptake in neurons (Skattebol and Triggle, 1987), which increases linearly with the extracellular potassium concentration between 5 and 55 mM; this difference probably reflects the behavior of astrocytes as perfect potassium electrodes even at low extracellular potassium concentrations. The dihydropyridine calcium channel blocker nimodipine potently (IC_{50} 2.7 nM) inhibits the potassium-induced calcium uptake in astrocytes but has virtually no effect on non-stimulated calcium uptake. This strongly suggests that the potassium induced ^{45}Ca uptake into astrocytes occurs through voltage-dependent L-channels (Hertz et al., 1989a). It is generally assumed that increased calcium inward currents into neurons are responsible for the decreased extracellular calcium levels observed during increased CNS excitability. However, the findings described above suggest that during periods of increased neuronal activity, a large amount of calcium enters astrocytes, and that neuronally released potassium may act as the trigger for this effect.

Transmitter Effects on Potassium and Calcium Homeostasis

The astrocytic part of neuronal-astrocytic interactions in potassium and calcium homeostasis constitutes a major target for transmitter actions. Exposure to noradrenaline increases Na^+, K^+-ATPase activity in primary cultures of astrocytes (Narumi et al., 1978), an increase which is reflected by an almost doubling of Na^+, K^+-ATPase activity after culturing with dBcAMP (K.V. Subbarao and L. Hertz, unpublished experiments). Potassium permeability appears, on the other hand, to be decreased, as suggested by a lowering of membrane potential at unaltered potassium concentrations (Bowman and Kimelberg, 1988). In addition, potassium permeability in mouse astrocytes is drastically decreased by active phorbolesters (Hertz, 1989a), which stimulate proteinkinase C activity and thus mimic transmitters acting on the phosphoinositol second messenger system. A similar response is evoked by exposure of astrocytes to the highly specific $alpha_2$-adrenergic agonist dexmedetomidine but not by exposure to noradrenaline (W.E. Code and L. Hertz, unpublished experiments).

Transmitter effects on free intracellular calcium concentration in astrocytes have been studied in depth by McCarthy and coworkers. They found that noradrenergic agonists cause increases in intracellular calcium in over 80% of all astrocytes tested in primary cultures. One component of the response was sensitive to reductions in extracellular calcium concentrations and might thus reflect an effect on calcium uptake. Both $alpha_1$ and $alpha_2$ adrenergic agonists could induce an increase in free intracellular calcium concentration, although only a minor fraction of the cells reacted to the $alpha_2$ agonist clonidine (Salm and McCarthy, 1990). Such an increase in intracellular calcium after exposure to $alpha_2$ adrenergic agonists has not previously been demonstrated in cells from the nervous system, but smooth muscle contracts in response to $alpha_2$-adrenergic stimulation (Young et al., 1988), suggesting an increased intracellular calcium concentration.

An exciting recent development in knowledge about astrocytic physiology is the demonstration that application of glutamate to primary cultures of astrocytes evokes an increase in concentration of free intracellular calcium (Enkvist et al., 1989) and that this effect can spread through a syncytium of astrocytes (Cornell-Bell et al., 1990). That this is part of a neuronal-astrocytic interaction can be seen by the observation in brain slices that the glutamate receptor agonist NMDA (which by itself does not trigger an increase in free intracellular calcium concentration in astrocytes, but does have such an effect in neurons) leads to an initial increase in free intracellular calcium concentration in neurons which secondarily triggers a wave of increase in free intracellular calcium concentration in astrocytes (Dani et al., 1990). The rate at which this phenomenon propagates is similar to that of Leao's spreading depression (Leibowitz, 1990), a phenomenon in which a depression of EEG activity (occasionally preceded by a brief stimulation), which is accompanied by an increase in glycogenolysis, propagates over the entire cortical surface (Bures et al., 1974). The salient point, which does not seem to have been answered yet, is whether the wave of increase in free intracellular calcium concentration is able to trigger excitation in a different neuron, in other words whether excitation of one neuron is able to spread through a glial syncytium and excite neurons which have no direct synaptic contacts with the originally stimulated neuron. Such a neuronal-glial-neuronal impulse transmission system (which was envisaged to operate via a transfer of potassium and to be reflected by a potassium-induced increase in energy metabolism) has previously been suggested (Hertz, 1965). Since an elevated potassium concentration enhances glycogenolysis in astrocytes by a dihydropyridine sensitive mechanism (K.V. Subbaro and L. Hertz, unpublished experiments) the propagation of a wave of an elevated free intracellular calcium concentration which might lead to an increase in glycogen breakdown, is compatible with - but not proof of - such potassium movements.

NEURONAL-ASTROCYTIC INTERACTIONS IN BRAIN DISEASE

If neuronal-astrocytic interactions are of major importance in normal brain function it could be expected that impairment of these interactions might occur under pathological conditions. Evidence will be presented below that Alzheimer's Disease, Down's Syndrome (and maybe especially the premature development of Alzheimer's pathology in middle-aged patients with Down's Syndrome), hepatic encephalopathy, manic depressive disease and maybe some of the consequences of brain ischemia might constitute disease states where a major cause of the cerebral dysfunction is interruption of normal neuronal-astrocytic interactions.

Alzheimer's Disease: According to most general concepts the initial insult in Alzheimer's Disease occurs in the cortex and subsequently evokes a descending degeneration in different nuclei in the basal forebrain and in the brain stem. It has, however, been suggested by a few authors that the disease may begin in the brain stem and from here ascend towards nucleus basalis, the cerebral cortex and hippocampus (Rossor, 1981; Hertz, 1989b). The basis for this suggestion is that cell destruction in locus coeruleus or the raphe nuclei is very prominent in postmortem brains from persons with Alzheimer's Disease (Hertz, 1989b; Chan-Palay, 1989) and may occur early during the development of the disease. The serotonin receptor $5HT_{1A}$ which is partly localized on astrocytes and causes a release of S-100 (see "Neuronal-astrocytic Interactions During Development") is decreased in Alzheimer's Disease, whereas the amount of S-100 is increased, leading Azmitia and Whitaker-Azmitia (1990) to propose that a reduced release of S-100, a neuronal growth factor, could be a primary cause of the neurodegenerative process. Moreover, many of the abnormalities in Alzheimer's Disease reminds of phenomena which can be observed after locus coeruleus lesions in the rat (Hertz, 1989b). Thus, impaired energy metabolism has repeatedly been demonstrated by the 2-deoxyglucose method in man in vivo as a key characteristic in Alzheimer's Disease (e.g. Rapoport et al., 1986; Haxby et al., 1990), together with a reduced density of the glucose transporter (Kalaria and Harik, 1989), and metabolic abnormalities occur after locus ceruleus lesions (see Hertz, 1989b). In locus coeruleus-lesioned animals there is a reduced glycogenolysis, and the finding of peculiar glycogen bodies in brains from patients with Alzheimer's Disease suggests that this may also be the case in these patients. Moreover, a beta receptor upregulation has been described under both conditions. There might be abnormalities in potassium homeostasis at the cellular level in locus ceruleus-lesioned animals (Hertz, 1989b), and recent experiments by Harik et al. (1989) have demonstrated a dramatic reduction in ouabain binding sites (i.e., in Na^+,K^+-ATP-ase activity) in brain parenchyma (but not in microvessels) from patients with Alzheimer's Disease. This might be secondary to a reduced adrenergic input. A reduced potassium content in hippocampus has also been reported (Gramsbergen et al., 1987) but the significance of such alterations in post-mortem material is difficult to establish. Whether or not abnormal cortical histology can be brought about by locus ceruleus lesion is unknown. We have examined rat brain cortices nine months after locus ceruleus lesion and at that stage there were no obvious abnormalities (A. Paterson, L.F. Eng, A.C.H. Yu and L. Hertz, unpublished experiments). However, evidence is accumulating that astrocytes may play a major role in the development of plaques (i.e. not only react secondarily to plaque formation). Thus, mRNA for alpha1-antichymotrypsin which is tightly associated with the beta protein in amyloid deposits, characteristic for Alzheimer's Disease, may predominantly or exclusively be synthesized by astrocytes (Pasternack et al. 1990).

It is an indication of dysfunction in glutamatergic transmission in Alzheimer's Disease that sodium dependent binding of D-aspartate, a compound which may bind specifically to a glutamate uptake site, is decreased (Procter et al., 1986; Hertz, 1989b; Cowburn et al., 1990). Ordinarily this decrease has been taken as an indication of a decreased glutamate uptake into glutamatergic neurons but in view of the predominant location of glutamate uptake sites in astrocytes it may at least equally well suggest a reduction of astrocytic glutamate uptake, a process enhanced by adrenergic stimulation (see also the note, added in proof, at the end of the paper). In contrast, glutamate receptor sites are often little affected, although a decrease may occur in severe cases. It should be specifically mentioned that there is no indication of any upregulation of glutamatergic receptors as could be expected if primarily presynaptic glutamatergic terminals were destroyed. There is, however, a drastic decrease of glutaminase activity (Palmer and Gershon, 1990), maybe reflecting an impaired return of glutamine from astrocytes to neurons. Moreover, glutamate content is reduced in biopsy samples from the brains of Alzheimer's patients (Lowe and Bowen, 1990).

The monoamineoxidase inhibitor, deprenyl, which is specific for MAO B, has been found to cause a significant improvement of cognitive function in patients with Alzheimer's Disease (Tariot et al., 1987; Piccinin et al., 1990). In confirmation of the suggestion that MAO inhibitors may be specially valuable during early stages of Alzheimer's Disease where they might be able to counteract further deterioration of neuronal-glial interactions (Hertz, 1989b), a tendency was found for deprenyl to be especially effective in patients with early disturbances of memory (Piccinin et al., 1990). In view of the low toxicity of this drug it appears that further drug trials with deprenyl are urgently needed in *early* Alzheimer's cases, including patients with Down's syndrome approaching middle age (see below). It would also be interesting to know if such medication could counteract the impairment of the phosphoinositol-protein kinase C second messenger system known to occur in Alzheimer's Disease (Fowler et al., 1990).

Down's Syndrome: Patients with Down's syndrome invariably develop Alzheimer's Disease with all characteristic symptoms and histological alterations prematurely if they live to middle age (for references, see Hertz, 1989b). In addition there is an early massive increase in S-100, indicating an early astrocytic response (Griffin et al., 1989). In view of the effect of serotonergic agonists on S-100 during development this might reflect an impairment of normal serotonergic function.

Hepatic encephalopathy: The typical histological abnormality in hepatic encephalopathy is a conversion of normal astrocytes to the so-called Alzheimer type 2 astrocytes. Many of the clinical abnormalities may be caused by an increase of the ammonia concentration in blood and brain (Cooper and Plum, 1987). Chronic exposure of astrocytes to an elevated ammonia concentration leads to an increase in glutamine synthesis and a concomitant decrease of both glutamate content and glutamate oxidation in astrocytes in primary cultures (Hertz et al., 1987; Ch. R.K. Murthy, G. Kala and L. Hertz, unpublished experiments). This astrocytic dysfunction might secondarily affect astrocytic-neuronal interactions, although chronic and acute exposure of neurons to ammonia also affects neuronal metabolism in the absence of astrocytes (Hertz et al., 1987 ; Yudkoff et al., 1990).

Manic-depressive illness: The subtype of beta-adrenergic receptor which is down-regulated after chronic exposure to antidepressant drugs in vivo is the $beta_1$-receptor. This receptor appears to be located primarily on astrocytes. Chronic, but not acute, exposure of astrocytes in primary cultures leads to down-regulation of beta-adrenergic function (Hertz and Richardson, 1983) but, at least in glioma cells, the reaction to the drug is not stereospecifically correlated with the active isomer (Manier and Sulser, 1990).

Ischemia: Although most emphasis has been placed on neuronal reactions to ischemia (severe hypoxia and substrate-deprivation), astrocytic reactions occur both in vivo (Petito and Babiak, 1982), in brain slices (Moller et al. 1974), and in astrocytes in primary cultures. These responses include morphological alterations in mitochondria after re-oxygenation (Petito et al., 1991) and severe metabolic impairment (Yu et al., 1989). It is unknown whether they reflect metabolic abnormalities which could contribute to delayed neuronal death. However, astrocytes do show indications of damage not only during hypoxia but also after re-oxygenation, including cell membrane leakage (Sochocka et al., 1991).

CONCLUDING REMARKS

The main theme of this review has been the concept that monoaminergic neurons exert a large part of their effects on astrocytes (although some monoamines clearly affect also neurons), and that the astrocytes in turn are capable of modifying their interaction with neurons. During development, these effects are morphogenic and include release of compounds (e.g., S-100) from astrocytes which enhance neuronal development. During normal function in the differentiated CNS the effects of monoamines to a large extent focus on astrocytic (and neuronal) functions regulating the contents of neuroactive amino acid transmitters and of potassium in the extracellular space. This will secondarily affect the output of other neurons, i.e., there is a neuronal-astrocytic-neuronal chain of events.

On the basis of the evidence presented it can be safely concluded that neuronal-astrocytic interactions are, indeed, important in CNS development and function. It is also obvious that any interruption of a *chain* of events, be it at a neuronal step or at an astrocytic step, can lead to catastrophic dysfunction under pathological conditions. However, the functional input of, respectively, astrocytic and neuronal events remains undetermined. On one hand, there is the possibility that astrocytes, dispersed over large brain areas, in unison respond to a neuronal signal in order to attend to neuronal demands as suppliers of nutrients and precursors, and as removers of waste. Such a *global* response might be connected with phenomena like sleep-wakefulness, mood, etc. On the other hand, there is also the possibility that astrocytes contribute actively to CNS output at a *local* level and play a role in deciding neuronal pathways. Such a situation may be compared with a neuronal network of roads and an astrocytic input into which roads are actually followed and maybe also with an astrocytic capability to make shortcuts between existing roads. This astrocytic input may be exerted in at least two different manners: an astrocytic release of glutamate and/or potassium might lead to a direct excitation of adjacent neurons in an non-synaptic manner (Galambos, 1961; Hertz, 1965; Schmitt, 1984) and/or astrocytes might possess the ability to promote some neuronal pathways by delivering transmitter precursors and to paralyze other neuronal pathways by depriving neurons of transmitter precursors. The extent to which astrocytes could affect CNS function in these manners is unknown but also unlimited.

REFERENCES

Albrecht, J., Hilgier, W., Lazarewicz, J.W., Rafalowska, U., and Wysmyk-Cybula, U., 1988, Astrocytes in acute hepatic encephalopathy: Metabolic properties and transport functions. in: "The Biochemical Pathology of Astrocytes," Norenberg M.D., Hertz, L., and Schousboe, A., eds., New York: Alan R. Liss, 465-476.

Aoki, C., Joh, T.H., and Pickel, V.M., 1987, Ultrastructural localization of immunoreactivity for beta-adrenergic receptors in cortex and neostriatum of rat brain, Brain Res. 437: 264-282.

Azmitia, E.C., Dolan, K. and Whitaker-Azmitia, P.M., 1990, S-100ß but not NGF, EGF, insulin or calmodulin is a CNS serotonergic growth factor. Brain Res. 516: 354-6.

Azmitia, E.C., and Whitaker-Azmitia, P.M., 1990, CNS 5-HT neurons and glial S-100_{β}. Clin. Neuropharmacol. 13: 633-634.

Balazs, R., Gallo, V. and Kingsbury, A., 1988, Effect of depolarization on the maturation of cerebellar granule cells in culture. Brain Res. 468: 269-276.

Balazs, R., Hack, N., and Jorgensen, O.S., 1990, Interactive effects of different classes of excitatory amino acid receptors and the survival of cerebellar granule cells in culture, Int. J. Dev. Neurosci., 8: 347-359.

Barres, B.A. Chun, L.L., and Corey, D.P., 1989, Calcium current in cortical astrocytes: induction by cAMP and neurotransmitters and permissive effect of serum factors. J. Neurosci, 9: 3169-3175.

Beaudet, A., and Descarries, L., 1984, Fine structure of monoamine axon terminals in cerebral cortex, in: "Monoamine Innervation of Cerebral Cortex," L. Descarries, T.R. Reader, and H.H. Jasper, eds., Alan R. Liss, New York, 77-93.

Benjamin, A.M. and Quastel, J.H., 1975, Metabolism of amino acids and ammonia in rat brain cortex slices in vitro: A possible role of ammonia in brain function. J. Neurochem. 25: 197-206.

Bowman, C.L., and Kimelberg, H.K., 1988, Adrenergic-receptor-mediated depolarization of astrocytes, in: "Glial Cell Receptors," H.K. Kimelberg, ed., Raven, New York, 53-76.

Brenneman, D.E., Neale, E.A., Foster, G.A., d'Autremont, S.W., and Westbrook, G.L., 1984, Non-neuronal cells mediate neurotrophic action of vasoactive intestinal peptide, J. Cell Biol., 104: 1603-1610.

Bures, J., Buresova, O., and Krivanek, J., 1974, The mechanism and applications of Leao's spreading depression of electroencephalographic activity. Academic Press, New York.

Buznikov, G.A., and Shmukler, Y.B., 1981, Possible role of "prenervous" neurotransmitters in cellular interactions of early embryogenesis, Neurochem. Res., 6: 55-68.

Chan-Palay, V., and Asan, E., 1989, Alterations in catecholamine neurons of the locus coeruleus in senile dementia of the Alzheimer type and in Parkinson's disease with and without dementia and depression, J. Comp. Neurol. 287, 373-392.

Code, W.E., White, H.S., and Hertz, L., 1991, Midazolam effects on calcium signalling in astrocytes, submitted for publication.
Coles, J.A., 1989, Functions of glial cells in the retina of the honeybee drone, Glia 2, 1-9.
Cooper, A.J.L. and Plum, F., 1987, Biochemistry and Physiology of brain ammonia. Physiol. Rev. 67, 440-519.
Cornell-Bell, A.H., Finkbeiner, S.M., Cooper, M.S., and Smith, S.J., 1990, Glutamate induces calcium waves in cultured astrocytes: long-range glial signaling, Science, 247: 470-474.
Cowburn, R.F., Hardy, J.A., and Roberts, P.J., 1990, Glutamatergic neurotransmission in Alzheimer's disease. Biochem. Soc. Trans., 18: 390-392.
Dani, J.W., Chernjavsky, A., and Smith, S.J., 1990, Calcium waves propagate through astrocyte networks in developing hippocampal brain slices. Abstracts, Soc. Neurosci. 16: 970.
Dietzel, I., Heineman, U., and Lux, H.D., 1989, Relations between slow extracellular potential changes, glial potassium buffering, and electrolyte and cellular volume changes during neuronal hyperactivity in cat brain, Glia, 2: 25-44.
Dismukes, K., 1977, New look at the aminergic neuron systems, Nature, 269: 557-558.
Enkvist, M.O., Holopainen, I., and Akerman, K.E., 1989, Glutamate receptor-linked changes in membrane potential and intracellular Ca^{2+} in primary rat astrocytes. GLIA, 2: 397-402.
Fowler, C.J., O'Neill, C., Garlind, A., and Cowburn, R.F., 1990, Alzheimer's disease: is there a problem beyond recognition? Trends Pharmacol. Sci., 11: 183-184.
Fonnum, F., 1984, Glutamate: a neurotransmitter in mammalian brain, J. Neurochem., 42: 1-11.
Galambos, R., 1961, A glial-neuronal theory for brain function, Proc. Nat. Acad. Sci. USA, 47: 129-136.
Grafe, P. and Ballanyi, K., 1987, Cellular mechanisms of potassium homeostasis in the mammalian nervous system. Can. J. Physiol. Pharmacol. 65: 1038-1042.
Gramsbergen, J.B., Mountjoy, C.Q., Rossor, M.N., Reynolds, G.P., and Korf, J., 1987, A correlative study on hippocampal cation shifts and amino acids and clinicopathological data in Alzheimer's Disease, Neurobiol. Aging, 88: 487-494.
Griffin, W.S., Stanley, L.C., Ling, C., White, L., MacLeod, V., Perrot, L.J., White, C.L. and Araoz, C., 1989, Brain interleukin 1 and ß 100 immunoreactivity are elevated in Down's Syndrome and Alzheimer's Disease. Proc. Nat. Sci. US Acd., 86: 7611-7615.
Hansson and Ronnback, 1988a, Neurons from substantia nigra increase the efficacy and potency of second messenger arising from striatal astroglia dopamine receptor, Glia, 1: 393-397.
Hansson, E., and Ronnback, L., 1988b, Regulation of glutamate and GABA transport by adrenoceptors in primary astroglial cell cultures, Life Sci., 44: 27-34.
Harik, S.I., Mitchell, M.J. and Kalaria, R.N., 1989, Ouabain binding in the human brain. Effects of Alzheimer's disease and aging. Arch. Neurol., 46: 951-954.
Harris, R.J., and Symon, L., 1984, Extracellular pH, potassium and calcium activities in progressive ischemia of rat cortex, J. Cereb. Blood Flow & Met., 4: 178-186.
Hartman, B.K., Swanson, L.W., Raichle, M.E., Preskorn, S.H., and Clark, H.B., 1979, Central adrenergic regulation of cerebral microvascular permeability and blood flow: anatomic and physiologic evidence, Adv. Exp. Med. Biol., 131: 113-126.
Hatten, M.E., 1985, Neuronal regulation of astroglial morphology and proliferation in vitro. J. Cell. Biol. 100, 384-396.
Haxby, J.V., Grandy, C.L., Koss, E., Horwitz, B., Heston, L., Schapiro, M., Friedland, R.P., and Rapoport, S.I., 1990, Longitudinal study of cerebral metabolic asymmetries and associated neuropsychological patterns in early dementia of the Alzheimer type. Arch. Neurol., 47: 753-760.
Hertz, L., 1965, Possible role of neuroglia: A potassium-mediated neuronal-neuroglial-neuronal impulse transmission system, Nature, 206: 1091-1094.
Hertz, L., Neuronal-glial interactions, 1989a, in: "Regulatory Mechanisms of Neurons to Vessel Communication in Brain," S. Govoni, G., Battaini and M.S. Mangoni, eds., Springer, Heidelberg, 271-305.
Hertz, L., 1989b, Is Alzheimer's Disease an anterograde neuronal-glial degeneration, originating in the brain stem, and disrupting metabolic and functional interactions between neurons and glial cells? Brain Res.Rev., 14: 335-353.

Hertz, L., 1990a, Regulation of potassium homeostasis by glial cells, in: "Development and Function of Glial Cells," G. Levi, ed., Alan R. Liss, N.Y., 225-234.

Hertz, L., 1990b, Dibutyryl cyclic AMP treatment of astrocytes in primary cultures as a substitute for normal morphogenic and "functiogenic" transmitter signals, in: "Molecular Aspects of Development and Aging in the Nervous System," A. Privat, E. Giacobini, P. Timiras and A. Vernadakis, eds., Plenum, NY, 227-243.

Hertz, L., and Franck, G., 1978, Effect of increased potassium concentrations on potassium fluxes in brain slices and in glial cells, in: "Dynamic Properties of Glial Cells, " E. Schoffeniels, G. Franck, L. Hertz, and D.B. Tower, eds., Pergamon Press, Oxford, 383-388.

Hertz, L., and Richardson, J.S., 1983, Acute and chronic effects of antidepressant drugs on ß-adrenergic function in astrocytes in primary cultures - an indication of glial involvement in affective disorders? J. Neurosci. Res., 9: 173-183.

Hertz, L., and Schousboe, A., 1986, Role of astrocytes in compartmentation of amino acid and energy metabolism, in: "Astrocytes," S. Fedoroff, and A. Vernadakis, eds., Academic Press, New York, 2: 179-208.

Hertz, L. and Schousboe, A., 1987, Primary cultures of GABAergic and glutamatergic neurons as model systems to study neurotransmitter functions. I. Differentiated cells. in: "Model Systems of Development and Aging of the Nervous System", A. Vernadakis, A. Privat, J.M. Lauder, P.S.Timiras and E. Giacobini, eds., Martinus Nijhoff Publishers, Mass., 19-31.

Hertz, L., and Schousboe, A., 1988, Metabolism of glutamate and glutamine in neurons and astrocytes in primary cultures, in: "Glutamine and Glutamate in Mammals," E. Kvamme, ed., CRC Press, Boca Raton, FL., 2: 39-55.

Hertz,L., Murthy, Ch.R.K., Lai, J.C.K., Fitzpatrick, S.M., and Cooper, A.J.L., 1987, Some metabolic effects of ammonia on astrocytes and neurons in primary cultures, Neurochem. Pathol., 6: 97-129.

Hertz, L., Bender, A.S., Woodbury, D., and White, H.S., 1989a, Potassium induced calcium uptake in astrocytes and its potent inhibition by a calcium channel blocker, J. Neurosci. Res., 22. 209-215..

Hertz, L., Peng, L., Hertz, E., Juurlink, B.H.J., and Yu, P.H., 1989b, Development of monoamine oxidase activity and monoamine effects on glutamate release in cerebellar neurons and astrocytes, Neurochem. Res., 1039-1096.

Hertz, L., Code, W.E., Shokeir, O., Shargool, M., Woodbury, D.M., and White, M.S., 1991, Calcium signalling in astrocytes, in: "Neuroglial Function," A.I. Roitbak, ed., Tbilisi, USSR.

Hogstad, S., Svenneby, G., Torgner, I.Aa., Kvamme, E., Hertz, L., and Schousboe, A., 1988, Glutaminase in neurons and astrocytes cultured from mouse brain: Kinetic properties and effects of phosphate, glutamate and ammonia, Neurochem. Res., 13: 383-388.

Hyden, H., 1959, Quantitative assay of compounds in isolated, fresh nerve cells and glial cells from control and stimulated animals, Nature, 184: 433-435.

Hyden, H., and McEwen, B., 1966, A glial protein specific for the nervous system, Proc. Nat. Acad. Sci. USA, 55: 354-358.

Janis, R.A., Silver, P.J., and Triggle, D.J., 1987, Drug action and cellular calcium function, Adv. Drug Res., 16: 309-591.

Joo, F., 1983, The blood-brain barrier in vitro: Ten years of research on microvessels isolated from the brain, Neurochem., 7: 1-25.

Kalaria, R.N., Stockmeier, C.A., and Harik, S.I., 1989a, Brain microvessels are innervated by locus coeruleus noradrenergic neurons, Neurosci. Lett., 97: 203-208.

Kalaria, R.N., and Harik, S.I., 1989, Reduced glucose transporter at the blood-brain barrier and in cerebral cortex in Alzheimer's disease. J. Neurochem. 53: 1083-1086.

Kaufman, E.G. and Driscoll, B.F., 1990, The effect of $[K^+]$ on CO_2 fixation in cultured glial cells. Trans. Am. Soc. Neurochem. 21: 289.

Kihara, M., and Kubo, T., 1989, Aspartate aminotransferase for synthesis of transmitter glutamate in the medulla oblongata: effect of aminooxyacetic acid and 2-oxoglutarate, J. Neurochem., 52: 1127-1134.

Kuffler, S.W., 1967, Neuroglial cells: Physiological properties and a potassium mediated effect of neuronal activity on the glial membrane potential, Proc. R. Soc. Series B, 168: 1-21.

Lauder, J.M., 1987, Neurotransmitters as morphogenetic signals and trophic factors. in: "Model Systems of Development and Aging of the Nervous System," Vernadakis, A., Privat, A., Lauder, J.M., Timiras, P.S. and Giacobini, E., eds, Martinus Nijhoff, Boston, 219-237.
Lauder, J.M., 1988, Neurotransmitters as morphogens. Prog. Brain Res. 73: 365-87.
Lazarewicz, J.W., Kanje, M., Sellstrom, A., and Hamberger, A., 1977, Calcium fluxes in cultured and bulk isolated neuronal and glial cells. J. Neurochem. 29: 495-502.
Leibowitz, D.H., 1990, A glial cytocal wave is the conduction velocity-determining propagation mechanism of spreading depression, Abstracts, Soc. Neurosci., 16: 970.
Lowe, S.L. and Bowen, D.M., 1990, Glutamic acid concentration in brains of patients with Alzheimer's Disease, Biochem. Soc. Trans., 18, 443-444.
MacVicar, B.A., 1984, Voltage-dependent calcium channels in glial cells, Science, 226: 1345-1347.
Manier, D.H., Sulser, F., 1990, Chronic exposure of rat glioma C cells to oxaprotiline reduces the density of beta adrenoceptors, Abstracts, Soc. Neurosci., 16: 385.
Mattson, M.P., 1988, Neurotransmitters in the regulation of neuronal cytoarchitecture, Brain Res., Rev., 13: 179-212.
Meier, E., Hertz, L., and Schousboe, A., 1991, Neurotransmitters as developmental signals, Neurochem. Int., in press.
Miller, R.J., 1987, Multiple calcium channels and neuronal function, Science, 235: 46-52.
Mobley, P.L., Scott, S.L., and Cruz, E.G., 1986, Protein kinase C in astrocytes: a determinant of cell morphology, Brain Res., 398: 366-369.
Moller, M., Mollgard, K., Lund-Andersen, H., and Hertz, L., 1974, Concordance between morphological and biochemical estimates of fluid spaces in rat brain cortex slices, Exp. Brain Res., 21: 299-314.
Narumi, S., Kimelberg, H.K., and Bourke, R.S., 1978, Effects of norepinephrine on the morphology and some enzyme activities of primary monolayer cultures from rat brain, J. Neurochem., 31, 1479-1490.
Norenberg, M.D., and Martinez-Hernandez, A., 1979, Fine structural localization of glutamine synthetase in astrocytes of rat brain, Brain Res., 161: 303-310.
Orkand, R.K., Nicholls, J.G., and Kuffler, S.W., 1966, Effect of nerve impulses on the membrane potential of glial cells in the central nervous system of amphibia, J. Neurophysiol., 29: 788-806.
Palaiologos, G., Hertz, L., and Schousboe, A., 1989, Role of aspartate amino-transferase and mitochondrial dicarboxylate transport for release of endogenously and exogenously supplied neurotransmitter in glutamatergic neurons, Neurochem. Res., 14: 359-366.
Palmer, A.M. and Gershon, S., 1990, Is the neuronal basis of Alzheimer's disease cholinergic or glutamatergic? FASEB J., 4: 2745-2752.
Pasternack, J.M., Abraham, C.R., Van Dyke, B.J., Potter, H., and Younkin, S.G., 1989, Astrocytes in Alzheimer's disease gray matter express alpha 1-antichymotrypsin mRNA, Am. J., Pathol., 135: 827-834.
Patel, A.J., Hunt, A., Gordon, R.D., and Balazs, R., 1982, The activities in different neural cell types of certain enzymes associated with the metabolic compartmentation of glutamate. Dev. Brain. Res. 4, 3-11.
Peng, L., Schousboe, A., and Hertz, L., 1990, Utilization of alpha-ketoglutarate as a precursor for transmitter glutamate in cultured cerebellar granule cells. Neurochem. Res., in press.
Peng, L., Juurlink, B.H.J., and Hertz, L., 1991, Development of cerebellar granule cells in the presence and absence of excess extracellular potassium - Do the two culture system provide a means of distinguishing between events in transmitter-related and non-transmitter-related glutamate pools?, Brain Res., submitted for publication.
Petito, C.K., and Babiak, T., 1982, Early proliferative changes in astrocytes in postischemic noninfarcted rat brain. Ann. Neurol. 11: 510-518.
Petito, C., Juurlink, B.H.J., and Hertz, L., 1991, An in vitro model differentiating between direct and indirect effects of ischemia on astrocytes, Exp. Neurol., in press.
Piccinin, G.L., Finali, G., and Picirilli, M., 1990, Neuropsychological effects of L-deprenyl in Alzheimer's type dementia. Clin. Neuropharmacol., 13: 147-163.
Procter, A.W., Palmer, A.M., Stratman, G.C., and Bowen, D.M., 1986, Glutamate aspartate-releasing neurons in Alzheimer's Disease, N. Eng. J. Med., 314: 1711-1712.
Ramon y Cajal, S., 1909, Histologie du systeme nerveux de l'homme et des vertébrés.

Rapoport, S.I., Horwitz, B., Haxby, J.V., and Grady, C.L., 1986, Alzheimer's Disease: metabolic uncoupling of associative brain regions, Can. J. Neurol. Sci., 13: 540-545.
Reichelt, W., Dettmer, D., Bruckner, G., Brust, P., Eberhardt, W., and Reichenbach, A., 1989, Potassium as a signal for both proliferation and differentiation of rabbit retina (Muller) glia growing in cell culture. Cell Signal, 1: 187-94.
Rossor, M.N., 1981, Parkinson's disease and Alzheimer's disease as disorders of the isodendritic core, Br. Med. J., 283: 1588-1590.
Salm, A.K. and McCarthy, K.D. Norepinephrine-evoked calcium transients in cultured cerebral type 1 astroglia. Glia 3: 529-538 (1990).
Savaki, H.E., Kadekaro, M., McCulloch, J., and Sokoloff, L., 1982, The central noradrenergic system in the rat: metabolic mapping with alpha-adrenergic blocking agents. Brain Res., 234: 65-79.
Schlue, W.R. and Wuttke, W., 1983, Potassium activity in leech neuropile glial cells changes with external potassium concentration. Brain Res., 270, 368-372.
Schmitt, F.O., 1984, Molecular regulators of brain function: a new view, Neurosci., 13: 991-999.
Schousboe, A., Hertz, L., Svenneby, G., and Kvamme, E., 1979, Phosphate activated glutaminase activity and glutamine uptake in astrocytes in primary cultures, J. Neurochem., 32: 943-950.
Schousboe, A., Drejer, J., and Hertz, L., 1988, Uptake and release of glutamate and glutamine in neurons and astrocytes in primary cultures, in: "Glutamine and Glutamate in Mammals," E. Kvamme, ed., CRC Press, Boca Raton, Fl, 2: 21-38.
Shank, R.P., and Aprison, M.H., 1988, Glutamate as a neurotransmitter, in: "Glutamine and Glutamate in Mammals," E Kvamme, ed., CRC Press, Boca Raton, Fl, 2: 3-19.
Shank, R.P., and Campbell, G.leM., 1984, α-Ketoglutarate and malate uptake and metabolism by synaptosomes: Further evidence for an astrocyte to neuron metabolic shuttle, J. Neurochem., 42: 1153-1161.
Shank, R.P., Bennett, G.S., Freytag, S.D., and Campbell, G.L., 1985, Pyruvate carboxylase: an astrocyte-specific enzyme implicated in the replenishment of amino acid neurotransmitter pools, Brain Res., 329: 364-367.
Skattebol, A., and Triggle, D.J., 1987, $^{45}Ca^{2+}$ uptake in rat brain neurons: absence of sensitivity to the Ca^{2+} channel ligands nitrendipine and Bay K 8644. Can. J. Physiol. Pharmacol., 65: 344-347.
Sochocka, E., Code, W.E., Shuaib, A., and Hertz, L., 1991, Effects of ischemia on cultured neurons and astrocytes, Trans. Am. Soc. Neurochem., 22, in press.
Stone, E.A., and Ariano, M.A., 1989, Are glial cells targets of the central noradrenergic system? A review of the evidence, Brain Res. Rev., 14: 297-309.
Subbarao, K., and Hertz, L., 1990, Effects of adrenergic agonists on glycogenolysis in primary cultures of astrocytes, Brain Res., in press
Subbarao, K.V., and Hertz, L., 1991, Stimulation of energy metabolism in astrocytes by adrenergic agonists, J. Neurosci.Res., in press.
Sykova, E., 1983, Extracellular K^+ accumulation in the central nervous system, Prog. Biophys. Molec. Biol. 42: 135-189.
Tariot, P.N., Sunderland, T., Weingartner, H., Murphy, D.L., Welkowitz, J.A., Thompson, K., and Cohen, R.M., 1987, Cognitive effects of L-deprenyl in Alzheimer's disease. Psychopharmacol., 91: 489-495.
Tas, P.W.L., Massa, P.T., Kress, H.G. and Koschel, K., 1987, Characterization of a $Na^+/K^+/Cl^-$ co-transport in primary cultures of rat astrocytes. Bichim. Biophys. Acta. 903, 411-416.
Turner, T.J., and Goldin, S.M., 1988, Do dihydropyridine-sensitive calcium channels play a role in neurosecretion in the central nervous system? Ann. NY Acad, Sci., 522: 278-283.
Van den Berg, C.J. and Garfinkel, D., 1971, A simulation study of brain compartments. Metabolism of glutamate and related substances in mouse brain. Biochem. J. 123, 211-218.
Walz, W.,1989, Role of glial cells in the regulation of the brain ion microenvironment. Progress in Neurobiology 33, 309-333.
Walz, W., and Hertz, L., 1983, Functional interactions between neurons and astrocytes. II. Potassium homeostasis at the cellular level, Progr. Neurobiol., 20: 133-183.

Walz, W. and Hertz, L., 1984, Intense furosemide-sensitive potassium accumulation into astrocytes in the presence of pathologically high extracellular potassium levels. J. Cerebr. Blood Flow & Metab. 4: 301-304.
Whitaker-Azmitia, P.M. and Azmitia, E.C., 1989, Stimulation of astroglial serotonin receptors produces culture media which regulates growth of serotonergic neurons. Brain Res. 497: 80-85.
Young, M.A., Vatner, D.E., Knight, D.R., Graham, R.M., Homey, C.J., and Vatner, S.F., 1988, α-Adrenergic vasoconstriction and receptor subtypes in large coronary arteries of calves. Am. J. Physiol., 255: H1452-H1459.
Yu, A.C.H., and Hertz, L., 1983, Metabolic sources of energy in astrocytes, in: "Glutamine, Glutamate and GABA in the Central Nervous System," L. Hertz, E. Kvamme, E.G. McGeer, and A. Schousboe, eds., Alan R. Liss, NY, 431-439.
Yu, A.C.H., Schousboe, A., and Hertz, L., 1982, Metabolic fate of (^{14}C)-labelled glutamate in astrocytes, J. Neurochem., 39: 954-966.
Yu, A.C.H., Drejer, J., Hertz, L., and Schousboe, A., 1983, Pyruvate carboxylase activity in primary cultures of astrocytes and neurons, J. Neurochem., 41: 1484-1487.
Yu, A.C., Gregory, G.A., and Chan, P.H., 1989, Hypoxia-induced dysfunctions and injury of astrocytes in primary cell cultures. J. Cereb. Blood Flow & Metab. 9: 20-28.
Yudkoff, M., Nissim, I., Hummeler, K., Medow, M., and Pleasure, D., 1986, Utilization of (^{15}N)-glutamate by cultured astrocytes, Biochem. J., 234: 185-192.
Yudkoff, M., Nissim, I., and Hertz, L., 1990, Precursors of glutamic acid nitrogen in primary neuronal cultures: studies with 15N. Neurochem. Res., in press.

Note added in proof.

Direct evidence that sodium-dependent binding of D-aspartate does not represent glutamate uptake into presynaptic neurons has recently been obtained by Greenamyre et al. (1990).

Greenamyre, J.T., Higgins, D.S., and Young, A.B., 1990, Sodium-dependent D-aspartate binding in not a measure of presynaptic neuronal uptake sites in an autoradiographic assay, Brain Res. 511: 310-318.

STRUCTURE AND FUNCTION OF GLIA MATURATION FACTOR BETA

Ramon Lim and Asgar Zaheer

Division of Neurochemistry and Neurobiology,
Department of Neurology, University of Iowa
College of Medicine, Iowa City, IA 52242

In 1972 this laboratory observed the ability of brain extracts to promote the phenotypic expression of cultured astrocytes and named the active agent "glia maturation factor" (GMF) (Lim et al., 1972, 1973; Lim and Mitsunobu, 1974). Attempts at purifying this factor encountered a number of difficulties. First, it exists in low abundance. Second, the purer the protein is the more unstable it becomes. Third, as purification steps increase, the procedure becomes more and more laborious to carry through. These problems are common to many growth factors and therefore not unique to GMF. However, the purification of GMF was further complicated by the fact that the brain is a source of several other potent growth factors, some of which interfered with the monitoring of GMF.

A major advance was made in 1985 when we obtained a monoclonal antibody against GMF, designated G2-09, which adsorbed the activity of GMF (Lim et al., 1985). This antibody served as a useful chemical handle for tracking down GMF even in the presence of interfering growth factors and led to the isolation of GMF in 1988 (Lim and Miller, 1988; Lim et al., 1989). The purified protein, designated GMF-beta, is a 17 kDa acidic protein containing 141 amino acid residues. The complete sequence of the protein was obtained by automated Edman degradation and tandem mass spectrometry of overlapped peptide fragments generated by cyanogen bromide cleavage and enzymatic digestion with trypsin, chymotrypsin, and endoproteinases Asp-N and Lys-C (Lim et al., 1990a). The protein contains three cysteines, three methionines and one tryptophan and has no potential N-glycosylation sites. The amino-terminal serine residue is blocked by N-acetylation. The complete sequence of GMF-beta, which shows no significant homology with any other proteins reported, is depicted in Fig. 1.

From the sequence of the protein a series of oligonucleotide probes were synthesized to screen a human cDNA library. The cDNA was cloned and eventually expressed in E. coli. The deduced amino acid sequence of human GMF-beta turned out to be identical to that of the natural protein purified from bovine brain (Lim et al., 1990b). The high degree of conservation across diverse species suggests structural stringency and the fundamental importance of the protein.

GMF-beta exhibits a strong antiproliferative function on cultured tumor cells. With a few exceptions, both neural and non-neural tumors are inhibited (Lim et al., 1990c). The cells are arrested at the G_0/G_1

Plasticity and Regeneration of the Nervous System
Edited by P.S. Timiras *et al.*, Plenum Press, New York, 1991

```
        10         20         30         40         50
SESLVVCDVA EDLVEKLRKF RFRKETNNAA IIMKIDKDKR LVVLDEELEG

        60         70         80         90        100
ISPDELKDEL PERQPRFIVY SYKYQHDDGR VSYPLCFIFS SPVGCKPEQQ

       110        120        130        140
MMYAGSKNKL VQTAELTKVF EIRNTEDLTE EWLREKLGFF H
```

Fig. 1. Amino acid sequence of bovine glia maturation factor beta (GMF-beta). From Lim et al. (1990a). The one-letter abbreviations for the amino acids are: A, Ala; C, Cys; D, Asp; E, Glu; F, Phe; G, Gly; H, His; I, Ile; K, Lys; L, Leu; M, Met; N, Asn; P, Pro; Q, Gln; R, Arg; S, Ser; T, Thr; V, Val; W, Trp; and Y, Tyr.

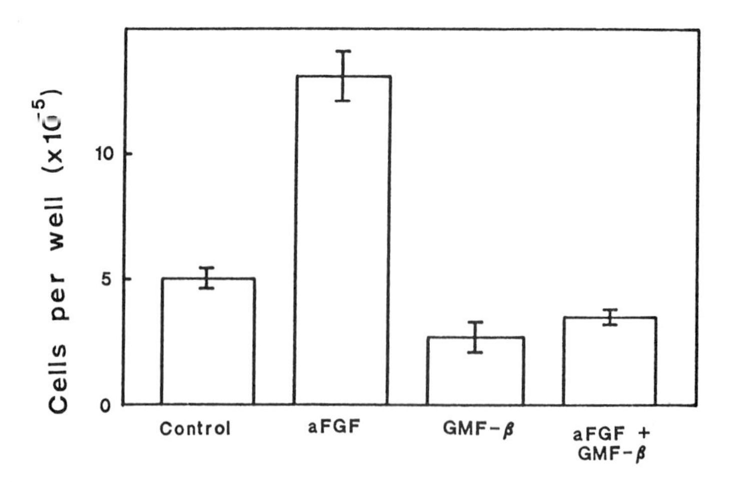

Fig. 2. GMF-beta inhibits the mitogenic effect of acidic fibroblast growth factor. C6 glioma cells were seeded at 2 x 10^5 cells per well in 24-well trays and grown for 3 days in serum-free defined medium in the presence of either acidic FGF (100 ng/ml), GMF-beta (250 ng/ml), or both. The cells were harvested and counted with a Coulter counter. The defined medium consisted of F12/DMEM (1:1; v/v) containing 10 nM hydrocortisone, 30 nM sodium selenite, 50 μg/ml transferrin, 10 ng/ml biotin, 5 μg/ml insulin and 1.5 μg/ml fibronectin. All values are means (± S.D.) of 4 wells. (From Lim et al., 1990c.)

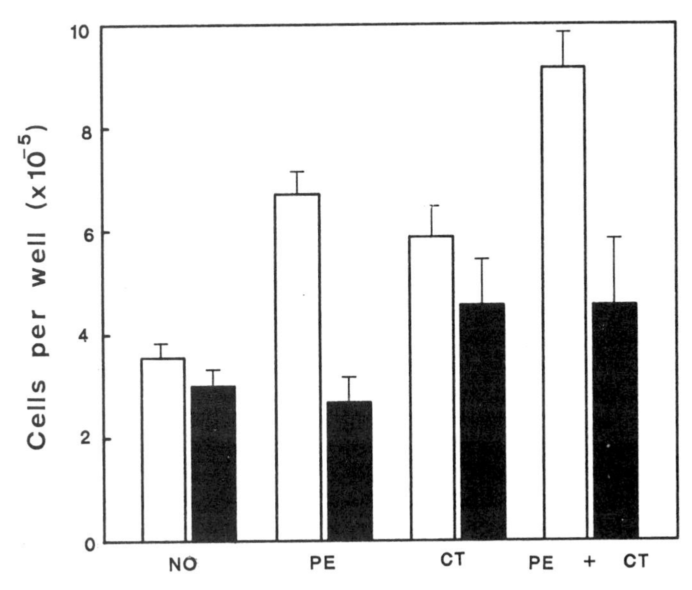

Fig. 3. GMF-beta inhibits the mitogenic effect of pituitary extract and cholera toxin on Schwann cells. Purified Schwann cells from primary cultures were seeded at 2 x 10^5 cells per well in 24-well trays and grown for 3 days in F12/DMEM containing 5% fetal calf serum under no stimulation (NO), or stimulation by pituitary extract (PE), cholera toxin (CT), or a combination of the two (PE + CT). Empty bars, cells cultured in the absence of GMF-beta; solid bars, in the presence of 250 ng/ml of GMF-beta. Pituitary extract was used at 500 μg/ml; cholera toxin (Sigma Chem. Co.), 1 μg/ml. All values are means (± S.D.) of 4 wells. (From Lim et al., 1990c).

phase and the inhibition is reversible. On C6 glioma cells, GMF-beta suppresses the mitogenic effect of acidic fibroblast growth factor (Fig. 2). On Schwann cells, GMF-beta antagonizes the proliferative action of pituitary extract and cholera toxin (Fig. 3).

GMF-beta adds to a small but growing list of proteins which may be products of tumor suppressor genes, as opposed to the mitogenic growth factors which are products of the protooncogenes. The characterization and understanding of the two distinct types of intercellular messages should help us define some of the decisive influences in development, regeneration and malignant transformation.

Acknowledgment: We thank Peggy Harris for the typing of this manuscript.

REFERENCES

Lim, R., Li, W.K.P., and Mitsunobu, K., 1972, Morphological transformation of dissociated embryonic brain cells in the presence of brain extracts. Abstracts, 2nd Meeting, Soc. Neuroscience, p. 181, Houston.

Lim, R., Mitsunobu, K., and Li, W.K.P., 1973, Maturation-stimulating effect of brain extract and dibutyryl cyclic AMP on dissociated embryonic brain cells in culture, Exp. Cell Res., 79:243-246.

Lim, R., and Mitsunobu, K., 1974, Brain cells in culture: morphological transformation by a protein, Science, 185:63-66.

Lim R., Miller J.F., Hicklin D.J., and Andresen A.A., 1985, Purification of bovine glia maturation factor and characterization with monoclonal antibody, Biochem., 24:8070-8074.

Lim, R., and Miller, J.F., 1988, Isolation and sequence analysis of GMF-beta: a neural growth factor, J. Cell Biol., 107:270a.

Lim, R., Miller, J.F., and Zaheer, A., 1989, Purification and characterization of glia maturation factor-β: a growth regulator for neurons and glia, Proc. Natl. Acad. Sci. USA, 86:3901-3905.

Lim, R., Zaheer, A., and Lane, W.S., 1990a, Complete amino acid sequence of bovine glia maturation factor β, Proc. Natl. Acad. Sci USA, 87:5233-5237.

Lim, R., Zaheer A., Kaplan, R. Jaye, M. and Lane, W.S., 1990b, Complete amino acid sequence of glia maturation factor β, Trans. Am. Soc. Neurochem., 21:295.

Lim, R., Zhong, W., and Zaheer, A., 1990c, Antiproliferative function of glia maturation factor β, Cell Regulation, 1:741-746.

NEUROMODULATORY ACTIONS OF GLUTAMATE, GABA AND TAURINE: REGULATORY ROLE OF ASTROCYTES

Arne Schousboe[1], Orla M. Larsson[1], Aase Frandsen[1], Bo Belhage[2], Herminia Pasantes-Morales[3] and Povl Krogsgaard-Larsen[4]

PharmaBiotec Research Center, Dept. of [1]Biology and [4]Organic Chemistry, Royal Danish School of Pharmacy, DK-2100 Copenhagen and Dept. of [2]Biochemistry A, Panum Institute, DK-2200 Copenhagen, Denmark. [3]Inst. of Cellular Physiology, National Univ. of Mexico, Mexico D.F., Mexico

INTRODUCTION

In addition to being actively involved in the general metabolism of the central nervous system, the amino acids glutamate and γ-aminobutyrate (GABA) serve as respectively excitatory and inhibitory neurotransmitters (Schousboe, 1990). This function is mediated by pharmacologically distinct receptors for each of the two amino acids. Receptors for glutamate are generally subdivided into 3 major classes exhibiting different pharmacological profiles, i.e. N-methyl-D-aspartate-(NMDA), kainate-, and quisqualate/AMPA (RS-α-amino-3-hydroxy-5-methyl-4-isoxazolo-propionate)-receptors (Watkins and Olverman, 1987). Likewise, GABA receptors are subdivided into two major classes referred to as $GABA_A$ and $GABA_B$ receptors (Johnston et al., 1984). It should, however, be emphasized that diversities exist within these respective classes of receptors as binding sites with different affinities for the amino acids have been observed (Johnston et al., 1984; Drejer & Honoré, 1988). Such a diversity is consistent with the discovery of a large number of subunits of these receptors with different amino acid sequences (Schofield et al., 1987; Hollmann et al., 1989). This, in turn, may be reflected by the multiple functions of these receptors such as involvement in modulation of neuro-

Plasticity and Regeneration of the Nervous System
Edited by P.S. Timiras *et al.*, Plenum Press, New York, 1991

transmitter release and mediation of trophic and toxic actions. These functional aspects will be discussed below. Since the amino acid taurine is able to mimic a number of events mediated by GABA (Huxtable, 1989), this amino acid will also be included in the discussions.

During recent years it has become evident that astrocytes which are localized in close proximity to synapses (Guldner and Wolff, 1973) may be actively involved in the control and regulation of a number of these neuromodulatory actions of glutamate, GABA and taurine. This is based on the repeated demonstrations that astrocytes have very active high affinity transport systems for these amino acids (Schousboe, 1981; 1982) which enable these cells to regulate the extracellular synaptic concentration of the amino acids. Since there may also be a release of these amino acids from astrocytes under a variety of physiological and pathological conditions, such processes could conceivably also contribute to the maintenance and balance of the extrasynaptic concentration of these amino acids. The present review will emphasize these astrocytic functions.

NEURONAL ACTIONS OF GLUTAMATE

1. Receptor-mediated GABA release

Using different preparations of cultured neurons it has been demonstrated that glutamate is able to stimulate GABA release (Drejer *et al.*, 1987; Weiss, 1988). At least part of this GABA release is Ca^{++} dependent but release occurs also *via* a Ca^{++} independent mechanism which presumably involves a reversal of the GABA carrier secondary to a depolarization coupled disruption of the sodium gradient (Pin & Bockaert, 1989; Dunlop *et al.*, 1989). It is, however, quite clear that glutamate receptors are involved regardless of the precise mechanism for GABA release. This conclusion is based on a pharmacological characterization of the action of glutamate. In cerebral cortex neurons, release of GABA can be stimulated not only by glutamate or sulphur containing excitatory amino acids (Dunlop *et al.*, 1989) but also by the glutamate receptor subtype selective agonists quisqualate, kainate and NMDA (Drejer *et al.*, 1987). Moreover, this excitatory amino acid mediated GABA release can be blocked by either NMDA receptor antagonists like APV (5-aminophosphonovaleric acid) or Mg^{++} (Drejer *et al.*, 1987) or non-NMDA antagonists such as DNQX (6,7-dinitroquinoxaline-2,3-dione) (Dunlop *et al.*, 1989). A similar pharmacological profile has been reported for excitatory amino acid stimulated GABA release in striatal neurons (Pin et al., 1988).

2. Receptor-mediated glutamate release

Using glutamatergic cerebellar granule cells in culture it was first demonstrated by Drejer et al. (1986) that release of transmitter glutamate can be stimulated by glutamate acting on glutamate receptors exhibiting a somewhat unusual pharmacology, i.e. they are only sensitive to selective glutamate subtype agonists at very high concentrations. Similar, although pharmacologically somewhat different results have been reported by Gallo et al. (1987). Such an atypical glutamate receptor subtype has also been described in other preparations both in vitro and in vivo (Davies et al., 1988; Frandsen et al., 1989b; Greenamyre et al., 1990). Again, these glutamate receptors are sensitive to the sulphur containing excitatory amino acids homocysteic acid and cysteine sulphonic acid (Dunlop et al., 1989; 1990).

3. Neurotoxicity

Excessive exposure of neurons to glutamate may lead to irreversible damage and finally cell death (Rothman & Olney, 1986). Since the extracellular glutamate concentration in the brain increases dramatically under pathological conditions such as ischemia and hypoglycemia (Benveniste et al., 1984; Sandberg et al., 1986) it has repeatedly been proposed that neuronal degeneration associated with such disturbances may involve glutamate induced neurotoxicity (cf. Choi, 1988). The mechanism for this excitotoxic action has been extensively studied during the past few years and cultured neurons have proven to be useful as a model system (Rothman, 1984; Frandsen & Schousboe, 1987; 1990; Choi et al., 1987; 1988). While it is clear that at least the toxic effects seen after prolonged exposure of the cells to glutamate are mediated independently by the different subtypes of glutamate receptors (Table I) the exact mechanism is largely unknown. It appears, however, that a persistent elevation of the intracellular Ca^{++} concentration plays a prominent role (Wahl et al., 1989). This increase in the intracellular Ca^{++} concentration is brought about by an increase in Ca^{++}-influx as well as a release of Ca^{++} from intracellular stores (Nicoletti et al., 1986; Bouchelouche et al., 1989; Wahl et al., 1989).

While the availability of a relatively large number of competitive and non-competitive antagonists specific for the NMDA receptor has facilitated the investigation of the involvement of this glutamate receptor subtype

Table I. Excitatory amino acid induced leakage of lactate dehydrogenase (LDH) from cultured cerebral cortex neurons.

Excitatory amino acid (μM)		None	LDH activity in media (% of total) with antagonists (μM) APV (100)	CNQX (10)	DNQX (10)	APV+CNQX	APV+DNQX
L-glutamate	(100)	61	53	48	49	14	0
Kainate	(300)	57	60	20	3	ND	ND
AMPA	(10)	59	52	15	9	ND	ND
NMDA	(100)	62	5	60	59	ND	ND

Cerebral cortex neurons were exposed for 4 h to 100 μM L-glutamate (plus 500 μM L-aspartic acid ß-hydroxamate), 300 μM KA, 10 μM AMPA, or 100 μM NMDA or to the agonists plus the EAA antagonists APV (100 μM), CNQX (10 μM), and DNQX (10 μM) alone or in combination as indicated. The cytotoxicity was evaluated by measurement of LDH activity in the culture media, expressed as a percentage of the total activity per culture. Fifty-five percent activity was equivalent to cell death (Frandsen and Schousboe, 1987). Values, corrected for the appropriate controls, are averages of six experiments with an SEM of ≤6%. Exposure of the cultures to 10 μM CNQX alone resulted in LDH activity of 15% after 4 h of incubation, whereas 100 μM APV, 10 μM DNQX, or 500 μM L-aspartic acid ß-hydroxamate alone induced no LDH activity in the media. ND, not determined. From Frandsen et al. (1989a).

in excitotoxicity, it has been difficult to study separately the individual importance of kainate and quisqualate/AMPA receptors. The development of the quinoxalinediones CNQX (6-cyano-7-nitroquinoxaline-2,3-dione) and DNQX (Honoré et al., 1988) has played a major role for the understanding of these non-NMDA receptors. These compounds have, however, been unable to distinguish between the non-NMDA receptor subtypes. In this context it should be mentioned that a newly developed drug with a similar structure, NBQX (2,3-dihydroxy-6-nitro-7-sulphamoyl-benzo(F)-quinoxaline) apparently is able to distinguish between kainate and AMPA receptors (Sheardown et al., 1990). This appears also to be the case for a completely different group of glutamate antagonists (Krogsgaard-Larsen et al., 1990; Frandsen et al., 1990a). As seen in Table II these compounds protect selectively against kainate induced toxicity in cultured neurons since AMPA induced toxicity could not be prevented by AMOA (2-amino-3-[3-(carboxymethoxy)-5-methylisoxazol-4-yl]propionic acid) and AMNH (2-amino-3-[2-(3-hydroxy-5-methyl-isoxazol-4-yl)methyl-5-methyl-3-oxoisoxazolin-4-yl]propionic acid). Interestingly, phenobarbital has also been reported to protect neurons selectively against kainate toxicity (Frandsen et al., 1990b).

Table II. Effects of AMOA and AMNH on excitatory amino acid induced neurotoxicity in cultured cerebral cortex neurons

Agonist (μM)	LDH activity in media (% of total activity) with antagonist (μM): none	AMOA (300)	AMNH (500)
None	13	8	14
KA (500)	72	10*	19*
AMPA (100)	66	65	71
NMDA (500)	97	70	80
L-Glu(10)	85	81	79
L-Asp(10)	90	93	88

Cerebral cortex neurons cultured for 9 days were exposed for 24 h to various EAAs with or without the addition of AMOA (300 μM) or AMNH (500 μM) respectively. To ensure a constant concentration of L-Glu and L-Asp, 500 μM of the uptake inhibitor L-Aspartate-β-hydroxamate was added in these experiments. Results are means of 9-12 experiments with SEM values <10%. Asterisks indicate statistically significant differences from the KA-treated cultures ($P < 0.001$; Student's t-test). From Frandsen et al. (1990a).

NEURONAL ACTIONS OF GABA AND TAURINE

1. GABA receptor expression

From in vivo and in vitro studies it has been documented that expression of $GABA_A$ receptors is influenced by GABA and GABA agonists (Redburn & Schousboe, 1987). Cerebellar granule cells in culture which in the absence of GABA express only high affinity $GABA_A$ receptors (Meier & Schousboe, 1982), have proven useful for studies of the mechanism of this action of GABA. It has been shown that this action is mediated by high affinity GABA receptors (Belhage et al., 1986) and that induction of low affinity GABA receptors requires de novo protein synthesis (Belhage et al., 1990a). Moreover, it has been shown that chloride channels coupled to the high affinity GABA receptors are involved (Belhage et al., 1990b). It should also be mentioned that taurine shares this ability of GABA to induce low affinity GABA receptors. This can be explained by an action of taurine on high affinity GABA receptors which are sensitive to taurine at concentrations exceeding 10 μM (Abraham & Schousboe, 1989). The functional importance of the low affinity GABA receptors is related to glutamatergic activity since they mediate the attenuating effect of GABA on evoked glutamate release from cerebellar granule cells (Meier et al., 1984; Belhage et al., 1990a). This means that GABA is not only directly involved in the regulation of excitatory activity mediated by glutamate but also regulates the expression of the receptors necessary for this function.

2. Seizure activity

Any imbalance between excitatory and inhibitory processes is likely to result in seizure activity (Schousboe, 1990). Such imbalances may be the result of increased glutamatergic activity due to excessive release of glutamate and/or a dysfunction of the inactivation processes involving high affinity uptake into astrocytes (cf. below). Alternatively, there may be a deficit in GABAergic activity secondary to diminished synthesis and release of GABA. In agreement with this, it is well documented that drugs inhibiting the GABA synthesizing enzyme glutamate decarboxylase are normally potent convulsants (Tapia, 1975) whereas drugs which inhibit GABA degradation *via* GABA-transaminase often act as anticonvulsants (Wood, 1975). Since GABA is inactivated by high affinity uptake into neurons and glia (cf. below), drugs acting on these processes are also of interest as anticonvulsants (Schousboe, 1990).

REGULATORY ROLE OF ASTROCYTES

1. High affinity uptake

a. Glutamate

From studies of glutamate uptake into cultured astrocytes and different types of neurons including glutamatergic cerebellar granule cells it can be concluded that astrocytes and glutamatergic neurons have a high capacity for glutamate uptake (Schousboe *et al.*, 1988). The astrocytic uptake is likely to be somewhat higher than that into neurons which means that the astrocytes play a key role in the maintenance of an extracellular glutamate concentration, a notion compatible with normal function of glutamate mediated excitatory activity. In agreement with this, it has been found that astrocytic glutamate uptake is impaired during ischemia (Drejer *et al.*, 1985) where the extracellular glutamate concentration greatly exceeds normal levels and reaches a value (Benveniste *et al.*, 1984) at which excitotoxic damage to neurons is likely to occur (cf. above). Another indication for the regulation of glutamatergic activity by astrocytes comes from the observation that glutamate uptake is particularly efficient in astrocytes originating from brain areas with a high level of glutamatergic activity (Schousboe & Divac, 1979). Moreover, expression of glutamate carriers in the astrocytic plasma membrane is influenced by macromolecules synthesized and released from neurons (Drejer *et al.*, 1983) indicating that this astrocytic function may be tightly coupled to neuronal activity.

b. GABA and taurine

From numerous studies of GABA and taurine uptake in different preparations of astrocytes it is clear that these cells possess an efficient high affinity uptake system for these amino acids (cf. Schousboe, 1982). However, in contrast to glutamate, astrocytic GABA and taurine uptake does not appear to be more efficient than the corresponding uptake systems in neurons (Schousboe _et al._, 1988). It has been suggested (Holopainen & Kontro, 1986) that taurine and GABA may share the same carrier in astrocytes. However, detailed kinetic analyses of GABA and taurine uptake in neurons, astrocytes or synaptosomes (Larsson _et al._, 1986; Debler & Lajtha, 1987) have led to the conclusion that taurine and GABA utilize distinct and selective carriers. Taurine and GABA are likely to be involved in different functional processes in the brain (cf. Schousboe _et al._, 1990a). Therefore, it seems likely that their removal from the extracellular space is subject to different regulatory processes.

Table III. Inhibition of neuronal and glial GABA uptake by GABA analogues of restricted conformation.

GABA analogue	Ki (μM) Neuron	Ki (μM) Glia
(_R_)-nipecotic acid	11	15
Guvacine	31	28
N-methyl-nipecotic acid	74	94
Cis-4-OH-nipecotic acid	53	148[a]
SKF-89976 A (DPB-Nip.)	1	2
SKF-100330 A (DPB-Guv.)	5	4
THPO	--	550
THAO	--	600
N-DPB-THPO	38	26
N-DPB-THAO	9	3

[a]Indicates a non-competetive inhibition. In all other cases competitive inhibition was observed. Modified from Schousboe _et al._ (1990b).
DPB-Nip. ((_RS_)-N-(4,4-diphenyl-3-butenyl)nipecotic acid)
DPB-Guv. (N-(4,4-diphenyl-3-butenyl)guvacine)
THPO (4,5,6,7-tetrahydroisoxazolo[4,5-_c_]pyridin-3-ol)
THAO (5,6,7,8-tetrahydro-4H-isoxazolo[4,5-_c_]azepin-3-ol)
N-DPB-THPO (N-(4,4-diphenyl-3-butenyl)4,5,6,7-tetrahydroisoxazolo[4,5-_c_]pyridin-3-ol)
N-DPB-THAO (N-(4,4-diphenyl-3-butenyl)5,6,7,8-tetrahydro-4H-isoxazolo[4,5-_c_]azepin-3-ol

While little is known about possible differences between neurons and glia with regard to the substrate specificity of the taurine high affinity carrier, detailed studies of the substrate specificity of the corresponding GABA carriers are available (Krogsgaard-Larsen _et al._, 1987; Schousboe _et al._, 1990b; Schousboe, 1990). As can be seen from Table III a number of GABA analogues are selective inhibitors of astrocytic GABA uptake. Since astrocytic GABA uptake may prevent synaptically released GABA from being reutilized as a neurotransmitter (Schousboe, 1979) such GABA uptake inhibitors may act as anticonvulsant drugs. By preventing GABA uptake into astrocytes such drugs would facilitate GABA reuptake into nerve endings where GABA would enter the neurotransmitter pool.

As seen from Table IV such compounds are indeed anticonvulsants and at least in some chemical seizure models they are more efficacious than GABA analogues acting as potent inhibitors of both neuronal and glial GABA uptake (Gonsalves _et al._, 1989a). These results also indirectly indicate that astroglial GABA uptake indeed participates in the control and regulation of the extracellular (synaptic) concentration of GABA. That this is the case is further substantiated by the demonstration that neuronally released factors are able to induce GABA carriers in the astrocytic plasma membrane (Drejer _et al._, 1983). A neuronal protein with these properties is currently being purified from conditioned media of cultured cerebellar granule cells (J. Nissen, I. Schousboe and A. Schousboe, unpublished).

Table IV. Semiquantitative summary of antiseizure effects of GABAmimetics.

Treatment	Seizure model		Proconvulsant activity
	Max PTZ	INH	
THPO	++++	+	No
THAO	++++	+	No
Cis-4-OH-nipecotic acid	0	0	No
DABA	++++	0	Yes

For maximal PTZ seizures, only compounds protecting at least 50% of animals against the tonic extensor component were considered anticonvulsant. Degree of anticonvulsant activity is indicated as follows: + = ≤25% increase over CSF control; ++ = >25%, ≤50% increase; +++ = >50%, ≤100% increase; ++++ = >100% increase. Proconvulsant activity was recorded as present or absent. Original data are from Gonsalves _et al._ (1989a,b).
DABA: 2,4-Diaminobutyric acid
PTZ: Pentylenetetrazol
INH: Isonicotinic acid hydrazide

2. Release from astrocytes

a. Glutamate

A potassium-stimulated release of glutamate from astrocytes has been reported (Schousboe et al., 1989; Westergaard et al., 1991) but it does not appear to be Ca^{++}-dependent. Since exposure of astrocytes to K^+ induces cell swelling (Pasantes-Morales & Schousboe, 1989) and such swelling to some extent may elicit glutamate release (Pasantes-Morales & Schousboe, 1988) it is possible that this process may be related to volume regulation (cf. below). In any case, since glutamate will exert neurotoxic actions at elevated extracellular concentrations (cf. above) an excessive release from astrocytes could have an adverse effect on neuronal function.

b. GABA

Under normal conditions, there is only a limited release of GABA from astrocytes (Schousboe et al., 1988). However, since excitatory amino acids acting on astrocytic glutamate receptors have been reported to stimulate GABA release (Gallo et al., 1989) it is possible that GABA could be released from astrocytes during excitatory activity. Such a GABA-release may, in turn, produce an attenuation of this excitatory activity since GABA is able to inhibit glutamate release (cf. above).

c. Taurine

Astrocytes have a very high content of taurine (Pasantes-Morales & Schousboe, 1988) which may be a consequence of an efficient high affinity uptake (cf. above) and a very slow spontaneous release (Schousboe et al., 1988; Pasantes-Morales & Schousboe, 1988; 1989). However, as can be seen

Table V. Potassium and hyposmolarity-stimulated release of ^{3}H-taurine from cultured cerebral cortex astrocytes.

Condition	Taurine release (%)
Isosmotic, 5mM KCl	4.7 ± 0.4 (16)
Isosmotic, 56mM KCl	15.6 ± 1.6 (7)
Hyposmotic (50%)	65.3 ± 2.2 (16)

Cells were preloaded with ^{3}H-taurine for 30 min and subsequently release of ^{3}H-taurine was followed using a superfusion system (Drejer et al., 1987). Results are expressed as fractional release, i.e. percent of total taurine released during a 4 min period. Results represent averages ± SEM of the number of experiments shown in parentheses.
From Pasantes-Morales & Schousboe (1988;1989).

from Table V, release of taurine from astrocytes is dramatically stimulated during swelling of the cells elicited by exposure to either hyposmotic conditions or elevated potassium concentrations (Pasantes-Morales & Schousboe, 1988; 1989). Since this release of taurine is sensitive to even small changes in osmolarity (Pasantes-Morales et al., 1990) or extracellular potassium (Pasantes-Morales & Schousboe, 1989) it could play a functional role under normal neuronal activity. Moreover, under pathological conditions such as ischemia and hypoglycemia where the extracellular concentration of potassium is markedly increased (Hansen, 1978), such a volume sensitive release of taurine is likely to be of a considerable magnitude. This is consistent with the observation that the extracellular concentration of taurine in the brain measured by the microdialysis technique increases significantly under these conditions (Benveniste et al., 1984; Sandberg et al., 1986). Since taurine may have a depressant action on neuronal activity and a protective effect on neuronal membranes (Lehmann et al., 1984) these astrocytic release processes may be important as a neuroprotective mechanism. It should also be mentioned that not only astrocytes but also neurons may be able to release large quantities of taurine subsequent to increases in cell volume (Schousboe et al., 1990c).

REFERENCES

Abraham, J.H. and Schousboe, A., 1989, Effects of taurine on cell morphology and expression of low-affinity GABA receptors in cultured cerebellar granule cells. Neurochem. Res. 14:1031.

Belhage, B., Meier, E., and Schousboe, A., 1986, GABA-agonists induce the formation of low-affinity GABA-receptors on cultured cerebellar granule cells via preexisting high affinity GABA receptors. Neurochem. Res. 11:599.

Belhage, B., Hansen, G.H., Meier, E. and Schousboe, A., 1990a, Effects of inhibitors of protein synthesis and intracellular transport on the GABA-agonist induced functional differentiation of cultured cerebellar granule cells. J. Neurochem. 55:1107.

Belhage, B., Hansen, G.H. and Schousboe, A., 1990b, GABA agonist induced changes in ultrastructure and GABA receptor expression in cerebellar granule cells is linked to hyperpolarization of the neurons. Int. J. Devl. Neurosci. 8:473.

Benveniste, H., Drejer, J., Schousboe, A., and Diemer, N.H., 1984, Elevation of the extracellular concentrations of glutamate and aspartate in rat hippocampus during transient cerebral ischemia monitored by intracerebral microdialysis. J. Neurochem. 43:1369.

Bouchelouche, P., Belhage, B., Frandsen, A., Drejer, J., and Schousboe, A., 1989, Glutamate receptor activation in cultured cerebellar granule cells increases cytosolic free Ca^{2+} by mobilization of cellular Ca^{2+} and activation of Ca^{2+} influx. Exp. Brain Res. 76:281.

Choi, D.W., 1988, Glutamate neurotoxicity and diseases of the nervous system. Neuron 1:623.

Choi, D.W., Maulucci-Gedde, M.A., and Kriegstein, A.R., 1987, Glutamate neurotoxicity in cortical cell culture. J. Neurosci. 7:357.

Choi, D.W., Koh, J., and Peters, S., 1988, Pharmacology of glutamate neurotoxicity in cortical cell culture: Attenuation by NMDA antagonists. J. Neurosci. 8:185.

Davies, S.N., Fletcher, E.J., and Lodge, D., 1988, Evidence for a fourth glutamate receptor subtype on rat central neurones in vivo and in vitro. J. Physiol. 406:15P.

Debler, E.A., and Lajtha, A., 1987, High-affinity transport of γ-aminobutyric acid, glycine, taurine, L-aspartic acid, and L-glutamic acid in synaptosomal (P_2) tissue: A kinetic and substrate specificity analysis. J. Neurochem. 48:1851.

Drejer, J., and Honoré, T., 1988, Excitatory amino acid receptors, in: "Glutamine and Glutamate in Mammals," E. Kvamme, ed., Volume 2, CRC Press, FL, 89.

Drejer, J., Meier, E., and Schousboe, A., 1983, Novel neuron-related regulatory mechanisms for astrocytic glutamate and GABA high affinity uptake. Neurosci. Lett. 37:301.

Drejer, J., Benveniste, H., Diemer, N.H., and Schousboe, A., 1985, Cellular origin of ischemia-induced glutamate release from brain tissue in vivo and in vitro. J. Neurochem. 45:145.

Drejer, J., Honoré, T., Meier, E., and Schousboe, A., 1986, Pharmacologically distinct glutamate receptors on cerebellar granule cells. Life Sci. 38:2077.

Drejer, J., Honoré, T., and Schousboe, A., 1987, Excitatory amino acid-induced release of ^{3}H-GABA from cultured mouse cerebral cortex interneurons. J. Neurosci. 7:2910.

Dunlop, J., Grieve, A., Schousboe, A., and Griffiths, R., 1989, Neuroactive sulphur amino acids evoke a calcium-dependent transmitter release from cultured neurones that is sensitive to excitatory amino acid receptor antagonists. J. Neurochem. 52:1648.

Dunlop, J., Grieve, A., Schousboe, A., and Griffiths, R., 1990, Characterization of the receptor-mediated sulphur amino acid-evoked release of [^{3}H]D-aspartate from primary cultures of cerebellar granule cells. Neurochem. Int. 16:119.

Frandsen, A., and Schousboe, A., 1987, Time and concentration dependency of the toxicity of excitatory amino acids on cerebral neurones in primary culture. Neurochem. Int. 10:583.

Frandsen, A. and Schousboe, A., 1990, Development of excitatory amino acid induced cytotoxicity in cultured neurones. Int. J. Devl. Neurosci., 8:209.

Frandsen, A., Drejer, J., and Schousboe, A., 1989a, Direct evidence that excitotoxicity in cultured neurons is mediated via N-methyl-D-aspartate (NMDA) as well as non-NMDA receptors. J. Neurochem. 53:297.

Frandsen, A., Drejer, J., and Schousboe, A., 1989b, Glutamate-induced $^{45}Ca^{2+}$ uptake into immature cerebral cortex neurons shows a distinct pharmacological profile. J. Neurochem. 53:1959.

Frandsen, A., Krogsgaard-Larsen, P. and Schousboe, A., 1990a, Novel glutamate receptor antagonists selectively protect against kainic acid neurotoxicity in cultured cerebral cortex neurons. J. Neurochem. 55: in press.

Frandsen, A., Quistorff, B. and Schousboe, A., 1990b, Phenobarbital protects cerebral cortex neurones against toxicity induced by kainate but not by other excitatory amino acids. Neurosci. Lett. 11:233.

Gallo, V., Suergiu, R., Giovannini, C., and Levi, G., 1987, Glutamate receptor subtypes in cultured cerebellar neurons: Modulation of glutamate and γ-aminobutyric acid release. J. Neurochem. 49: 1801.

Gallo, V., Suergiu, R., Giovannini, C., and Levi, G., 1989, Expression of excitatory amino acid receptors by cerebellar cells of the type-2 astrocyte cell lineage. J. Neurochem. 52:1.

Gonsalves, S.F., Twitchell, B., Harbaugh, R.E., Krogsgaard-Larsen, P., and Schousboe, A., 1989a, Anticonvulsant activity of intracerebroventricularly administered glial GABA uptake inhibitors and other GABAmimetics in chemical seizure models. Epilepsy Res. 4:34.

Gonsalves, S.F., Twitchell, B., Harbaugh, R.E., Krogsgaard-Larsen, P., and Schousboe, A., 1989b, Anticonvulsant activity of the glial GABA uptake inhibitor, THAO, in chemical seizures. Eur. J. Pharmacol. 168:265.

Greenamyre, J.T., Higgins, D.S., Young, A.B., and Penney, J.B., 1990, Regional ontogeny of a unique glutamate recognition site in rat brain: An autoradiographic study. Int. J. Devl. Neurosci. 8:437.

Guldner, F.H., and Wolff, J.R., 1973, Neuron-glial synaptoid contacts of the median eminence of the rat: Ultrastructure, staining properties, and distribution on tanycytes. Brain Res. 61:217.

Hansen, A.J., 1978, The extracellular potassium concentration in brain cortex following ischemia in hypo- and hyperglycemic rats. Acta Physiol. Scand. 102:324.

Holopainen, I., and Kontro, P., 1986, High-affinity uptake of taurine and ß-alanine in primary cultures of rat astrocytes. Neurochem. Res. 11:207.

Hollmann, M., O'Shear-Greenfield, A., Rogers, S.W., and Heinemann, S., 1989, Cloning by functional expression of a member of the glutamate receptor family. Nature 342:643.

Honoré, T., Davis, S.N., Drejer, J., Fletcher, J.E., Jacobsen, P., Lodge, D., and Nielsen, F.E., 1988, Quinoxalinediones: Potent competitive non-NMDA glutamate receptor antagonists. Science 241:701.

Huxtable, R.J., 1989, Taurine in the central nervous system and the mammalian actions of taurine. Prog. Neurobiol. 32:471.

Johnston, G.A.R., Allan, R.D., and Skerritt, J.H., 1984, GABA receptors, in: "Handbook of Neurochemistry, 2nd. Edition, Vol. 6," A. Lajtha, ed., Plenum Press, New York, 213.

Krogsgaard-Larsen, P., Falch, E., Larsson, O.M., and Schousboe, A., 1987, GABA uptake inhibitors: Relevance to antiepileptic drug research. Epilepsy Res. 1:77.

Krogsgaard-Larsen, P., Ferkany, J.W., Nielsen, E.Ø., Madsen, U., Ebert, B., Johansen, J.S., Diemer, N.H., Bruhn, T., Beattie, D.T., and Curtis, D.R., 1990, Novel class of antagonists at non-N-methyl-D-aspartic acid excitatory amino acid receptors. Synthesis, in vitro and in vivo pharmacology and neuroprotection. J. Med. Chem., in press.

Larsson, O.M., Griffiths, R., Allan, I.C., and Schousboe, A., 1986, Mutual inhibition kinetic analysis of γ-aminobutyric acid, taurine and ß-alanine high affinity transport into neurons and astrocytes: evidence for similarity between the taurine and ß-alanine carriers in both cell types. J. Neurochem. 47:426.

Lehmann, A., Hagberg, H., and Hamberger, A., 1984, A role for taurine in the maintenance of homeostasis in the central nervous system during hyperexcitation? Neurosci. Lett. 52:341.

Meier, E., and Schousboe, A., 1982, Differences between GABA receptor binding to membranes from cerebellum during postnatal development and from cultured cerebellar granule cells. Dev. Neurosci. 5:546.

Meier, E., Drejer, J., and Schousboe, A., 1984, GABA induces functionally active low-affinity GABA receptors on cultured cerebellar granule cells. J. Neurochem. 43:1737.

Pasantes-Morales, H., and Schousboe, A., 1988, Volume regulation in astrocytes: A role for taurine as an osmoeffector. J. Neurosci. Res. 20:505.

Pasantes-Morales, H., and Schousboe, A., 1989, Release of taurine from astrocytes during potassium-evoked swelling. Glia 2:45.

Pasantes-Morales, H., Morán, J., and Schousboe, A., 1990, Volume-sensitive release of taurine from cultured astrocytes: Properties and mechanism. Glia 3: in press.

Pin, J.-P., and Bockaert, J., 1989, Two distinct mechanisms, differentially affected by excitatory amino acids, trigger GABA release from fetal mouse striatal neurons in primary culture. J. Neurosci. 9:648.

Pin, J.-P., Van-Vliet, B.J., and Bockaert, J., 1988, NMDA- and Kainate-evoked GABA release from striatal neurones differentiated in primary culture: Differential blocking by phencyclidine. Neurosci. Lett. 87:87.

Rothman, S., 1984, Synaptic release of excitatory amino acid neurotransmitter mediates anoxic neuronal death. J. Neurosci. 4:1884.

Rothman, S., and Olney, J.W., 1986, Glutamate and the pathophysiology of hypoxic ischemic brain damage. Ann. Neurol. 19:105.

Sandberg, S., Butcher, S.P., and Hagberg, H., 1986, Extracellular overflow of neuroactive amino acids during severe insulin-induced hypoglycemia in vivo dialysis of the rat hippocampus. J. Neurochem. 47:178.

Schofield, P.R., Darlison, M.G., Fujita, N., Burt, D.R., Stephenson, F.A., Rodriguez, H., Rhee, L.M., Ramachandran, J., Reale, V., Glencorse, T.A., Seeburg, P.H., and Barnard, E.A., 1987, Sequence and functional expression of the $GABA_A$ receptor show a ligand-gated receptor superfamily. Nature 328:221.

Schousboe, A., 1979, Effects of GABA-analogues on the high-affinity uptake of GABA in astrocytes in primary cultures, in: "GABA-Biochemistry and CNS Functions," P. Mandel, and DeFeudis, F.V., eds., Plenum Publ. Corp., New York, 219.

Schousboe, A., 1981, Transport and metabolism of glutamate and GABA in neurons and glial cells. Int. Rev. Neurobiol. 22:1.

Schousboe, A., 1982, Metabolism and function of neurotransmitters, in: "Neuroscience Approached through Cell Culture, Vol. 1," S.E. Pfeiffer, ed., CRC Press, Boca Raton, FL, 108.

Schousboe, A., 1990, Neurochemical alterations associated with epilepsy or seizure activity, in: "Comprehensive Epileptology," M. Dam, and Gram, L., eds., Raven Press, New York, 1.

Schousboe, A., and Divac, I., 1979, Differences in glutamate uptake in astrocytes cultured from different brain regions. Brain Res. 177:407.

Schousboe, A., Larsson, O.M., Krogsgaard-Larsen, P., Drejer, J., and Hertz, L., 1988, Uptake and release processes for neurotransmitter amino acids in astrocytes, in: "The Biochemical Pathology of Astrocytes," M.D. Norenberg, Hertz, L., and Schousboe, A., eds., Alan R. Liss, New York, 381.

Schousboe, A., Frandsen, A., and Drejer, J., 1989, Evidence for evoked release of adenosine and glutamate from cultured cerebellar granule cells. Neurochem. Res., 14:871.

Schousboe, A., Sánchez-Olea, R., and Pasantes-Morales, H., 1990a, Depolarization induced neuronal release of taurine in relation to synaptic transmission: Comparison with GABA and glutamate, in: "Functional Neurochemistry of Taurine," H. Pasantes-Morales, Shain, W., Martin D.L., and del Rio, R.M., eds., Alan R. Liss, New York, 289.

Schousboe, A., Krogsgaard-Larsen, P., Larsson, O.M., Gonsalves, S.F., Harbaugh, R.E., and Wood, J.D., 1990b, GABA uptake inhibitors: Possible use as anti-epileptic drugs, in: "Amino Acids, Chemistry, Biology and Medicine," G. Lubec, and Rosenthal, G.A., eds., ESCOM Science Publishers B.V., Leiden, 345.

Schousboe, A., Moran, J., and Pasantes-Morales, H., 1990c, Potassium-stimulated release of taurine from cultured cerebellar granule neurons is associated with cell swelling. J. Neurosci. Res., in press.

Sheardown, M.J., Nielsen, E.Ø., Hansen, A.J., Jacobsen, P., and Honoré, T., 1990, 2,3-Dihydroxy-6-nitro-7-sulphamoyl-benzo(F)quinoxaline: A neuroprotectant for cerebral ischemia. Science 247:571.

Tapia, R., 1975, Biochemical Pharmacology of GABA in CNS, in: "Handbook of Psychopharmacology, Vol. 4," L.L. Iversen, Iversen, S.D., and Snyder S.H., eds., Plenum Publ. Corp., New York, 1.

Wahl, P., Schousboe, A., Honoré, T., and Drejer, J., 1989, Glutamate-induced increase in intracellular Ca^{2+} in cerebral cortex neurons is transient in immature cells but permanent in mature cells. J. Neurochem. 53:1316.

Watkins, J.C., and Olverman, H.J., 1987, Agonists and antagonists for excitatory amino acid receptors. Trends Neurosci. 10:265.

Weiss, S., 1988, Excitatory amino acid-evoked release of gamma-[^{3}H]-aminobutyric acid from striatal neurones in primary culture. J. Neurochem. 51:435.

Westergaard, N., Fosmark, H., and Schousboe, A., 1991, Metabolism and release of glutamate in cerebellar granule cells cocultured with astrocytes from cerebellum or cerebral cortex. J. Neurochem. 56: in press.

Wood, J.D., 1975, The role of γ-aminobutyric acid in the mechanism of seizures. Prog. Neurobiol. 5:79.

C-6 GLIOMA CELLS OF EARLY PASSAGE HAVE PROGENITOR PROPERTIES IN CULTURE

Antonia Vernadakis, Susan Kentroti, Chaya Brodie, Dimitra Mangoura* and Nikos Sakellaridis**

Departments of Psychiatry and Pharmacology
University of Colorado School of Medicine
Denver, Colorado 80262, USA

INTRODUCTION

Although considerable progress has been made in the last decade in our understanding of the role of glial cells in neuronal development and function, the factors which regulate glia cell growth and function are only recently being investigated (see refs in review Vernadakis, 1988). C-6 glioma cells have provided a useful model to study glial cell properties, glial factors and sensitivity of glial cells to various substances and conditions. In an early study, we reported (Parker et al, 1980) that C-6 glioma cells, 2B clone, exhibited differential enzyme expression with cell passage: the activity of cyclic nucleotide phosphohydrolase (CNP) an enzyme marker for oligodendrocytes (Poduslo and Norton, 1972; Poduslo, 1975) was markedly high and that of glutamine synthetase (GS), an enzyme marker for astrocytes (Martinez-Hernandez et al, 1977; Norenberg and Martinez-Hernandez, 1979) was low in early passages (up to passage 26) and this relationship was reversed in the late passages (beyond passage 70). We have also found that early passage cells express low level GFA immunoreactivity in contrast to late passage cells which express a high intensity of immunoreactive staining (Brodie and Vernadakis, 1990). In this chapter, we will discuss findings which provide supportive evidence that early passage cells have progenitor glia properties and can be geared towards either astrocytic or oligodendrocytic expression whereas the late passages are more committed to astrocytic expression.

*Present address: Dept. of Pediatrics, University of Chicago, Chicago, IL. 60637, USA.
**Present address: Dept. of Pharmacology and Toxicology, University of Indiana School of Medicine, Northwest Center for Medical Education, Gary, Indiana 46408, USA.

FACTORS INFLUENCING EARLY PASSAGE C-6 GLIAL CELLS TO EXPRESS OLIGODENDROCYTIC PHENOTYPE

Studies in vivo and in vitro indicate that the lineages of astroglia and oligodendroglia originally thought to be totally independent may have a common origin. The utilization of immunocytochemical techniques at the electron microscopic level have revealed the existence, at early developmental stages, of radial glial cells in human spinal cord exhibiting a phenotype that could be considered as intermediate between that of astrocytes and oligodendrocytes (Choi and Kim, 1985; Choi et al, 1983). Studies in culture have shown that environmental conditions could affect the differentiation of glial precursors into cells exhibiting astroglial or oligodendroglial phenotypes (Eccleston and Silberberg, 1984; Goldman et al, 1986; Saneto and de Vellis, 1985). The reports by Raff and colleagues (Raff et al, 1983; Temple and Raff, 1985; Ffrench-Constant and Raff, 1986), that rat optic nerve fibrous astrocytes and oligodendrocytes derive from a common bipotential cell that can differentiate into an oligodendrocyte when cultured in media not supplemented with serum and into a type-2 astrocyte when cultured in the presence of serum, have had a strong impact in glial research. Levi et al (1986) have also shown that in the rat cerebellum, putative fibrous astrocytes and oligodendrocytes derive from a common precursor whose differentiation route in vitro is determined by the presence or absence of serum in the culture medium. The same investigators (Levi et al 1987) have reported a detailed analysis of the developmental profile of a set of surface antigens, recognized by different monoclonal antibodies, which are expressed by bipotential cerebellar glial precursors during their differentiation into oligodendrocytes. Trotter and Schachner (1989) have reported that glial cells expressing surface antigen 04 (0 antigen expressed by murine oligodendrocyte) are bipotential precursor cells able to differentiate into astrocytes or oligodendrocytes depending on the culture medium. Moreover, these authors claim that serum factors induce the expression of GFAP in 04-positive cells rather than selecting for the survival of cells expressing the two markers, since 04-positive cells were never seen to express GFAP prior to sorting. More importantly are their findings that bipotentiality of these glial precursor cells is retained up to a later developmental stage. We are now proposing that early passage C-6 glial cells exhibit progenitor properties in that they can be geared to express predominantly the oligodendrocytic phenotype. Some of the factors which can influence early passage C-6 glial cells to express this phenotype are shown in Fig. 1. In this chapter, we will discuss only the evidence that neuronal factors influence oligodendrocytic phenotype.

C-6 glial cells, 2B clone, were originally provided to us by Dr. Jean de Vellis, University of California at Los Angeles at passage 12. We currently have several passages (from 16-130) frozen in liquid nitrogen. In this study (Mangoura et al, 1989), C-6 glial cells at passage 24 were used. Culture medium was Dulbecco's Modified Eagle's medium (DMEM) containing 10% fetal calf serum (FCS). For neuron conditioned medium (NCM) we used either NCM derived from neuronal cultures prepared from 8-day-old chick embryo cerebral hemispheres or neuroblastoma cell cultures (mouse neuroblastoma NBP_2) which express both tyrosine hydroxylase and choline acetyltransferase activity (Prasad et al, 1973). In the presence of these conditioned media the expression of GS, the astrocytic marker, is markedly reduced (Figures 2,3) whereas activity of CNP, the oligodendrocytic marker, is enhanced (Figure 4). It is of interest that FCS did not completely counteract the CM-produced increase in CNP whereas FCS reversed the decline in GS activity produced by CM. We have also found that in primary glial cell cultures derived from 15-day-old chick embryo cerebral hemispheres astrocytic expression is reduced and oligodendrocytic expression is increased in cultures grown in the presence of neuronal factors (neuron conditioned medium) (Sakellaridis et al, 1986). These data suggest the possibility that a glial mitogen is present in CM, which would unilaterally favor only glioblastic proliferation, opposing differentiation and thus resulting in reduced expression. That the neuronal input favors oligodendrocyte expression has been reported extensively (Bologa et al, 1982, 1983). Hunter and Bottenstein (1989) have recently reported that conditioned medium from the B104 CNS neuronal cell line produced large numbers of oligodendrocytes and multipolar glial progenitors after 8 to 12 day treatment of glial cultures derived from rat brain. In another study reported by the same group (Frost et al, 1989) C6 glioma cells exhibited a dose-dependent increase in number when treated with CM, paralleling the increase observed in primary glial cell cultures. These findings support our view that C6 glial cells from early passages can be utilized as in vitro models of glial progenitor cells.

FACTORS INFLUENCING C-6 GLIAL CELLS TO EXPRESS ASTROCYTIC PHENOTYPE

Some of the factors which promote astrocyte expression in progenitor cells are illustrated in Figure 5. In this chapter we will only discuss the data derived from our laboratory using platelet-activating factor and muscle-derived factors.

Platelet-activating factor

In order to examine the potential role of glial cells in neuro-inflammation and reactive gliosis, we examined the response of C6 glioma cells to platelet-activating factor (PAF). PAF (1-0 alkyl-2-acetyl-sn-glycero-3-phosphocholine) is a potent biologically active phospholipid found in various cells including: endothelial, mast, and kidney cells as well as platelets and macrophages. It is normally released during inflammation and immune reactions to modulate cellular functions (Braquet, 1987; Braquet et al, 1987). In the central nervous system, PAF exerts modulatory effects on neuronal differentiation and calcium fluxes (Kornecki and Ehrlich, 1988) and its synthesis is enhanced by certain neurotransmitters (Bussolino et al, 1986).

In this study (Kentroti et al, 1990) C-6 glial cells from early (20-25) passage were grown in serum free chemically defined medium described by Aizenman et al (1986). Cultures were treated with PAF on day 1 and harvested on day 4. We found that 200 mM PAF markedly increases (5-fold) GS activity (Fig. 6).

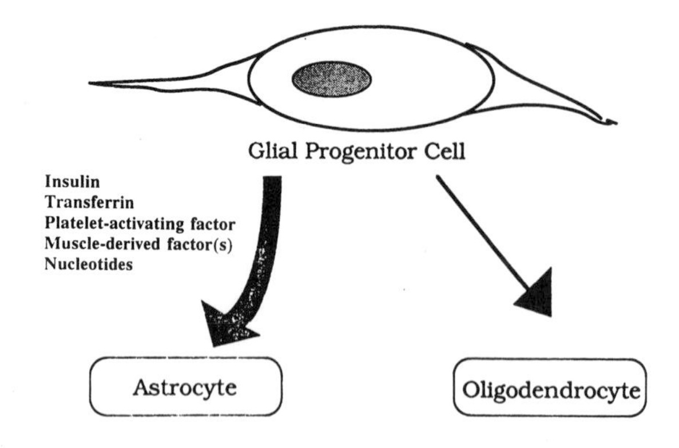

Fig. 1. Schematic representation of factors affecting oligodendrocytic expression in glial progenitor cells.

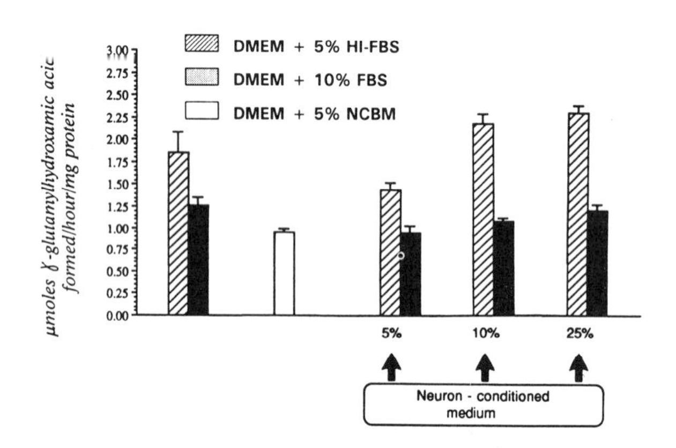

Fig 2. Glutamine Synthetase activity in C-6 glial cell cultures, Early passage. Cells were plated on culture dishes at $1x10^6$ cells/100 mm culture dish in the following media: DMEM plus 10% Fetal Bovine Serum (FBS); + 10% heat inactivated fetal bovine serum (HI-FBS); + 5% NCM; + 5% NCM with either 5% HI-FBS or 10% FBS; + 10% NCM with either 5% HI-FBS or 10% FBS; and + 25% NCM with either 5% HI-FBS or 10% FBS; Cultures were harvested at culture day 7. Barograms represent means ± SEM. Experiments were repeated at least twice, with an n of 4 each time. Relevant Statistical comparisons: 5% NCM versus 5% NCM + 5% HI-FBS, $P<0.05$; 5% NCM versus 10% NCM + 5% NCM + 5% HI-FBS, $P<0.001$; 5% NCM versus 25% NCM + 5% HI-FBS, $P<0.001$; 5% NCM versus 5% HI-FBS, $P<0.005$; 5% NCM versus 10% FBS, $P<0.01$.

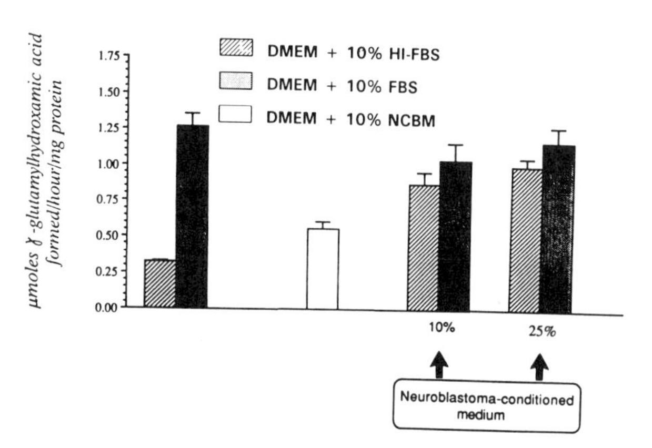

Fig 3. Glutamine synthetase activity in C-6 glial cell cultures from early passages. Experimental design similar to that of Figure 2, except that the cells were plated in DMEM plus; 10% FBS; 10% HI-FBS; 10% neuroblastoma-conditioned medium (NBCM); 10% NBCM with either 10% HI-FBS or 10% FBS; and 25% NBCM with either 10% HI-FBS or 10% FBS. Relevant statistical comparisons; NBCM treatment alone differs significantly from any sera + NBCM treatment and the 10% FBS alone ($P<0.005$).

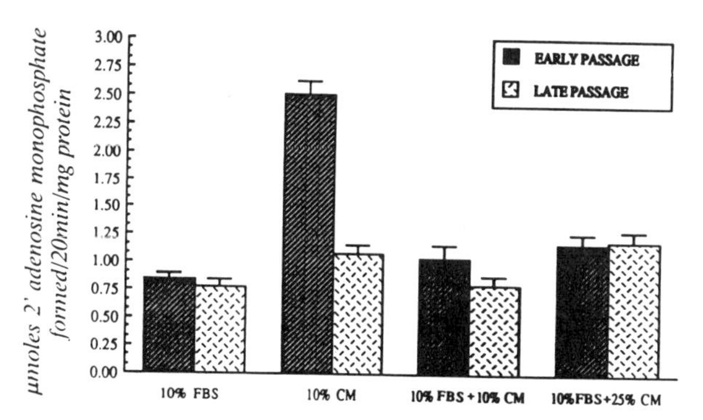

Fig 4. 2′3′-Nucleotide 3′-Phosphohydrolase activity in C-6 glial cells from early and late passages. Experimental design similar to that of Figures 2 and 3 (culture medium was supplemented with 10% FBS and 0-25% neuroblastoma-conditioned medium). Relevant statistical comparisons: Late passage 10% FBS versus 10% NBCM, $P<0.001$; Early passage, NBCM treatment alone differs significantly from the sera only or sera + NBCM treatments ($P<0.001$).

Platelet-activating factor is produced by a variety of cells primarily associated with the immune system. Among its many proinflammatory properties are: 1) a chemotaxic response by neutrophils; 2) promotion of platelet aggregation; 3) increased vascular permeability. PAF is generated by inflammatory cells and although it has not been isolated per se from the extracellular matrix, its participation in the immune response at the time of its release into the microenvironment is likely.

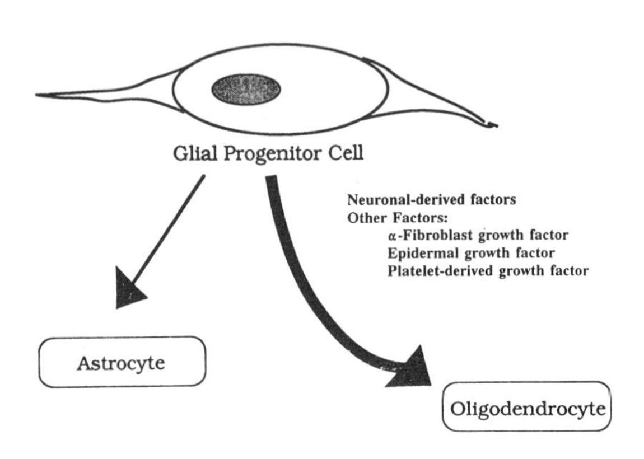

Fig. 5. Schematic representation of factors affecting astrocytic expression in glial progenitor cells.

The increase in GS activity in the early passage cells in response to PAF may reflect a promotion of astrocytic expression in these glioblastic cells. Evidence for the role of PAF as a growth-promoting factor during embryonic development is beginning to accumulate. Ryan et al (1989) have reported significant quantities of PAF produced by pre-implantation, morula stage mouse embryos in vitro. In addition, PAF has been shown to influence expression of c-fos and c-jun, two proto-oncogenes which are involved in the neuronal response to environmental stimuli (Greenberg et al, 1986). In this way, PAF may contribute to the long-term phenotypic responses of a cell to its environment (Squinto et al, 1989). PAF has been reported to increase inositol phosphate production and cytosolic free Ca^{++} concentrations in cultured rat Kupffer cells (Fisher et al, 1989). By expanding the category of actions of PAF to include glial cells one can hypothesize as to the role of Ca^{++} in glial differentiation as it has been shown for other cells (See review in Lipton and Kater, 1989). In our model system, the effect of PAF on GS in glioblasts could be partially mediated through these second messenger signals.

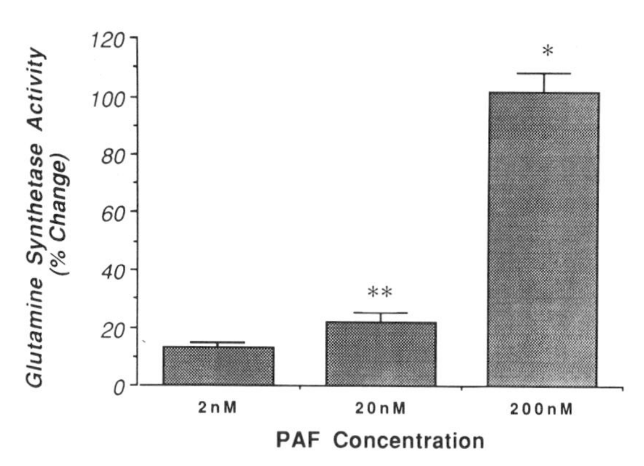

Fig. 6. Glutamine synthetase (GS) activity in early passage C-6 glioma cells following treatment with platelet activating factor (PAF). Cells were grown for 4 days (C1-4) in chemically defined medium supplemented with PAF. Cultures were then harvested at C4 for assay of GS activity. Barograms represent the mean percent change in GS activity as compared to controls. Bracketed lines represent the S.E.M. of 5-7 samples. *P<0.001 and **P<0.02 as compared with controls.

Muscle-derived factor(s)

Muscle-derived factors have been shown to influence neuronal survival, neuronal sprouting and cholinergic neurotransmitter expression (Bennett and Nurcombe, 1979; Davies, 1986; Flanigan et al., 1985; Oh et al., 1988; Smith and Appel, 1983). Recently, we have found that muscle-derived factors influence cholinergic neuronal phenotypic expression, as assessed by choline acetyltransferase activity (ChAT), when administered to chick embryos in ovo during early neuroembryogenesis (day 1-3), and also in neuron-enriched cultures derived from chick embryonic brain (unpublished observations). We have concluded that muscle-derived factors may have universal cholinotrophic effects and are not only confined to target-derived effects such as reported for spinal motorneurons. Based on these findings and in view of the recent report that neurons and glial cells may have a common progenitor cell in the chicken (Galileo et al, 1990), we tested the possibility of muscle-derived factors influencing glial phenotypes.

In this study (Brodie and Vernadakis, 1990) C-6 glial cells, passages (19-20 were grown in chemically defined medium (CDM) described in the above section and were treated with limb muscle extract (LME) or breast muscle extract (BME), 1-10% at day 1 and harvested at day 5. LME-treated cultures exhibit a dose-dependent increase in GS activity (Fig 7A). In addition, LME treatment produces a 171% increase in cell number (Fig 8) which is similar to the increase observed in protein content (Fig 7B). The marked cell proliferation of the LME-treated cultures is also evident morphologically (Fig 9). Thus the increase in GS activity reflects the marked astrocytic proliferation and differentiation of these glioblastic cells.

Reports concerning the nature of neurotrophic factors derived from muscle extracts are not in agreement and range

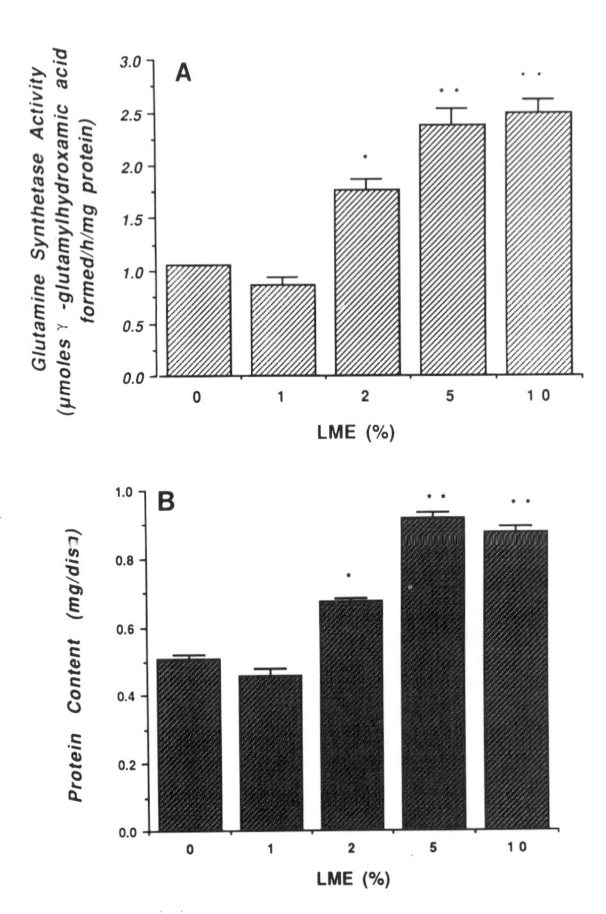

Fig. 7 Dose effect of LME on glutamine synthetase activity (A) and protein content (B) of early passage C-6 glial cells. Cells were plated in DMEM + 10% FBS as described in Figure 1, treated with LME on culture day 1 and harvested at culture day 5. Barograms with bracketed lines represent the mean ± S.E.M. of 2-3 experiements each one consists of 5 samples. *P <0.002 **P <0.001 as compared to control.

from a muscle cell NGF (Murphy et al., 1977) to a heat-labile, trypsin sensitive, non-dialyzable factor functionally and immunologically different from NGF (Davies, 1986), to an unidentified factor (Dohrmann et al., 1987), a protein not related to laminin or type II cyclic adenosine 3′,5′ monophosphate-dependent protein kinase (cAMP-dPK) (Oh et al, 1988). Recently Pernaud et al. (1988) have reported two fibroblast growth factors named acidic and basic FGF (aFGF and bFGF) which stimulate proliferation of rat astroblasts and also the expression of glutamine synthetase activity in cells grown in primary culture. In a subsequent report, the same group (Loret et al., 1989) compared aFGF, bFGF and thrombin in a variety of pleiotropic responses on rat astroblasts, proliferation, morphology, GS activity and phenotypic expression. Astroblast proliferation was stimulated transiently by these growth factors while GS activity significantly increased. Acidic FGF and EGF but not thrombin modified the cell morphology. They have proposed a model in which cell maturation is characterized by the modulation of the synthesis of many proteins which can be grouped into classes. Each class appears to be under the control of one regulatory element. In our study, we cannot conclude whether the effects we observed on glial cell proliferation and differentiation represent the effect of one factor or two different factors. We are currently attempting to characterize the nature of these muscle-derived factors.

The influence of muscle-derived factors on early glioblastic cells leads us to propose that cells of non-ectodermal origin may play a role in glia cell differentiation. Recently, Andres and Salopok (1989) reported that meningeal cells (another mesodermal origin cell, like muscle) increase astrocytic gap junction communication in vitro. They report that the presence of meningeal cells are necessary for the normal development of the glia limitans. They speculated that the prevalence of gap junctions in the glia limitans is due to the meningeal-glial interactions. In our paradigm, glial cells appear to be influenced by factors derived by another mesodermal-origin cell, muscle. These few examples introduce a new relationship which may be important during early glial phenotypic expression, that is the interaction of glial cells with mesodermal cells. The recent report by Fontaine-Perus et al (1989) that muscle-containing tissues can increase the rate of proliferation of neuroepithelial cells when these tissues are placed together, provide further support to our view that during early neuroembryogenesis signals provided by non-ectodermal cells may play a vital role in neuronal and glial differentiation.

CONCLUSIONS

Using C-6 glial cells of early passage, we have presented evidence that glial phenotypes are influenced by various microenvironmental factors including neuronal-derived factors, platelet-activating factor and muscle-derived factors. That glial phenotypic expression is regulated by such factors during early gliogenesis remains speculative. There is abundant evidence that glial cells influence neuronal differentiation and growth (see review Vernadakis, 1988). However, the influence of somatic cells in the development of neurons and

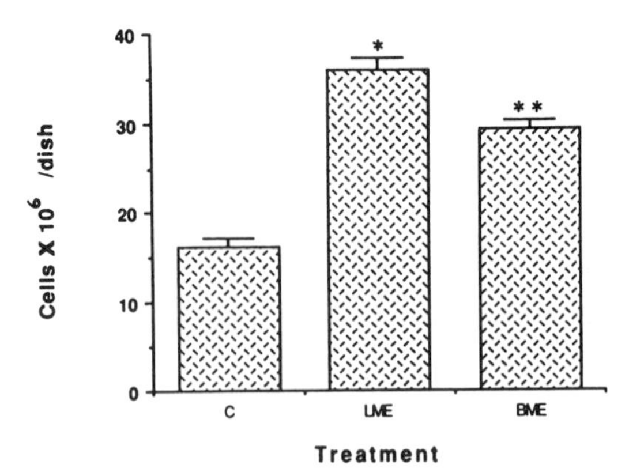

Fig 8. Cell number of early passage C-6 glial cells in cultures treated with either 5% ME or 5% BME. Experimental design as in Figure 2. Barograms with bracketed lines represent the mean ± S.E.M. of 2 experiments each one consists of 4 samples. The increase in cell number appears to be equivalent to the enhanced morphological proliferation. *p<0.02 **p<0.005.

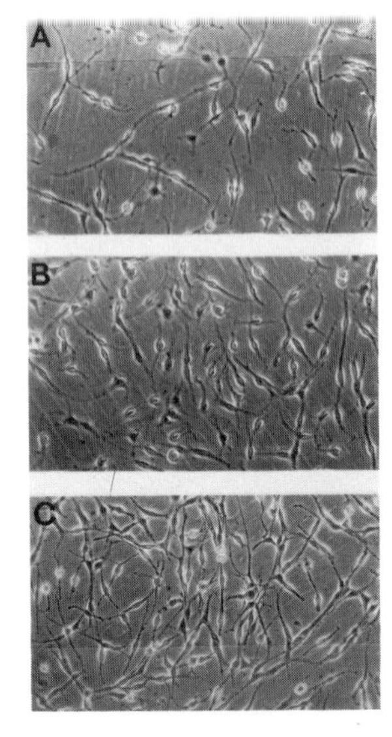

Fig 9. Photomicrographs of early passage (19-20) C-6 glial cells (2B-clone) at culture day 5 (C5). Cells were plated in DMEM supplemented with 10% FBS. After 24h media was replaced with Basal Nutrient Medium (BNM) (A); BNM + 5% BME (B) or BNM + 5% LME (C). Controls exhibit their characteristic spindle shape pattern whereas both groups of treated cells exhibited proliferation and astrocytic-like differentiation (magnification 590x).

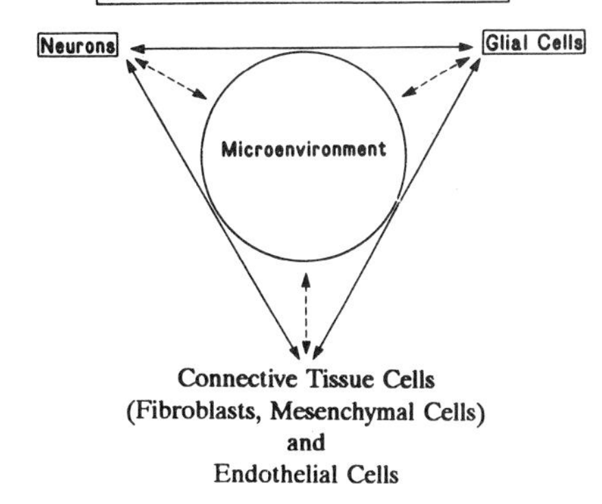

Fig. 10. Schematic representation of cell-cell interactions in the central nervous system tissue.

glial cells is also recently receiving attention. Our findings provide some insight into possible relationships among the various cell types in the CNS as illustrated in Figure 10.

REFERENCES

Aizenman, Y., Weischel, M.E. and De Vellis, J., 1986. Changes in insulin and transferrin requirements of pure brain neuronal cultures during embryonic development. Proc. Nat. Acad. Sci., USA. 83:2263-2266.

Andres, J.J. and Salopik, M., 1989. Meningeal cells increase in vitro astrocytic gap junctional communication as measured by fluorescence recovery after laser photobleaching. J. Neurocytol. 18:257-264.

Bennett, M.R. and Nurcombe, V., 1979. The survival and development of cholinergic neurons in skeletal muscle conditioned media. Brain Res. 173:543-548.

Bologa, L., Z'Graggen, A., and Herschkowitz, N. 1983. Proliferation rate of oligodendrocytes in culture can be influenced by extrinsic factors. Devel. Neurosci. 6:26-31.

Bologa, L., Bisconte, J.C., Joubert, R., Marangos, P.J., Derbin, C., Rioux, F. and Herschkowitz, N. 1982. Accelerated differentiation of oligodendrocytes in neuronal rich mouse brain cell cultures. Brain Res. 252:129-136.

Braquet, P. 1987. The ginkgolides: Potent platelet-activating factor antagonists isolated from Ginkgo biloba L.: Chemistry, pharmacology and clinical applications. Drugs Future 12:643-699.

Braquet, P., Touqui, L., Shen, T.Y. and Vargaftig, B.B. 1987. Perspectives in platelet-activating factor research. Pharmacol. Rev. 39:97-145, 1987.

Brodie, C. and Vernadakis, A. 1990. Muscle-derived factors induce proliferation and astrocytic phenotypic expression in C-6 glial cells. GLIA (in press).

Bussolino, F., Gremo, F., Tetta, C., Pescarmona, G.P. and Camussi, G. 1986. Production of platelet-activating factor by chick retina. J. Biol. Chem. 261:16502-16508.

Choi, B.H. and Kim, R.C. 1985. Expression of glial fibrillary acidic protein by immature oligodendroglia and its implications. J. Neuroimmunol. 8:215-235.

Choi, B.H., Kim, R.C. and Lapham, L.W. 1983. Do radial glia give rise to both astroglial and oligodendroglial cells? Dev. Brain Res. 8:119-130.

Davies, A.M. 1986. The survival and growth of embryonic proprioceptive neurons is promoted by a factor present in skeletal muscle. Dev. Biol. 115:56-67.

Dohrmann, U., Edgar, D. and Theonen, H. 1987. Distinct neurotrophic factors from skeletal muscle and the central nervous system interact synergistically to support the survival of cultured embryonic spinal cord neurons. Dev. Biol. 124:145-152.

Eccleston, P.A., Silberberg, D.H., 1984. The differentiation of oligodendrocytes in a serum-free hormone-supplemented medium. Dev. Brain Res. 16:1-9.

Fisher, R.A., Sharma, R.V. and Bhalla, R.C., 1989. Platelet-activating factor increases inositos phosphate production and cytosolic free Ca^{++}concentrations in cultured rat Kupffer cells. FEBS Lett. 251:22-26.

Flanigan, T.P., Dickson, J.G., and Walsh, F.S. 1985. Cell survival characteristics and choline acetlytransferase activity in motor neuron-enriched cultures from chick embryo spinal cord. J. Neurochem. 45:1323-1326.

Fontaine-Perus, J.C., Chancaine, M., Le Douerin, N.M., Gershon, M.D., and Rothman, T.P. 1989. Mitogenic effect of muscle on the neuroepithelium of the developing spinal cord. Development 107:413-422.

Ffrench-Constant, C. and Raff, M.C. 1986. Proliferating bipotential glial progenitor cells in adult rat optic nerve. Nature 319:499-502.

Frost, G.H., Thangnipon, W. and Bottenstein, J.E. 1989. Glial progenitor growth factor assay using rat glima cells. Trans. Amer. Soc. Neurochem. 20:188.

Galileo, D.S., Graq, G.E., Owens, G.C., Majors, J. and Sanes, J.R. 1990. Neurons and glia arise from a common progentor in chicken optic tectum: Demonstration with two retroviruses and cell type-specific antibodies. Proc. Natl. Acad. Sci. USA 87:458-462.

Goldman, J.E., Geier, S.S. and Hirano, M. 1986. Differentiation of astrocytes and oligodendrocytes from germinal matrix cells in primary culture. J. Neurosci 6:52-60.

Greenberg, M.E., Ziff, E.B., and Greene, L.A. 1986. Stimulation of neuronal acetylcholine receptors induces rapid gene transcription. Science 234:80-83.

Hunter, S.F. and Bottenstein, J.E. Bipotential glial progenitors are targets of neuronal cell line-derived factors. Dev. Brain Res. 49:33-49.

Kentroti, S., Baker, K., Bruce, C., and Vernadakis, A. 1990. Platelet-activating factor increases glutamine synthetase activity in early and late passage C-6 glioma cells. J. Neurosci. Res. (in press).

Kornecki, E. and Ehrlich, Y.H., 1988: Neuroregulatory and neuropathological actions of the ether-phospholipid platelet-activating factor. Science 240:1792-1794.

Levi, G., Aloisi, F. and Wilkins, G.P. 1987. Differentiation of cerebellar bipotential glial precursors into oligodendrocytes in primary culture: Developmental profile of surface antigens and mitotic activity. J. Neurosci. Res. 18:407-417.

Levi, G., Gallo, V., and Ciotti, M.T. 1986. Bipotential precursors of putative fibrous astrocytes and oligodendrocytes in rat cerebellar cultures express distinct surfact featurs and "neuron-like" α-aminobutyric acid transport. Proc. Nat. Acad. Sci. USA 83:1504-1508.

Lipton, S.A. and Kater, S.B. 1989. Neurotransmitter regulation of neuronal growth, plasticity and survival. TINS 12:265-270.

Loret, C., Sensenbrenner, M. and Labourdette, G. 1989. Differential phenotypic expression induced in cultured rat astrocytes by acidic fibroblast growth factor, epidermal growth factor and thrombin. J. Biol. Chem. 264:8319-8327.

Mangoura, D., Sakellaridis, N., Jones, J., and Vernadakis, A. 1989. Early and late passage C-6 glial cell growth: Similarities with primary glial cells in culture. Neurochem. Res. 4:941-947.

Martinez-Hernandez, A., Bell, K.P., and Norenberg, M.D. 1977. Glutamine synthetase-glial localization in the brain. Science 195:1356-1358.

Murphy, R.A., Singer, R.H., Sarde, J.D., Pantazis, N.J., Blanchard, M.H., Byron, K.S., Arnason, B.G.W., and Young, M. 1977. Synthesis and secretion of a high molecular weight form of a nerve growth factor by skeletal muscle cells in culture. Proc. Natl. Acad. Sci. 74:4496-4500.

Norenberg, M.D., and Martinez-Hernandez, A. 1979. Fine structural localization of glutamine synthetase in astrocytes of rat brain. Brain Res. 161:303-310.

Oh, T.H., Markelanis, G.J., Dion, T.L, and Hobbs, S.L. 1988. A muscle-derived substrate-bound factor that promotes neurite outgrowth from neurons of the central and peripheral nervous system. Dev. Biol. 127:88-98.

Parker, K.K., Norenberg, M.D., and Vernadakis, A. 1980. "Transdifferentiation" of C-6 glial cells in culture. Science 208:179-181.

Pernaud, F., Bensand, F., Pettman, R., Sensenbrenner, M. and Labourdette, L. 1988. Effects of acidic and basic growth factors αFGF and βFGF) on the proliferation of rat astroblasts in culture. Glia 1:124-131.

Poduslo, S.E. 1975. The isolation and characterization of a plasma membrane and myelin fraction derived from oligodendroglia of calf brain. J. Neurochem. 24:647-664.

Poduslo, S.E. and Norton, W.T. 1972. Isolation and some chemical properties of oligodendroglia from calf brain. J. Neurochem. 19:727-736.

Prasad, K.N., Mandal, B., Waymire, J.C., Lees, G.J., Vernadakis, A. and Weiner, N. 1973. Basal level of neurotransmitter synthesizing enzymes and effect of cyclic AMP agents on the morphologcal differentiation of isolated neuroblastoma clones. Nature 241:117-119.

Raff, M.D., Miller, R.H. and Noble, M. 1983. A glial progenitor cell that develops in vitro into an astrocyte or an oligodendrocyte depending on culture medium. Nature 303:390-396.

Ryan, J.P., Spinks, N.R., O'Neill, C., Ammit, A.J., Wales, R.G. 1989. Platelet-activating factor (PAF) production by mouse embryos in vitro and its effects on embryonic metabolism. J. Cell Biochem. 40:387-395.

Sakellaridis, N., Mangoura, D., and Vernadakis, A. 1986. Effects of neuron-conditioned medium and fetal calf serum content on glial growth in dissociated cultures. Dev. Brain Res. 27:31-41.

Saneto, R.P. and de Vellis, J. 1985. Characterization of cultured rat oligodendrocytes proliferating in a serum-free, chemically defined medium. Proc. Natl. Acad. Sci. USA 82:3509-3513.

Squinto, S.P., Block, A.L., Braquet, P. and Bazan, N.G. 1989. Platelet-activating factor stimulates a Fos/Jun/AP-1 transcriptional signaling system in human neuroblastoma cells. J. Neurosci. Res. 24:558-556.

Smith, R.G. and Appel, S.H. 1983. Extracts of skeletal muscle increase neurite outgrowth and cholinergic activity of fetal rat spinal motor neurons. Science 219:1079-1081.

Temple, S., and Raff, M.C. 1985. Differentiation of a bipotential glial progentor cell in single cell microculture. Nature 313:223-225.

Trotter, J. and Schachner, M. 1989. Cells positive for the O4 surface antigen isolated by cell sorting are able to differentiate into astrocytes or oligodendrocytes. Dev. Brain Res. 115-122.

Vernadakis, A. 1988. Neuron-glia interrelations. Int. Neurobiol. Rev. 30:149-223.

BRAIN EXTRACELLULAR MATRIX AND NERVE REGENERATION

Amico Bignami, Richard Asher and George Perides

Department of Pathology, Harvard Medical School and Spinal Cord Injury Research Laboratory, Department of Veterans Affairs Medical Center, Boston, MA 02132

INTRODUCTION

Over the past five years, we have made some progress in our studies on the composition of brain extracellular matrix. As in previous work on GFA protein, a major component of glial scars, the motivation for these studies was to find out why axons do not regenerate in mammalian CNS. In fact, we started doing research on brain extracellular matrix because the experimental evidence suggested that the glial scar per se, could not explain the riddle of CNS regeneration (Bignami et al., 1986).

In our studies of GFA and neurofilament proteins, the basic assumption was that such proteins would turn out to be cell specific and this because of the existence of specific neurohistological stains, i.e., Weigert method for glial fibrils and Cajal method for neurofibrils (bundles of 10-nanometer filaments appear as fibrils under the light microscope). Based on this assumption, the specificity of antibodies to GFA and neurofilament proteins could be determined in tissue sections by immunohistological methods. As expected, antibodies specific for GFA protein stained as Weigert's method for glial fibrils (Bignami and Dahl, 1974) and antibodies specific for neurofilament proteins stained as Cajal's method for neurofibrils (Dahl and Bignami, 1977).

For the isolation of brain extracellular matrix proteins, the expectation was that in white matter these proteins would share some properties with cartilage proteins, and this in view of the fact that myelinated white matter and cartilage are among the few tissues of the body where axons do not grow. In cartilage, extracellular matrix proteins (proteoglycans and link proteins), form macromolecular complexes with hyaluronate due to non-covalent binding (Hascall and Hascall, 1981). Hyaluronate binding is thus a characteristic property of cartilage extracellular matrix proteins.

The purpose of this presentation is to summarize our findings and to suggest some ideas as to the reason why the extracellular matrix in myelinated white matter may not be conducive to axonal growth.

Glial Hyaluronate-binding Protein (GHAP)

GHAP is a major protein of myelinated white matter. The yields are 8.2 mg protein and 5.2 mg protein/100g wet tissue in human cerebral white matter

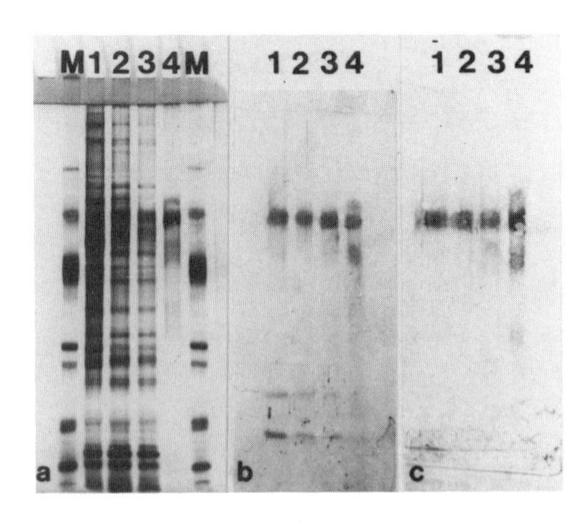

Fig. 1. Different stages during the isolation of bovine GHAP from acid extracts of spinal cord. a, silver stained gels; b and c, corresponding immunoblots with rabbit antiserum against human (b) and bovine (c) GHAP. Lane 1, supernatant at pH 2.2, lane 2, supernatant at pH 5.6, lane 3, supernatant at pH 7.5 which was applied to the hyaluronate-Sepharose column, and lane 4, isolated GHAP after elution with 4 M Guanidine-HCl. M, molecular weight standards. From top to bottom: phosphorylase b, 97,000 bovine serum albumin, 67,000; ovalbumin, 43,000; carbonic anhydrase, 30,000; trypsin inhibitor, 20,100; α-lactalbumin, 14,400.

and bovine spinal cord, respectively (Perides et al., 1989; Perides and Bignami, in preparation). The lower yield in bovine is probably due to the fact that spinal cord contains a sizable amount of gray matter. On SDS gels, GHAP isolated from human, dog and bovine tissues, migrates as a diffuse band with different molecular weights depending on the species: 60,000 (human), 70,000 (dog), and 64,000 (bovine). In bovine brain, two minor species at 76,000 and 54,000 dalton (Fig. 1) were shown to derive from the same protein by peptide mapping and partial amino acid sequencing. The pI of the 3 bovine polypeptides was very similar to that of human GHAP (4.3). As indicated by concavalin A blots, GHAP is a glycoprotein. After chemical and enzymatic deglycosylation, GHAP still migrates as a diffuse band, but with apparent molecular weights of 47,000 (human) and 43,000 (bovine). The incubation with N-Glycanase and O-Glycanase did not have any effect on the isoelectric points of human and bovine GHAP. After treatment with neuraminidase both proteins appeared to shift to a more basic one, that is 4.7-4.8 (Fig. 2).

Partial amino acid sequences revealed up to 83% identity with the hyaluronate binding domain of cartilage proteoglycans and link proteins. Furthermore, the complete cDNA sequence of versican, a large fibroblast proteoglycan, showed identical sequences within the hyaluronate binding domain, thus suggesting that GHAP is a proteolytically processed form of versican (Zimmerman and Ruoslahti, 1989). Another possibility is that GHAP is generated from the versican gene by alternative splicing.

The interactions of GHAP with hyaluronate were recently studied using oligomers of hyaluronate (Perides and Bignami, in preparation). Hyaluronate is formed by repeating disaccharide units (N-acetylglycosamine-glucuronic

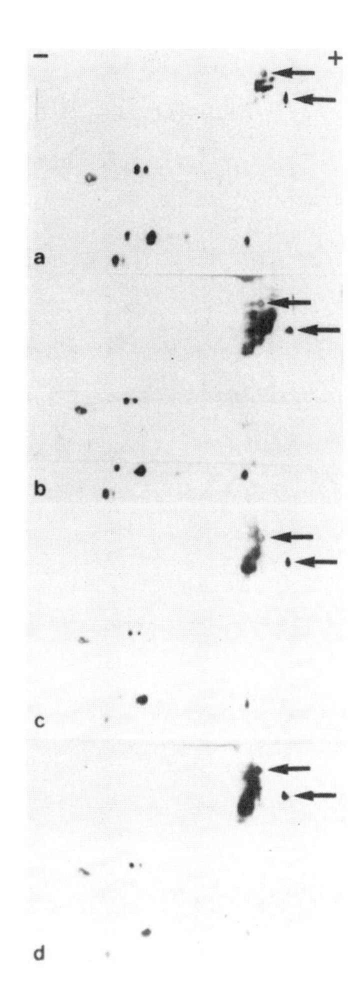

Fig. 2. Two-dimensional electrophoresis of deglycosylated bovine GHAP with pI standards 3.5-9.7. a, GHAP control; b, GHAP after incubation with N-Glycanase; c, GHAP after incubation with N-Glycanase and neuraminidase; d, GHAP after incubation with N-Glycanase, neuraminidase and O-Glycanase. Arrows in a-d point to the position of two pI standards.

acid) and it was originally shown by Hascall and Heinegard (1974), that the binding of hyaluronate to cartilage proteins was inhibited by 5 (but not 4) repeating units. Conversely, the hexasaccharide (3 repeating units) significantly inhibited the binding of hyaluronate to its cell receptor (Underhill et al., 1983), which is now believed to be identical to CD44 (Aruffo et al., 1990). CD44 is a membrane glycoprotein that in cooperation with other lymphocyte receptors and counter-receptors on the endothelium, allows lymphocyte circulation between specialized venules and lymph nodes (Springer, 1990). GHAP appeared to be different from cartilage proteins and the hyaluronate receptor, in that the binding was inhibited by 4 disaccharide repeat units but not by the hexasaccharide (Fig. 3). Interestingly, the GHAP-hyaluronate complex was dissociated at pH 3.0, probably explaining why GHAP extraction from brain tissues requires an acid pH. As we will discuss in the next section, GHAP forms an insoluble complex with hyaluronate in the extracellular matrix of myelinated white matter.

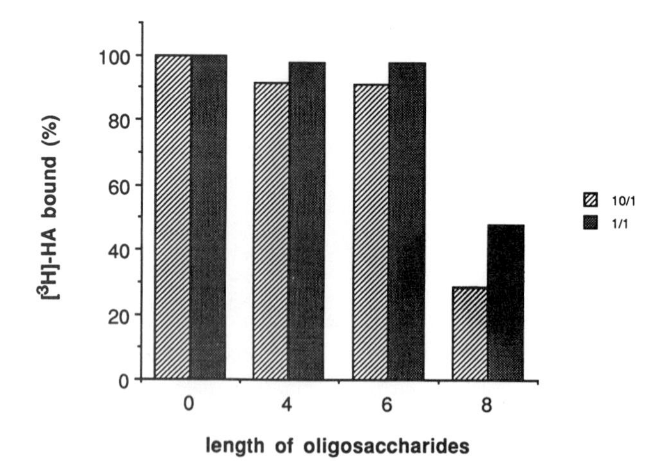

Fig. 3. Effect of hyaluronate (HA) oligosaccharides on the binding of bovine GHAP onto [^{3}H]HA. Note the inhibition of the binding in the presence of octosaccharide at a 10/1 and 1/1 oligosaccharide/[^{3}H]HA ratio. Tetrasaccharide and hexasaccharide do not inhibit the binding.

GHAP-hyaluronate Complex in CNS White Matter

The presence of an GHAP-hyaluronate complex in the extracellular space of myelinated white matter has been recently demonstrated in this laboratory (Asher et al., 1990). As noted before, GHAP is extracted from brain at acid pH, i.e. under conditions that dissociate the GHAP-hyaluronate complex *in vitro*. The demonstration that such complex also exists *in vivo* was prompted by the observation that the staining of tissue sections with GHAP antibodies was abolished by hyaluronidase digestion. It was then shown that intact GHAP was released from tissue homogenates by hyaluronidase digestion under conditions that did not allow the extraction of GHAP in the absence of the enzyme. Direct demonstration of the extracellular localization of GHAP in dog spinal cord white matter was obtained by immunoelectron microscopy (Asher et al., 1990). Although such extracellular localization was expected (hyaluronate is the only glycoaminoglycans synthesized on the plasma membrane and extruded in the extracellular space during elongation) an *intracellular* localization of hyaluronate and hyaluronate-binding proteins in mature brain has been reported (Ripellino et al., 1988, 1989).

It is of interest that the presence of hyaluronate in the extracellular matrix of myelinated white matter was first reported by Angelo Bairati, then Professor of Normal Human Anatomy at the University of Bari, Italy (Bairati, 1953). In Bairati's experiment, India ink was injected together with hyaluronidase in the spinal cord. Under these conditions, India ink spread in the extracellular space of spinal cord white matter and was deposited around myelinated axons, a virtually identical histological appearance being obtained with antibodies to GHAP (Fig. 4).

Why Hyaluronate-binding Proteins Inhibit Axonal Growth?

In primary cultures derived from newborn murine brain astrocytes and oligodendrocytes do not produce GHAP. However, astrocytes (but not oligodendrocytes) are surrounded by a hyaluronate coat (Asher and Bignami, in preparation; Figures 5 and 6). These findings may suggest some ideas as to the reason why hyaluronate-binding proteins may inhibit cell migration and axonal growth. The hypothesis is that immature astrocytes produce hyaluronate and that, due to the presence of hyaluronate receptors on their surface, migratory cells as well as axonal growth cones are capable of adhering

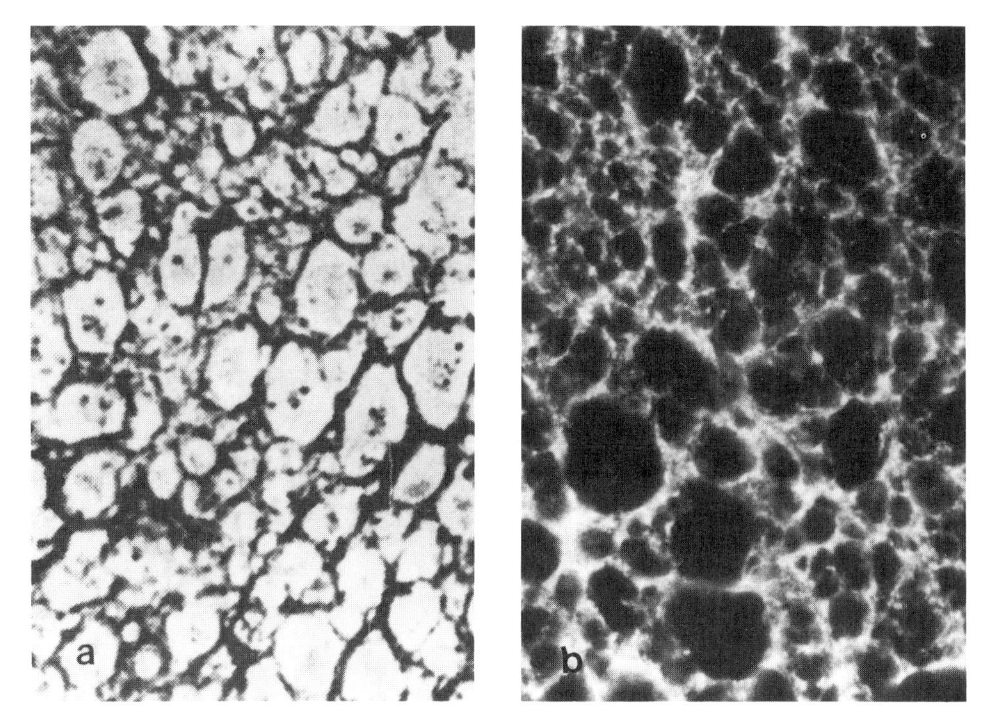

Fig. 4. Demonstration by two different methods of an extracellular matrix filling the space between myelinated axons in transverse sections of bovine spinal cord. In (a), the spinal cord has been injected with hyaluronidase and India ink (Bairati, 1953). Following the digestion of hyaluronate and the solubilization of GHAP (Asher et al., 1990), India ink can diffuse into the extracellular space. In (b), GHAP is stained with monoclonal antibodies. From Bairati, 1953 (a) and from Bignami and Dahl, 1986 (b).

to hyaluronate retained as a coat on the glial surface or secreted in the extracellular space. At a later stage, white matter astrocytes produce GHAP. The presence of a hyaluronate-binding protein in the extracellular space interferes with cell adhesion. Since GHAP appears relatively late in development i.e., after the onset of myelination (Bignami and Dahl, 1988), its expression will not interfere with cell migration and axonal elongation in the immature brain. However, it will prevent axonal growth in mature brain, or more specifically, in myelinated white matter where the protein is located. According to the hypothesis, reactive astrocytes at the brain-graft interface do not prevent axonal growth into peripheral nerve grafts because they do not produce GHAP (Mansour et al., 1990). These grafts cannot bridge the gap produced by a CNS lesion due to the persistence of GHAP in white matter tracts undergoing Wallerian degeneration (Bignami et al., 1989).

A few observations *in vitro* are compatible with this hypothesis. As shown in Fig. 7, and repeatedly demonstrated in several laboratories, neurons are capable of migrating and extending neurites on the surface of astrocytes in monolayer cultures derived from late fetal or neonatal murine brain (Baehr and Bunge, 1990; Bignami and Dahl, 1989; Fawcett et al., 1989a; Hatten, 1990; Noble et al., 1984; Rousselet et al., 1990; Schwab and Caroni, 1989; Smith et al., 1990). As shown before, many astrocytes are surrounded by a hyaluronate coat in these cultures. It remains to be seen whether neurons attach to hyaluronate or to other still unidentified components of the peri-astrocytic coat. Hyaluronate receptors have been reported in several cell types *in vitro* (Underhill and Toole, 1979), and it was recently suggested that CD44 is the principal cell surface receptor for hyaluronate (Aruffo et al., 1990). CD44, originally identified as a lymphocyte surface protein involved in endothelial cell recognition, exhibits sequence homology with the phylogenetically conserved, hyaluronate-binding region of cartilage proteins (Goldstein et al., 1989; Stamenkovic et al., 1989). We also note that axonal growth cones are unable to adhere to oligodendrocytes in primary monolayer cultures (Fawcett et al., 1989b; Schwab and Caroni, 1988), and that oligodendrocytes in these cultures lack a pericellular hyaluronate coat.

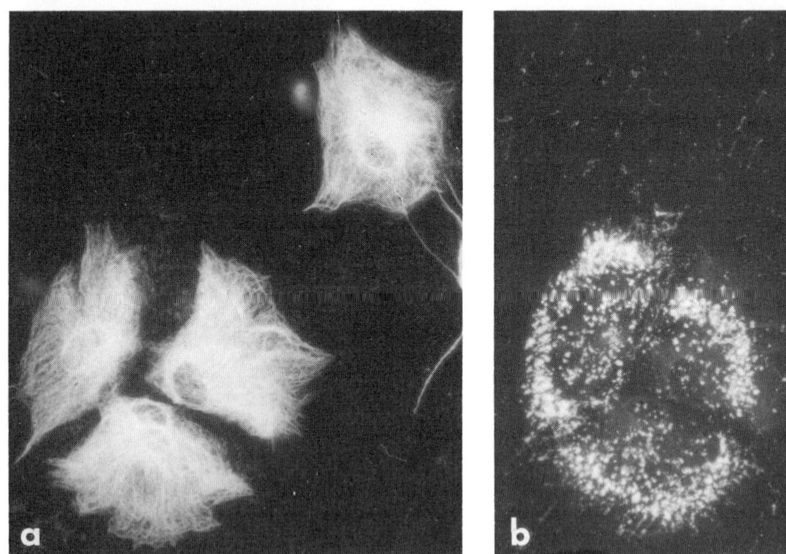
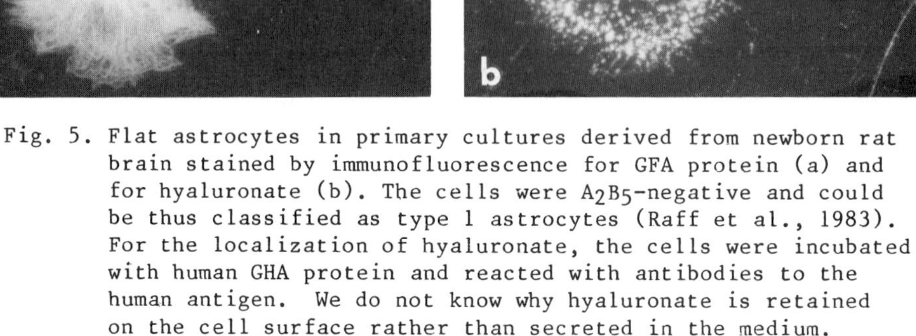

Fig. 5. Flat astrocytes in primary cultures derived from newborn rat brain stained by immunofluorescence for GFA protein (a) and for hyaluronate (b). The cells were A_2B_5-negative and could be thus classified as type 1 astrocytes (Raff et al., 1983). For the localization of hyaluronate, the cells were incubated with human GHA protein and reacted with antibodies to the human antigen. We do not know why hyaluronate is retained on the cell surface rather than secreted in the medium.

In conclusion, more should be known on brain extracellular matrix before we can solve the riddle of CNS regeneration. Although many years have passed since Van Harreveld et al. (1965) showed that the brain possess a considerable extracellular space, about 17-20% in volume according to current estimates (Nicholson and Rice, 1986), relatively little work has been done to fill the space.

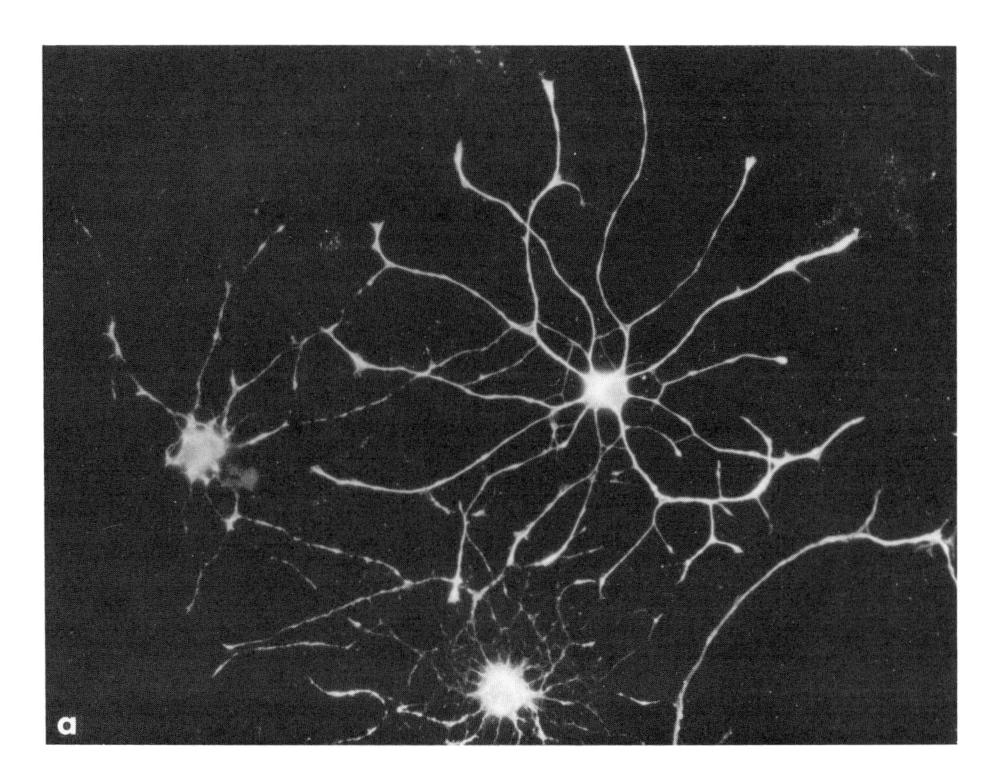

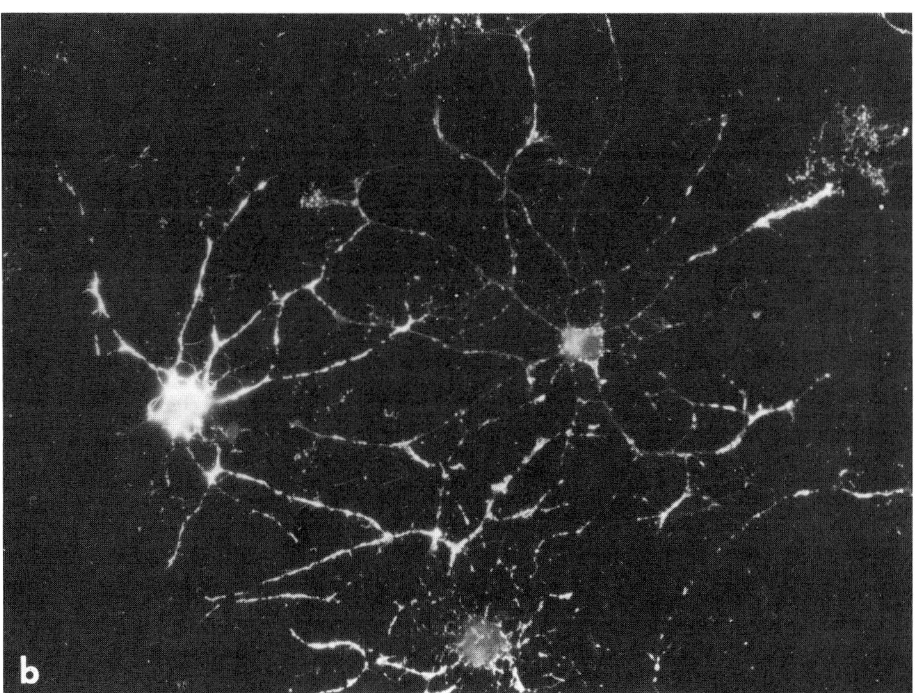

Fig. 6. Stellate astrocytes in primary cultures derived from newborn rat brain stained by indirect immunofluorescence for GFA protein (a) and for hyaluronate (b). The cells were A_2B_5-positive and could be thus classified as type 2 astrocytes (Raff et al., 1983).

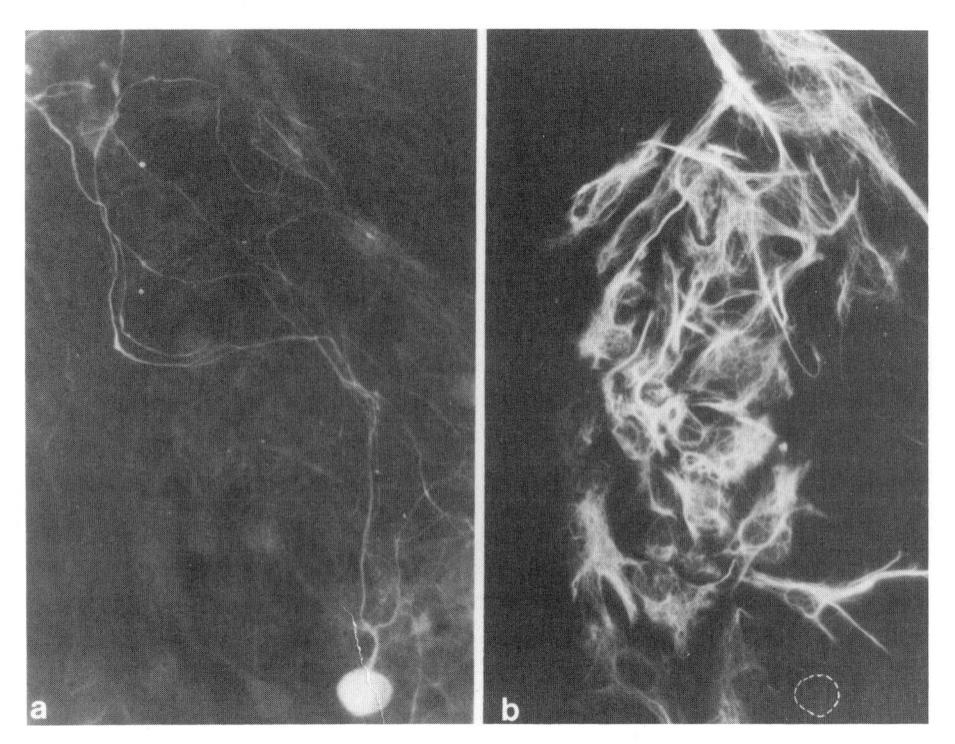

Fig. 7. Mixed spinal cord/dorsal root ganglion culture derived from 15 day-old embryos on laminin substrate (Bignami and Dahl, 1989). The culture is double stained with neurofilament polyclonal (a) and GFA monoclonal (b) antibodies by indirect immunofluorescence. A dorsal root ganglion neuron extends long neurites on GFA-positive cells. As shown in Figs. 5 and 6, astrocytes in culture are surrounded by an hyaluronate coat.

ACKNOWLEDGEMENTS

Supported by NIH Grant NS 13034 and by the Department of Veterans Affairs.

REFERENCES

Aruffo, A., Stamenkovic, I., Melnick, M., Underhill, C. B., and Seed, B., 1990, CD44 is the principal cell surface receptor for hyaluronate, Cell, 61:1303.

Asher, R., Perides, G., Vanderhaeghen, J.-J., and Bignami, A., 1990, The extracellular matrix of central nervous system white matter: demonstration of an hyaluronate-protein complex, J. Neurosci. Res., 27:in press.

Baehr, M., and Bunge, R. P., 1990, Growth of adult rat retinal ganglion cell neurites on astrocytes, Glia, 3:293.

Bairati, A., 1953, Spreading factor and mucopolysaccharides in the central nervous cystem of vertebrates, Experientia, 9:461.

Bignami, A., and Dahl, D., 1974, Astrocyte-specific protein and neuroglial differentiation. An immunofluorescence study with antibodies to the glial fibrillary acidic protein, J. Comp. Neurol., 153:27.

Bignami, A., and Dahl, D., 1986, Brain-specific hyaluronate-binding protein: an immunohistological study with monoclonal antibodies of human and bovine CNS, Proc. Nat. Acad. Sci. (USA), 83:3518.

Bignami, A., and Dahl, D., 1988, Expression of brain-specific hyaluronectin (BHN), a hyaluronate-binding protein, in dog postnatal development, Exp. Neurol., 99:107.

Bignami, A., and Dahl, D., 1989, Vimentin-GFAP transition in primary dissociated cultures of rat embryo spinal cord, Int. J. Dev. Neurosci., 7:343.

Bignami, A., Chi, N. H., and Dahl, D., 1986, The role of neuroglia in axonal growth and regeneration, in: "Neural Transplantation and Regeneration," Das and R. B. Wallace, eds., Springer-Verlag.

Bignami, A., Mansour, H., and Dahl, D., 1989, Glial hyaluronate-binding protein in Wallerian degeneration of dog spinal cord, Glia, 2:391.

Dahl, D., and Bignami, A., 1977, Preparation of antisera to neurofilament protein from chicken brain and human sciatic nerve, J. Comp. Neurol., 176:645.

Fawcett, J. W., Housden, E., Smith-Thomas, L., and Meyer, R. L., 1989a, The growth of axons in three-dimensional astrocyte cultures, Develop. Biol., 135:449.

Fawcett, J. W., Rokos, J., and Bakst, I., 1989b, Oligodendrocytes repel axons and cause axonal growth cone collapse, J. Cell Sci., 92:93.

Goldstein, L. A., Zhou, D. F. H., Picker, L. J., Minty, C. N., Bargatze, R. F., Ding, Jie F., and Butcher, E. C., 1989, A human lymphocyte homing receptor, the hermes antigen, is related to cartilage proteoglycan core and link proteins, Cell, 56:1063.

Hascall, V. C., and Hascall, G. K., 1981, Proteoglycans, in: "Cell Biology of Extracellular Matrix," E. D. Hay, ed., Plenum Publishing Corp., New York.

Hascall, V. C. and Heinegård, D., 1974, Aggregation of cartilage proteoglycans. II. Oligosaccharide competitors of the proteoglycan-hyaluronic acid interaction, J. Biol. Chem., 249:4242.

Hatten, M. E., 1990, Riding the glial monorail: a common mechanism for glial-guided neuronal migration in different regions of the developing mammalian brain, Trends Neurosci., 13:179.

Mansour, H., Asher, R., Dahl, D., Labkovsky, B., Perides, G., and Bignami, A., 1990, Permissive and non-permissive reactive astrocytes: immunofluorescence study with antibodies to the glial hyaluronate-binding protein, J. Neurosci. Res., 25:300.

Nicholson, C., and Rice, M. E., 1986, The migration of substances in the neuronal microenvironment, in: "The Neuronal Microenvironment," H. F. Cserr, ed., Ann. NY Acad. Sci.

Noble, M., Fog-Seang, J., and Cohen, J., 1984, Glia are a unique substrate for the in vitro growth of central nervous system neurons, J. Neurosci., 4:1892.

Perides, G., Lane, W. S., Andrews, D., Dahl, D., and Bignami, A., 1989, Isolation and partial characterization of a glial hyaluronate-binding protein, J. Biol. Chem., 264:5981.

Raff, M. C., Miller, R. H., and Noble, M., 1983, A glial progenitor cell that develops in vitro into an astrocyte or an oligodendrocyte depending on culture medium, Nature, 303:390.

Ripellino, J. A., Bailo, M., Margolis, R. U., and Margolis, R. K., 1988, Light and electron microscopic studies on the localization of hyaluronic acid in developing rat cerebellum, J. Cell Biol., 106:845.

Ripellino, J. A., Margolis, R. U., and Margolis, R. K., 1989, Immunoelectron microscopic localization of hyaluronic acid binding region and link protein epitopes in brain, J. Cell Biol., 108:1899.

Rousselet, A., Autillo-Touati, A., Araud, D., and Prochiantz, A., 1990, In vitro regulation of neuronal morphogenesis and polarity by astrocyte-derived factors, Develop. Biol., 137:33.

Schwab, M. E., and Caroni, P., 1988, Oligodendrocytes and CNS myelin are nonpermissive substrates for neurite growth and fibroblast spreading in vitro, J. Neurosci., 8:2381.

Smith, G. M., Rutrrishauser, U., Silver, J., and Miller, R. H., 1990, Maturation of astrocytes in vitro alters the extent and molecular basis of neurite outgrowth, Develop. Biol., 138:377.

Springer, T. A., 1990, Adhesion receptors of the immune system, Nature, 346:425.

Stamenkovic, I., Amiot, M., Pesando, J. M., and Seed, B., 1989, A lymphocyte molecule implicated in lymph node homing is a member of the cartilage link protein family, Cell, 56:1057.

Underhill, C. B., and Toole, B. P., 1979, Binding of hyaluronate to the surface of cultured cells, J. Cell Biol., 82:475.

Underhill, C. B., Chi-Rosso, G., and Toole, B. P., 1983, Effects of detergent solubilization on the hyaluronate-binding protein from membranes of simian virus 40-transformed 3T3 cells, J. Biol. Chem., 258:8086.

Van Harreveld, A., Crowell, J., and Malhotra, S. K., 1965, A study of extracellular space in central nervous tissue by freeze-substitution, J. Cell Biol., 25:117.

Zimmerman, D. R., and Ruoslahti, E., 1989, Multiple domains of the large fibroblast proteoglycan, versican, EMBO J., 8:2975.

HUMAN NERVE GROWTH FACTOR: BIOLOGICAL AND IMMUNOLOGICAL ACTIVITIES, AND CLINICAL POSSIBILITIES IN NEURODEGENERATIVE DISEASE

Ted Ebendal (1), Stine Söderström (1), Finn Hallböök (2), Patrik Ernfors (2), Carlos F. Ibáñez (2), Håkan Persson (2), Cynthia Wetmore (3), Ingrid Strömberg (3), and Lars Olson (3)

Department of Developmental Biology (1), Biomedical Center, Uppsala University, Uppsala, and Department of Medical Chemistry II, Laboratory of Molecular Neurobiology (2), and Department of Histology and Neurobiology (3), Karolinska Institute, Stockholm, Sweden

INTRODUCTION

During development of the vertebrate nervous system, an overproduction of neurons is often followed by naturally occurring neuronal death. Specific proteins, termed neurotrophic factors, are produced in limiting amounts in the target tissues and mediate the cell interactions regulating neuron survival. The release of these proteins is believed to regulate not only the survival of neurons but also the timing and and exent of innervation of the target tissues. Besides their role neuronal development, neurotrophic factors are of importance also in the function of the adult nervous system.

The best known neurotrophic factor is ß-nerve growth factor (NGF). NGF is a basic 118 amino acid protein acting as a trophic factor for many sensory and sympathetic neurons in the peripheral nervous system (Levi-Montalcini, 1966, 1987; Levi-Montalcini and Angeletti, 1968; Thoenen and Barde, 1980). In agreement with a trophic role for NGF also in adult sympathetic neurons, the levels of NGF mRNA and protein correlate with the density of sympathetic innervation (Korsching and Thoenen, 1983). Nerve growth factor has recently been found also in the brain (Korsching et al., 1985; Whittemore et al., 1986; Shelton & Reichardt, 1986; Goedert et al., 1986) where it serves a trophic function in the development and maintenance of cholinergic neurons situated in the basal forebrain (Korsching et al.,1985; Large et al., 1986; Richardson et al., 1986; Lärkfors et al., Ernfors et al., 1990; 1987; review by Ebendal, 1989a). NGF mRNA and protein have also been found in the brain with the highest levels in hippocampus and cerebral cortex, to which the major cholinergic pathways in the brain project (for reviews, see Thoenen et al., 1987; Ebendal, 1989a; Persson et al., 1990). In the brain, NGF supports the survival of basal forebrain cholinergic neurons. Thus, these neurons can be prevented from dying after axonal transection *in vivo* by addition of exogenous NGF (Hefti, 1986; Williams et al., 1986). NGF protein can be found in the brain of the rat and mouse with the use of a sensitive two-site enzyme immunoassay for mouse NGF. It would be of interest to study the distribution of NGF also in the human brain, especially against the background that cholinergic neurons deteriorate in Alzheimer's disease (Perry et al., 1978; Wilcock et al., 1982). The low level of antibody cross-reactivity between NGFs from different species has hampered attempts to determine levels of human NGF. We have now examined the biological activity and immunological properties of human recombinant NGF protein in an attempt to find conditions that will permit the detection of human NGF in tissue extracts. In particular, one monoclonal antibody to mouse NGF has been found to recognize human

Plasticity and Regeneration of the Nervous System
Edited by P.S. Timiras *et al.*, Plenum Press, New York, 1991

NGF as well as native NGF purified from the mouse submandibular gland (Söderström et al., 1990).

BIOLOGICALLY ACTIVE HUMAN RECOMBINANT NGF PRODUCED BY TRANSFECTED CELLS

The NGF protein has been purified and sequenced (Angeletti and Bradshaw, 1971; Angeletti et al., 1973) from the submandibular gland of the male mouse. In addition, NGF has later been purified from several other species (Server et al., 1976; Chapman et al., 1979; Hofmann and Unsicker, 1982; Harper et al., 1982). The amino acid sequence has been confirmed by analysis of complementary DNA (cDNA) and genomic clones for NGF. Nerve growth factor is synthesized as a 305 amino acid long prepro-NGF (Scott et al., 1983; Ullrich et al., 1983; Ebendal et al., 1986; Selby et al., 1987a; Whittemore et al., 1988). The mature 118 amino acid long NGF protein is generated by proteolytic cleavage at dibasic amino acid residues located in the carboxy-terminal region of prepro-NGF.

We have established and extensively used an *in vitro* transient expression system, utilizing monkey kidney COS cells, that allows production of high levels of biologically active recombinant NGF (Hallböök et al. 1988; Ebendal and Persson, 1988; Ibáñez et al., 1990a,b; Ernfors et al., 1990a; Ebendal et al., 1990). This expression system is rapid and reproducibly gives 100-200 ng of biologically active NGF per ml of conditioned medium (i.e. 1 - 2 µg per transfection), enough for many assays for NGF activity on explanted sympathetic ganglia (Ebendal et al.,1984; Ebendal, 1989b). We have now also produced human recombinant NGF in this transient expression system (Ibáñes et al., 1990b). Thus, a clone containing the 3'exon of the human NGF gene, including the entire coding region for the smaller size prepro-NGF (Scott et al., 1983; Ullrich et al., 1983; Selby et al., 1987a), was isolated from a human genomic EMBL phage library in by cross-hybridization to a mouse NGF cDNA clone. Subsequently a 1.5 kb SmaI-SacI fragment containing the 3'exon of the human NGF gene was subcloned in expression vector pXM (Yang et al., 1986). COS cells (Gluzman, 1981) grown to about 70% confluence were then transfected with 20 µg of the plasmid constructs per 10-cm dish using the DEAE-Dextran-chloroquine method (Luthman and Magnusson, 1983). Transfected cells were grown for three days and the conditioned media collected for assay of NGF. The biological activity of NGF was measured in a nerve fibre outgrowth assay with sympathetic ganglia from the 9-day-old chicken embryo explanted to a collagen matrix (Ebendal et al., 1978, 1980, 1984). Details of the technique have recently been presented (Ebendal, 1989b).

A biologically active human NGF protein was secreted by the COS cells after transfection. A similar strategy to produce recombinant human NGF has recently been taken also by Heinrich and collaborators (Heinrich and Meyer, 1988; Bruce and Heinrich, 1989). For comparisons, the biological activity of purified mouse NGF (Mobley et al., 1976; Chapman et al., 1979; Ebendal et al., 1984) present at 5 ng/ml evoked a dense circular fibre halo regarded as corresponding to an activity of 1 biological unit (BU). A response scored as 0.5 BU was found at a mean NGF concentration of 1.0 ng/ml and the 0.3 BU score was found at 0.4 ng/ml. Inhibition by monoclonal antibody 27/21 on the fibre outgrowth evoked by NGF at 5 ng/ml was also examined and a 50% inhibition was found using 15 ng of anti-NGF Ig per ml (Söderström et al., 1990).

The biological activity of this human recombinant NGF protein (Fig. 1a) was identical to that previously obtained using the purified mouse NGF or rat recombinant NGF (Hallböök et al., 1988; Ibáñez et al., 1990a,b). The monoclonal antibody 27/21 to mouse NGF was shown to efficiently block the activity of both the human recombinant NGF (Fig. 1b) and mouse native NGF (Söderström et al., 1990).

ENZYME IMMUNOASSAY FOR HUMAN NGF USING THE MONOCLONAL ANTIBODY 27/21

The introduction of a sensitive two-site enzyme immunoassay (EIA; Furukawa et al.,

1983; Korsching and Thoenen, 1983; Lärkfors and Ebendal, 1987; Weskamp and Otten, 1987; Arumäe et al., 1989) has allowed for reliable determination of the low levels of NGF present in peripheral and CNS tissues (Korsching and Thoenen, 1983, 1987; Heumann et al., 1984; Lärkfors and Ebendal, 1987; Weskamp and Otten, 1987). For instance, these assays detect NGF in the brain and estimate the level of NGF in the rat hippocampus to 1-2 ng/g wet weight of tissue (Korsching et al., 1985; Whittemore et al., 1986; Lärkfors and Ebendal, 1987). The presence of NGF mRNA demonstrated by RNA blot techniques, and the fact that several laboratories using different antibodies in the EIA (monoclonal and polyclonal) have reached similar values for the NGF concentration in regions of the brain, strongly indicate that authentic NGF is being measured (Korsching et al., 1985; Whittemore et al., 1986; Large et al., 1986; Lärkfors and Ebendal, 1987; Weskamp and Otten, 1987). In contrast, the competitive RIAs used for NGF determinations prior to the introduction of the two-site EIAs falsely indicated high levels of NGF in many tissues (Suda et al., 1978). Sensitivity has often been a limiting factor in determining the levels of NGF present in peripheral and central nervous tissues, especially in studies of NGF from other species than mouse or rat. Moreover, the detection limits for NGF have differed markedly depending on the immunoassay used. Early two-site radioimmunoassays detected NGF down to around 10 pg/ml (Suda et al., 1978; Ebendal et al., 1983). The same level was reported for the first two-site EIA (Furukawa et al., 1983), whereas Korsching and Thoenen (1983) reached a detection limit of 5 pg/ml. Using optimized conditions it has been possible to detect the NGF protein even at 1 pg/ml or less (Lärkfors and Ebendal, 1987; Weskamp and Otten, 1987; Nagata et al., 1987; Hellweg et al., 1989; Söderström et al., 1990).

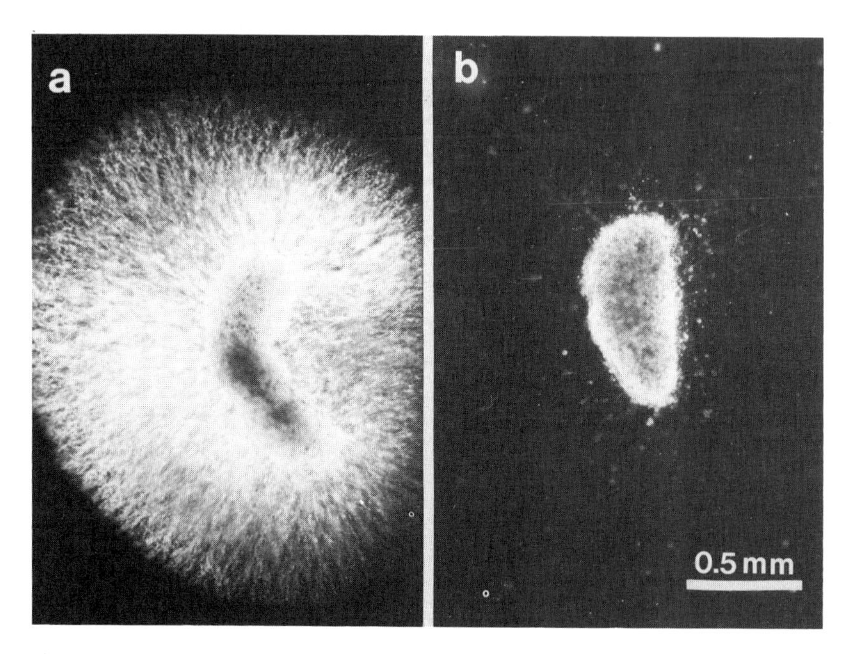

Fig. 1. Biological activity of human recombinant NGF determined as fibre outgrowth in a sympathetic ganglion assay (Ebendal et al., 1989b). (a): Fibre response to recombinant human NGF. Conditioned medium from transfected COS cells was added to a sympathetic ganglion. (b): The same but with monoclonal antibody 27/21 added (188 ng/ml). Dark-field microscopy of cultures incubated for two days.

We wanted to study this human recombinant NGF in a two-site EIA, comparing the result of the EIA with the response evoked by the human NGF protein in a sympathetic ganglion bioassay (Ebendal, 1989b). One major obstacle in assessing the function of NGF in neurodegenerative conditions such as Alzheimer's disease (Hefti and Weiner, 1986; Ebendal, 1989a) has been the lack of a sensitive EIA for the human NGF protein. It would be of interest to measure the NGF protein in the human brain, in particular since the cholinergic neurons are severely deteriorated in senile dementia of the Alzheimer type. The limited immunological cross-reactivity between NGFs from different species has previously hampered attempts to determine levels of the human NGF. From initial experiments we found that the monoclonal anti-mouse NGF antibody 27/21 (described by Korsching and Thoenen, 1983) recognized the human NGF with higher affinity than our previously used polyclonal sheep antibodies to mouse NGF (Lärkfors and Ebendal, 1987). A two-site EIA using monoclonal antibody 27/21 was then optimized (Söderström et al., 1990). Under the conditions used, this EIA detected the human recombinant NGF with the same sensitivity (1 pg/ml) as shown for the mouse NGF. It should now be possible to test the EIA on homogenized tissue to examine human NGF in brain samples from Alzheimer patients or age-matched controls.

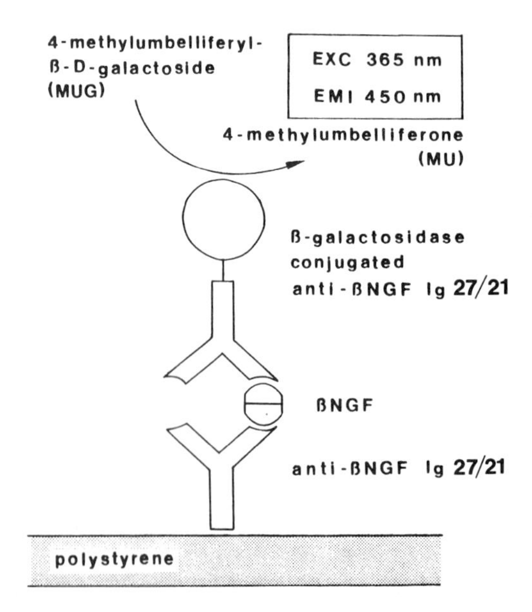

Fig. 2. Schematic illustration of a two-site EIA for NGF based on the monoclonal antibody 27/21 as capture antibody (coated to EIA multiwell dishes) and detection antibody (conjugated to ß-galactosidase). Details of the technique are described by Lärkfors and Ebendal (1987) and Söderström et al. (1990).

Details of the EIA have been given elsewhere (Söderström et al., 1990). In summary, immunoplates (black 96-well multidishes, Dynatech Microfluor) were coated with monoclonal anti-mouse-ßNGF antibody 27/21 (Korsching and Thoenen, 1983) obtained from Boehringer Mannheim Biochemica. Control wells were coated at 4 °C overnight with mouse normal immunoglobulin G (mouse IgG). After blocking and washing with Tris-buffered saline (TBS; comprising of 0.02 M Tris-HCl, pH 7.4, 0.5 M NaCl), conditioned medium with recombinant NGF was serially diluted in TBS containing 2 mg/ml BSA and 0.5% Tween 20. To serve as standards, samples of purified mouse NGF at 100, 10, 1, 0.1, 0.01, 0.001 ng/ml in the same buffer were also added, the plates sealed and incubated overnight at 4 °C. The plates were then washed extensively in TBS with Tween 20 before an antibody-27/21-ß-galactosidase-conjugate (Boehringer; 4 units of enzyme activity per ml) was added at a dilution of 1:200 for an overnight incubation at 4°C. Finally the dishes were washed for several hours in TBS and the enzyme reaction started by adding methylumbelliferyl-ß-galactoside. The accumulation of methylumbelliferone (excitation wavelength 365 nm, emission wavelength 450 nm) was followed in a microplate fluorometer and the result expressed as enzyme activity (used fmol substrate per min). The enzyme activity plotted as a function of the NGF concentration is shown in Fig. 3.

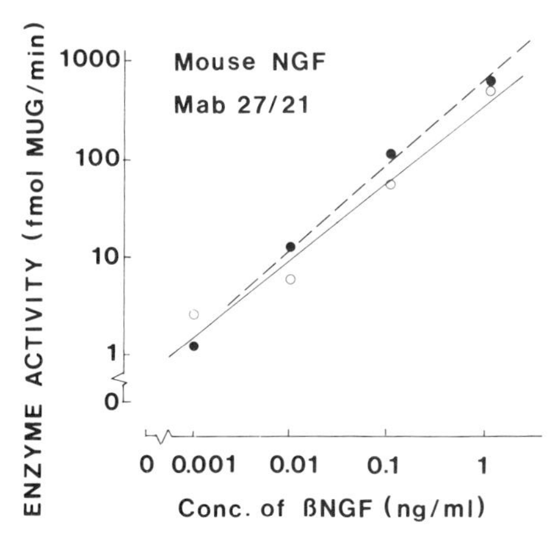

Fig. 3. The outcome of measuring standard known amounts of purified mouse ßNGF in the two-site EIA with monoclonal NGF antibody 27/21. The monoclonal antibody 27/21 served as capture antibody coated to EIA microplates. As detection antibody the same 27/21 IgG conjugated to ß-galactosidase was used. The result of two independent determinations (open and filled circles, respectively) and the resulting regression lines are shown as log-log plots. The sensitivity of the assay is in the range of 1 - 2 pg/ml under the conditions used (Söderström et al., 1990).

The ability of the monoclonal antibody 27/21 to detect the human recombinant NGF in the EIA was measured (Söderström et al., 1990). The resulting regression lines for the human recombinant NGF coincide well with that obtained for purified mouse NGF. It was recently indicated also in other studies that that the monoclonal antibody 27/21 to mouse ßNGF (Korsching and Thoenen, 1983) recognizes human NGF in a two-site EIA (Heinrich and Meyer, 1988; Bruce and Heinrich, 1989). The similarity in amino sequence between NGF from different species is high (66-98 %) with the variable regions located in a few hydrophilic domains (Meier et al., 1986; Whittemore et al., 1988; Ebendal et al., 1989), likely to correspond to antigenic epitopes. This variation among species, offers a plausible explanation for the limited immunological cross-reactivity between mouse NGF and NGF isolated from other species such as bull, rabbit, guinea pig, shrew and human (Harper and Thoenen, 1980; Harper et al., 1983; Ueyama et al., 1986; Nakata et al., 1988) as well as

```
                       V1                                     P3
NGF
 mouse     +1    S S T H P V F H M G E F S V C D S V S V W V G - - D K T
 rat       +1    S S T H P V F H M G E F S V C D S V S V W V G - - D K T
 Mastom    +1    S S T H P V F Q M G E F S V C D S V S V W V G - - D K T
 gui pig   +1    S S T H P V F H M G E F S V C D S V S V W V A - - D K T
 human     +1    S S S H P I F H R G E F S V C D S V S V W V G - - D K T
 bovine    +1    S S S H P V F H R G E F S V C D S I S V W V G - - D K T
 chicken   +1    T A - H P V L H R G E F S V C D S V S M W V G - - D K T
 cobra     +1    E - D H P V H N L G E H S V C D S V S A W V T - - - K T
HDNF/NT-3
 rat       +1    Y A E H K S - H R G E Y S V C D S E S L W V T - - D K S
 mouse     +1    Y A E H K S - H R G E Y S V C D S E S L W V T - - D K S
 human     +1    Y A E H K S - H R G E Y S V C D S E S L W V T - - D K S
BDNF
 pig       +1    H S D - P A - R R G E L S V C D S I S E W V T A A D K K
 rat       +1    H S D - P A - R R G E L S V C D S I S E W V T A A D K K
 mouse     +1    H S D - P A - R R G E L S V C D S I S E W V T A A D K K
 human     +1    H S D - P A - R R G E L S V C D S I S E W V T A A D K K

                     P3                        V2
NGF
 mouse    +27  T A T D I K G K E V T V L A E V N I N N S V F R Q Y F F E T
 rat      +27  T A T D I K G K E V T V L G E V N I N N S V F K Q Y F F E T
 Mastom   +27  T A T D I K G N E V T V L G E V N I N N S V F K Q Y F F E T
 gui pig  +27  T A T D I K G K E V T V L A E V N V N N N V F K Q Y F F E T
 human    +27  T A T D I K G K E V M V L G E V N I N N S V F K Q Y F F E T
 bovine   +27  T A T D I K G K E V M V L G E V N I N N S V F K Q Y F F E T
 chicken  +26  T A T D I K G K E V T V L G E V N I N N N V F K Q Y F F E T
 cobra    +26  T A T D I K G N T V T V M E N V N L D N K V Y K Q Y F F E T
HDNF/NT-3
 rat      +26  S A I D I R G H Q V T V L G E I K T G N S P V K Q Y F Y E T
 mouse    +26  S A I D I R G H Q V T V L G E I K T G N S P V K Q Y F Y E T
 human    +26  S A I D I R G H Q V T V L G E I K T G N S P V K Q Y F Y E T
BDNF
 pig      +27  T A V D M S G G T V T V L E K V P V S K G Q L K Q Y F Y E T
 rat      +27  T A V D M S G G T V T V L E K V P V S K G Q L K Q Y F Y E T
 mouse    +27  T A V D M S G G T V T V L E K V P V S K G Q L K Q Y F Y E T
 human    +27  T A V D M S G G T V T V L E K V P V S K G Q L K Q Y F Y E T

                       V3
NGF
 mouse    +57  K C R A S N P V E S G C R G I D S K H W N S Y C T T T H T F
 rat      +57  K C R A P N P V E S G C R G I D S K H W N S Y C T T T H T F
 Mastom   +57  K C R A R N P V E S G C R G I D S K H W N S Y C T T T H T F
 gui pig  +57  K C R D P S P V E S G C R G I D S K H W N S Y C T T T H T F
 human    +57  K C R D P N P V D S G C R G I D S K H W N S Y C T T T H T F
 bovine   +57  K C R D P N P V D S G C R G I D A K H W N S Y C T T T H T F
 chicken  +56  K C R D P R P V S S G C R G I D A K H W N S Y C T T T H T F
 cobra    +56  K C K N P N P E P S G C R G I D S S H W N S Y C T E T D T F
HDNF/NT-3
 rat      +56  R C K E A R P V K N G C R G I D D K H W N S Q C K T S Q T Y
 mouse    +56  R C K E A R P V K N G C R G I D D K H W N S Q C K T S Q T Y
 human    +56  R C K E A R P V K N G C R G I D D K H W N S Q C K T S Q T Y
BDNF
 pig      +57  K C N P M G Y T K E G C R G I D K R H W N S Q C R T T Q S Y
 rat      +57  K C N P M G Y T K E G C R G I D K R H W N S Q C R T T Q S Y
 mouse    +57  K C N P M G Y T K E G C R G I D K R H W N S Q C R T T Q S Y
 human    +57  K C N P M G Y T K E G C R G I D K R H W N S Q C R T T Q S Y
```

V4

	Position	Sequence
NGF		
mouse	+87	V K A L T T D E - K Q A A W R F I R I D T A C V C V L S R K
rat	+87	V K A L T T D D - K Q A A W R F I R I D T A C V C V L S R K
Mastom	+87	V K A L T T D D - R Q A A W R F I R I D T A C V C V L T R K
gui pig	+87	V K A L T T D N - K Q A A W R F I R I D T A C V C V L N R K
human	+87	V K A L T M D G - K Q A A W R F I R I D T A C V C V L S R K
bovine	+87	V K A L T M D G - K Q A A W R F I R I D T A C V C V L S R K
chicken	+86	V K A L T M E G - K Q A A W R F I R I D T A C V C V L S R K
cobra	+86	I K A L T M E G - N Q A S W R F I R I E T A C V C V - I T K
HDNF/NT-3		
rat	+86	V R A L T S E N N K L V G W R W I R I D T S C V C A L S R K
mouse	+86	V R A L T S E N N K L V G W R W I R I D T S C V C A L S R K
human	+86	V R A L T S E N N K L V G W R W I R I D T S C V C A L S R K
BDNF		
pig	+87	V R A L T M D S K K R I G W R F I R I D T S C V C T L T I K
rat	+87	V R A L T M D S K K R I G W R F I R I D T S C V C T L T I K
mouse	+87	V R A L T M D S K K R I G W R F I R I D T S C V C T L T I K
human	+87	V R A L T M D S K K R I G W R F I R I D T S C V C T L T I K

	Position	Sequence	End
NGF			
mouse	+116	A T R R G	+120
rat	+116	A A R R G	+120
Mastom	+116	A A R R G	+120
gui pig	+116	A A R R G	+120
human	+116	A V R R A	+120
bovine	+116	T G Q R A	+120
chicken	+115	S G - R P	+118
cobra	+114	- - K K G N	+117
HDNF/NT-3			
rat	+116	- I G R T	+119
mouse	+116	- I G R T	+119
human	+116	- I G R T	+119
BDNF			
pig	+117	- R G R	+119
rat	+117	- R G R	+119
mouse	+117	- R G R	+119
human	+117	- R G R	+119

Fig. 4. Comparison of the amino acid sequences for the members of the NGF family of neurotrophic factors. The mature protein sequences are shown. The amino acid residues most frequent at each position have been boxed. In the case that two amino acids are represented in equal number the residue shared by different members of the protein family is boxed. Four regions of variability are denoted V1 to V4. A region highly conserved within NGFs from different species is denoted P3 (our synthetic NGF peptide P3 used to raise antibodies as described previously (Ebendal et al., 1989). Data are based on the cloning of NGF from eight different species (mouse, rat, the African rat *Mastomys natalensis*, guinea pig, human, bovine, chicken, and cobra; Scott et al., 1983; Ullrich et al., 1983; Ebendal et al., 1986; Meier et al., 1986; Wion et al., 1986; Selby et al., 1987a, b; Fahnestock and Bell, 1988; Whittemore et al., 1988a, b; Schwarz et al., 1989). The sequences for the two additional members of the NGF family named brain-derived neurotrophic factor (BDNF) and hippocampus-derived neurotrophic factor or neurotrophin-3 (HDNF/NT-3) have recently been published for the indicated species by Leibrock et al. (1989), Hohn et al. (1990), Maisonpierre et al. (1990), Ernfors et al. (1990a), Kaisho et al. (1990), and Rosenthal et al. (1990).

cobra and viper (Bailey et al., 1976; Arumäe et al., 1987) despite the similarity in biological activity (Harper et al., 1983). Nakata et al. (1988) cross-checked NGF from mouse, guinea pig and cobra in two-site EIAs for each of the NGF from these species. Under the used conditions, there was in each case a preference for the homologous antigen. The mammalian NGFs showed estimated cross-reactivities of 57-71 %. Even NGF from the submandibular gland of the insectivore *Suncus murinus* (house musk shrew) has been reported to give 11 % cross-reactivity in the two-site EIA for mouse NGF (Ueyama et al., 1986). In contrast, the cobra NGF did not give a signal in the the two-site EIAs for mouse and guinea pig NGF. Similarly, the cobra NGF EIA, with a detection limit of 40 pg/ml of the homologous antigen, failed to detect mouse or guinea pig NGF even at 10 ng/ml (Nakata et al., 1988).

We thus have demonstrated the feasibility to determine human NGF in minute amounts using the monoclonal antibody 27/21 in the two-site EIA (Söderström et al., 1990). Further studies may, however, require antibodies with even higher affinity for the human NGF. It should be possible to use the recombinant human NGF as an immunogen for the production of both polyclonal and monoclonal antibodies. This approach requires purification of the recombinant human NGF from medium conditioned by transfected cells, and is furthermore likely to require the NGF to be preserved in a biologically active, and therefore assayable, form throughout the purification procedure.

MOLECULAR CLONING OF A NOVEL NEUROTROPHIC PROTEIN WITH SIMILARITIES TO ß-NERVE GROWTH FACTOR

The cloning of NGF (Scott et al., 1983; Ullrich et al., 1983; Ebendal et al., 1986; Meier et al., 1986; Wion et al., 1986; Selby et al., 1987a, b; Fahnestock and Bell, 1988; Whittemore et al., 1988a, b; Schwarz et al., 1989) has allowed deduction of the sequence for the mature NGF protein from eight different species (mouse, rat, the African rat *Mastomys natalensis*, guinea pig, human, bovine, chicken, cobra). In addition, another protein, termed brain-derived neurotrophic factor (BDNF) supports the survival of neural crest-derived embryonic sensory neurons in vitro (Barde et al., 1982; Leibrock et al., 1989), but nonoverlapping trophic activities are suggested by the finding that BDNF also supports placode-derived neurons from the nodose ganglion (Davies et al., 1986) and retinal ganglion cells, wich are less influenced by the activity of NGF. Regulation of neuronal survival in the brain by BDNF remains to be demonstrated.

Recently, a genomic clone was isolated for the porcine BDNF (Leibrock et al., 1989). Of considerable interest was the finding that the mature BDNF and NGF proteins show striking amino acid similarities, suggesting that they are structurally related and belong to a family of neurotrophic factors. In order to clone further members of the NGF family, a pool of degenerate oligonucleotides representing all possible codons in regions of homology between brain-derived neurotrophic factor (BDNF) and ß-nerve growth factor (NGF) was used to prime rat hippocampal cDNAs in the polymerase chain reaction (Ernfors et al., 1990a). Six separate mixtures of 28-mer oligonucleotides representing all possible codons corresponding to the amino acid sequence KQYFYET (5'-oligo) and WRFIRID (3'-oligo) were synthesized (see Fig. 4 for their positions). The 5'-oligo contained a synthetic EcoRI site and the 3'-oligo a synthetic HindIII site to facilitate subcloning. The PCR amplified products were cloned in pBS KS+ and sequencing revealed PCR products corresponding to both rat NGF and BDNF as well as a DNA sequence with 60 and 62% nucleotide sequence similarity to mouse NGF and pig BDNF, respectively. From the nucleotide sequence, a 282 amino acid long protein with approximately 45% amino acid similarity to both pig BDNF and rat NGF was deduced. The latter sequence was used to screen a rat hippocampus cDNA library and a 1020bp cDNA clone was isolated. Nucleotide sequence analysis of this showed an open reading frame encoding a 282 amino acid long protein. The carboxyterminal part of this protein contained a potential cleavage site for a 119 amino acid protein with 57% amino acid similarity to both rat mature NGF and pig mature BDNF. Included in this similarity were all six cysteine residues, involved in the formation of disulfide bridges in NGF (Angeletti et al., 1973). The prepro part of the protein showed weak, but significant, homology, to NGF. A potential N-glycosylation site located nine amino acids from the start of the mature protein was also conserved between the three proteins (Ernfors et al., 1990a).

The cDNA clone isolated by us (Ernfors et al., 1990a) encodes a protein, given the name hippocampus-derived neurotrophic factor (HDNF), having a remarkable sequence similarity to both NGF and BDNF and therefore represents a novel member of a family of neurotrophic proteins (Fig. 4). Recently, also other reports on the independent cloning of this protein, also termed neurotrophin-3 or NGF-2, from other species including human have been published (Rosenthal et al., 1990; Leibrock et al., 1990; Maisonpierre et al., 1990; Ernfors et al., 1990a, Kaisho et al., 1990). The distribution of HDNF mRNA in the adult brain showed remarkable regional specificity with high levels in hippocampus compared to other brain regions analyzed (Ernfors et al., 1990a,c). In the adult brain the mRNA for this protein was predominantly expressed in hippocampus, where it was confined to a subset of pyramidal and granule neurons. In fact, cerebellum was the only other region where HDNF mRNA was clearly detected, with the exception of cerebral cortex which showed a weak signal. BDNF mRNA was more widely distributed in the rat brain, although hippocampus also contained the highest amount, followed by cerebral cortex, pons and cerebellum. *In situ* hybridization histochemistry showed this factor to have a neuronal localization partly distinct from that of NGF and HDNF/NT-3 including e.g. many neurons in claustrum (Wetmore et al., 1990). Interestingly, these three neurotrophic proteins were maximally expressed at different times of brain development in the rat with a peak of HDNF mRNA shortly after birth, BDNF mRNA around two weeks and NGF mRNA three weeks after birth (Whittemore et al., 1986; Ernfors et al., 1990a). We have recently shown that the peak in HDNF mRNA expression after birth is due to a transient expression of HDNF mRNA in the cingulate cortex (Friedman et al., 1990).

The 1020bp HDNF cDNA insert (Ernfors et al., 1990a) was also inserted into the expression vector pXM (Yang et al., 1986). The construct was transfected into COS cells and three days later conditioned medium was tested for biological activity in bioassays measuring fibre outgrowth from various chicken embryo ganglia. A marked stimulation of neurite outgrowth, consistently resulting in dense fibre halos, was seen in the ganglion of Remak, a ganglionated nerve trunk in the mesorectum of the chicken embryo (Hedlund and Ebendal, 1978; Ebendal, 1979). Although NGF is known to stimulate the explantated ganglion of Remak (Ebendal, 1979) it was far less efficent than HDNF (Ernfors et al., 1990a). A modest stimulation of fiber outgrowth was also seen with HDNF in the nodose ganglion, consisting of neurons exclusively derived from an epidermal placode (Hedlund and Ebendal, 1978). Again, HDNF was superior to NGF in evoking this response. A weak but consistent fiber outgrowth in response to HDNF was seen in paravertebral symphathetic trunk ganglia (Ernfors et al., 1990) which, however, was much less pronounced compared to the massive response to NGF. In dorsal root spinal ganglia HDNF stimulated neurite outgrowth to the same extent as NGF. Antisera to NGF did not block the biological activity of HDNF, nor was HDNF detected in a protein blot using NGF antisera. We were also unable to detect HDNF with the 27/21 EIA for NGF (Söderström et al., 1990) in conditioned media from COS cells transfected with the HDNF gene (Ernfors et al., 1990a; and unpublished observations), in agreement with the well documented NGF-specificity of antibody 27/21, showing no cross-reactivity with other growth factors such as epidermal growth factor, fibroblast growth factor or insulin (Korsching and Thoenen, 1983; Donohue et al., 1989).

Thus, NGF, BDNF and HDNF/NT-3 each has unique biological activities and represent different members of a family of neurotrophic factors that may cooperate to support the development and maintenance of the vertebrate nervous system. It is worth noting that NGF seems to be the one in the family of neurotrophic factors having the largest interspecies variability (Fig. 4). Antibodies to BDNF and to HDNF/NT-3 have yet not been presented but it may be expected that when available they will show a high degree of cross-reactivity, a fact that should facilitate the development of EIAs detecting also the human mature BDNF and HDNF/NT-3 proteins, respectively, each showing amino acid identity to the corresponding amino acid sequences for the proteins from the rat (Fig. 4). The availability of antibodies for human NGF, BDNF and HDNF/NT-3 would make it possible to further study the functions of these factors in human brain development and their role in neurodegenerative disease.

SITE-DIRECTED MUTAGENESIS OF NGF

No information is yet available on the receptor-binding region of NGF. Little is also known about the amino acid sequences required for the neurotrophic activity of NGF. The mature NGF protein interacts with a specific NGF receptor *in vivo* and elicits the neurotrophic activity. Considering the wide expression of the NGF-R in development (Ernfors et al., 1988; Hallböök et al., 1990) it is also of interest to examine if any or several of the other members of the NGF family are using the same cell-surface receptor. Indeed recent findings indicate that the low affinity form of the NGF receptor may be used by both NGF and BDNF (Rodriguez-Tébar et al., 1990). Information about which part of the NGF protein that binds to the NGF receptor is also necessary for understanding the specific action of each of the members of the NGF family. Such data can be obtained by site-directed mutagenesis of the cloned NGF gene. Furthermore, specific deletions or exchanges in this gene, resulting in the synthesis of altered forms of the NGF protein or chimaeric NGF/BDNF/HDNF proteins, can also be used to obtain information on functionally important regions of these proteins and shed light on which differences between the members of the NGF family bring functional specificity to the different proteins. A prerequisite for such studies is a rapid and efficient system as described above in order to test mutated NGF genes for their ability to produce biologically active NGF protein. Attempts can also be directed towards finding a hyperactive NGF or an NGF with altered half-life for experimental purposes and possible therapeutic applications.

We have begun to use also the human NGF gene for site-specific mutagenesis and deletion studies. Truncated forms of the NGF protein have been produced by the insertion of in-frame translational stop codons in the human NGF gene using synthetic oligonucleotides (Taylor et al., 1985). Selected amino acid residues also in the chicken nerve growth factor (NGF) were replaced by site directed mutagenesis (Ibáñez et al., 1990a). Mutated NGF sequences were transiently expressed in COS cells using the plasmid pXM (Yang et al.,1986). The amount of each mutant polypeptide accumulated in conditioned medium was assessed by protein immunoblotting. This determination allowed us to correct for differences in the levels of protein, which varied over a 12-fold range. Binding of each mutant to NGF receptors on PC12 cells was evaluated in a competition assay (Ibáñes et al., 1990a). The biological activity was determined by measuring stimulation of neurite outgrowth from chick sympathetic ganglia (Ebendal, 1989b).

The structural and functional importance of highly conserved residues within the NGF polypeptide was thus determined by the systematic replacement of selected amino acids using site-directed mutagenesis (Ibáñez et al., 1990a). In most cases the mutant residues were chosen for having similar chemical characteristics to the replaced amino acids in the native sequence, thereby causing minimal conformational distortions. In addition, non-conservative changes were introduced to test further the structural importance of a particular residue. The three tryptophan residues at positions 20, 75 and 98 in chicken NGF are all conserved within the family of NGF-like polypeptides (Fig. 4), and may be essential for biological activity. To define the functional importance of these residues, we replaced them with phenylalanine, a residue that also has an aromatic side chain but lacks the indole ring. Predictions of the secondary structure and hydrophilicity of the mutants by computer analysis did not show any major differences when compared to those based on the wild type sequence. However, large differences were found in the yield of production and the W20F mutant accumulated only 15% as much NGF protein as the wild type (Ibáñez et al., 1990a). Both W20F and W98F mutants showed a somewhat reduced activity in the sympathetic ganglion bioassay. The modification of Trp 75 did not significantly affect the activity of the molecule. All three mutant proteins were detected by immunoprecipitation with mouse NGF antibodies, which indicates that the replacement by phenylalanine did not significantly change their characteristics. Trp 20 was also replaced by Gly, a residue which disrupts alpha and beta conformations, to probe for the structural importance of the aromatic side chain. However, this mutant did not accumulate in the medium at high enough concentrations to be detected by protein blotting.

Ser 18, Met 19, Val 21 and Asp 23 were systematically substituted by alanine, a residue that eliminates the side chain beyond the β-carbon, yet does not alter the main conformation or impose extreme electrostatic or steric effects. The mutant NGFs with either M19A or D23A substitutions showed levels similar to wild type in both biological activity and receptor binding, although they markedly differed in their yield of production. The S18A mutant exhibited about half the specific activity of the wild type, which correlated with a decrease of the same magnitude in receptor binding. Interestingly, the modification of Val 21 substantially reduced both binding and biological activity, suggesting a relative importance of this residue. The conserved Arg 99 and Tyr 51 were found to be of little essence for function but may play a structural role (Ibáñez et al., 1990a). We also investigated the structural and functional role of the highly conserved Arg 99 and Arg 102. The conservative replacement of Arg 99 by Lys did not affect the behaviour of the molecule in any of the assays performed. However, both binding and biological activity were greatly reduced when both Arg 99 and Arg 102 were simultaneously changed into Gly, probably due to conformational alterations.

A stretch of aromatic residues around the conserved Tyr 51 has been suggested as a potential site of contact with the NGF receptor. We therefore replaced Tyr 51 with either Phe or Gly to define the importance of this residue. The Y51F mutant displayed a similar binding affinity and biological activity to the wild type (Ibáñez et al., 1990a). Furthermore, the replacement did not significantly affect the main antigenic determinants of molecule. However, substitution of Tyr 51 by Gly prevented the accumulation of detectable levels of NGF protein in the medium, suggesting a structural importance of this position in determining a stable conformation of the protein.

The analysis presented here begins to unravel the importance of some structural and functional elements in the NGF protein. More insight into the structures responsible for receptor binding and biological activity of NGF will be gained by the systematic application of site-directed mutagenesis to additional regions of the NGF molecule. It may also, as noted above, be possible to design hybrid NGF/BDNF/HDNF factors with testable, novel neurotrophic activities for potential applications in neurodegenerative disease.

RESCUE OF AXOTOMIZED NEURONS WITH GRAFTED GENETICALLY MODIFIED, NGF-PRODUCING CELLS

We have used the mammalian expression vector OVEC (Westin et al., 1988), containing a synthetic oligonucleotide encoding a metal responsive element in conjunction with a rabbit ß-globin promoter to establish a mouse fibroblast cell line (3T3-3E) that produces large amounts of biologically active, recombinant rat NGF, continuously secreted at high levels (reaching approximately 5 ng/ml in the conditioned medium; Ernfors et al., 1989). Following transplantation of this cell line into both normal and lesioned rat and mouse brains, a marked increase in survival of basal forebrain cholinergic neurons has been documented (Ernfors et al., 1989; Strömberg et al., 1990). The 3E cells are stably expressing a 25-fold elevated level of NGF-release compared to the parent cell line and condition their growth medium so that the classical NGF-induced fibre halo around explanted embryonic ganglia is easily demonstrated as is the drastic increase of NGF mRNA in the 3E cells (Ernfors et al., 1989). The 3E were grafted under several different conditions, either as a cell suspension of entrapped in a collagen gel matrix (Ebendal, 1989b). When grafted in the collagen matrix to the intact striatum of adult rats, the NGF-producing cells but not a cell-free matrix or a collagen matrix with the parent 3T3 cells (Todaro and Green, 1963), evoked a markedly increased density of cholinergic nerve terminals around the implant as evaluted by immunohistochemistry for acetylcholine esterase (AChE; Ernfors et al., 1989), a good cholinergic marker in this part of the brain (Eriksdotter-Nilsson et al., 1989). In addition, AChE-positive fibres invaded the collagen implant only if containing 3E cells. As a model of the cortical cholinergic deficits of Alzheimer's disease, the NGF-producing 3E cells were also grafted together with NGF-dependent fetal cholinergic neurons from the basal forebrain into the cerebral cortex of adult rats that had previously received ibotenic acid injections to reduce the cholinergic projections to the cortical areas. Four weeks after the co-grafting, the survival of the fetal cholinergic neurons was significantly higher with the 3E cells than with the 3T3 parental cells. Cholinergic nerve fibre growth was also greatly enchanced. In addition to the effects on the grafted fetal neurons, the NGF-producing cells were found to stimulate intrinsic, local AChE fibres in the cortex (Ernfors et al. 1989; Olson et al., 1990).

We have also shown (Strömberg et al., 1990) that the genetically modified 3E cells can be grafted to the cavity after a fimbria-fornix lesion to provide a chronic source of NGF in order to rescue axotomized neurons. An unilateral aspiration lesion was made of the fimbria-fornix pathway in the brains of female Sprague-Dawley rats (200-220 g body weight) under deep anesthesia. The 3E cells or the parent 3T3 cells were first grown in a three-dimensional collagen gel matrix. The gels contained clusters of cells occupying approximatley 5% of the gel volume prior to grafting. Immediately after the lesion was made, pieces of gel measuring approximately 25 cubic millimeter were inserted into the cavity formed by aspiration of a portion of the fimbria-fornix pathway and the gels covered by gel foam. Some rats recevied grafts of pieces of collagen gel containing the stably transfected, NGF-producing cell line, and other rats the parental cell line. As the 3T3 cells are of mouse origin, host rats were immunosuppressed with daily injections of cyclosporine A (10 mg/kg i.p.) and protected from infection by Vibramycin (2 mg/kg i.p., daily). After 4-6 weeks rats were sacrificed under deep barbiturate anesthesia by transcardial perfusion of calcium-free Tyrode's solution, followed by chilled 4% paraformaldehyde in phosphate buffer, and the brains were processed for immunohistochemistry. Cryostat tissue sections (14 µm) were incubated with antibodies to acetylcholine esterase (AChE; generously provided by Drs. J. Grassi and J. Massoulié, France), followed by fluorescein-labeled swine anti-rabbit second antibodies. A mentioned above, a comparison between AChE histochemistry, AChE detected immunohistochemically, and choline acetyltransferase immunohistochemistry (Eriksdotter-Nilsson et al., 1989) showed that AChE immunohistochemistry is a reliable marker of cholinergic neurons in the septum. As a further test, the distrubution of AChE-immunoreactive (AChE-IR) neurons in the medial septum was compared with that of neurons immunoreactive with a monoclonal NGF-receptor (NGFR) antibody (MC192, kindly provided by Dr. Eugene Johnson, St. Louis, MO). All AChE-immunoreactive neurons in the medial septum-horizontal limb of the diagonal band of Broca region were counted on the lesioned as well as the unlesioned side in a series of approximately 20 sections taken from each brain. Completeness of the fimbria-fornix lesion was ascertained by examination of the fimbra-fornix area and, particulary, by the loss of AChE-immunoreactive fibers in the ipsilateral hippocampus.

The fimbria-fornix lesion caused complete or near complete loss of AChE-immunoreactive nerve terminales in the hippocampus on the lesioned side without affecting the density of such fibers on the contralateral side (Strömberg et al., 1990). Surviving fibroblasts were found in all gels 4-6 weeks after grafting, yet there was no indication that the cells had infiltrated the host brain tissue. Basal forebrain cholinergic neurons and processes in septum and the diagonal band were strongly labeled septal and diagonal band neurons also showed strong NGF receptor-like immunoreactivity on their cell surface. The distribution of AChE-and NGFR-immunoreactive neurons in the diagonal band and medial septal area were found to be very similar, if not identical. There was a clear loss of AChE-immunoreactive somata in the medial septum/vertical limb of the diagonal band area on the lesioned side in animales receiving grafts of 3T3 cell-containing gels. At the graft/host interface, AChE-IR nerve terminals from the surrounding cortex appeared to increase in density and sprout into the collagen gel in the presence of 3E cells. Little, if any, proliferative response was noted in host brain tissue adjacent to gels containing the parental cell line. The loss AChE-immunoreactive cells was much smaller on the lesioned side in animals receiving gels containing NGF-producing 3E cells (Strömberg et al., 1990). The results of the cell counts demonstrated that the percentage of AChE-immunoreactive cells lost in the medial septum on the lesioned side of the brain was significantly reduced in the group of animals that received the NGF-producing 3E cells. Thus, grafts of gels containing NGF-producing 3E cells resulted in a significantly higher percentage of cholinergic neurons surviving axotomy, compared with animals treated with the parental cell line (Strömberg et al., 1990).

These results are in line with the wealth of recent information showing that central cholinergic neurons depend on NGF for trophic support (for reviews, see Thoenen et al., 1987; Ebendal, 1989a). In particular, it has been shown that lesioned septal cholinergic neurons can be rescued by infusion or injection of exogenous NGF (Hefti, 1986; Williams et al., 1986; Kromer, 1987). In order to obtain these beneficial effects, NGF must be available for an extended period of time and must be obtained from a local source in order to circumvent the blood-brain barrier. Repeated local injections or chronic infusion using osmotic minipumps connected to implanted cannulas or dialysis fibres (Strömberg et al., 1985) can provide NGF effectively for extended periods. We and, independently, Rosenberg et al. (1988), who described a recombinant NGF-producing cell line produced by the use of a retroviral vector, have as described above devised an alternative means of local NGF delivery, based upon grafting genetically modified cell lines that secrete biologically active recombinant NGF. The fact that the NGF-producing 3E cell line gave marked neurotrophic effects after implantation into the rat brain, suggests that grafting of genetically modified cell lines that produce functional recombinant gene products (transgenes) is a feasible strategy to evoke growth responses in the host brain (Olson et al., 1990). Grafting of such cells to the brain can be used as a tool to unravel functional aspects of the product encoded by the transgene. The widespread expression of both NGF receptor mRNA and protein in different brain regions such as cerebellum (Ernfors et al.,1988) suggest that NGF may also exert an effect on non-cholinergic neurons. Further, implantation of the 3E, or similar NGF-producing cell lines, into intact or lesioned rat brain should reveal additional central nervous system neurons responsive to NGF *in vivo*. In addition, this approch may be used to treat central nervous system disease where the transgene product could be therapeutically benefical (Olson et al., 1990).

A similar approach is now being taken for the isolated human NGF gene in order to reach high and constitutive expression of the recombinant human NGF in human and mouse cell lines. We are currently isolating primary cell lines from different organs of rat and human fetuses to serve for transplantation purposes. It is well known that mouse and rat cells readily become immortalized upon passage in culture (Todaro and Green, 1963) whereas this is not the case for human primary cells that, without exceptions, have a finite lifespan in culture (Pontén, 1976). The growth properties of the several cell lines we have isolated are currently being studied. We also plan to genetically alter several of these cell lines to be in a position to compare production and secretion of recombinant NGF from different cell types. The resulting cells can be used in model experiments with cell grafts producing NGF in the brain and offer an alternative to NGF infusion into the brain (Strömberg et al., 1985; Fischer et al., 1987) to study the specific effects of NGF and the other members (BDNF, HDNF/NT-3) of the NGF family of proteins.

Magnocellular cholinergic neurons in nucleus basalis of Meynert undergo a profound and selective degeneration in patients with senile dementia of the Alzheimer type (SDAT). A strong correlation also exists between the reduction in cholinergic marker activity and severity of the dementia (Perry et al., 1978; Wilcock et al., 1982). The level of NGF mRNA is not significantly changed in hippocampal neurons in SDAT compared to aged-matched control samples (Goedert et al., 1986; Ernfors et al., 1990b). However, the recent finding of lowered levels of NGF protein in the brain of aged rats (Lärkfors et al., 1987, 1988) calls for a detailed study of the levels of NGF protein in the brain of SDAT patients compared to age-matched control brains. In this context, it is interesting that the surviving magnocellular cholinergic neurons in the nucleus basalis of Meynert from SDAT brains contain threefold higher levels of NGF receptor mRNA compared to their age-matched normal counterparts (Ernfors et al., 1990b), suggesting that a these neurons can respond to administered NGF, possibly retarding degeneration of basal forebrain cholinergic neurons in SDAT.

We suggest that implantation of genetically modified cells producing NGF (Ebendal et al., 1990; Olson et al., 1990) may have therapeutic applications in rescuing damaged central cholinergic neurons in senile dementia of the Alzheimer type as well as in providing trophic support for chromaffin tissue grafts in Parkinson's disease. Such efforts should make it possible to study the effects of human NGF on brain repair and in development after transplantation of the genetically modified cells. Furthermore, the availability of a cell line producing high levels of human NGF will allow for a homogenous preparation of biologically active, recombinant human NGF for clinical purposes.

ACKNOWLEDGMENTS

We thank Annika Jordell-Kylberg and Vibeke Nilsson for expert technical assistance. This work was supported by the Swedish Natural Science Reseach Council, the Swedish Medical Research Council, the Swedish Enviromental Protection Agency, Petrus and Augusta Hedlund Foundation, Magnus Bergvalls stiftelse and funds from the Karolinska Institute. Patrik Ernfors was supported by the Swedish Medical Research Council. The Genetics Institute, Boston, Massachussetts, provided expression vectors p91023(B) and pXM.

REFERENCES

Angeletti, R.H., and Bradshaw, R.A., 1971, Nerve growth factor from mouse submaxillary gland: Amino acid sequence, Proc. Natl. Acad. Sci. USA, 68: 2417-2420.

Angeletti, R.H., Hermodson, M.A., and Bradshaw, R.A. ,1973, Amino acid sequences of mouse 2.5 S nerve growth factor. II. Isolation and characterization of the thermolytic and peptic peptides and the complete covalent structure, Biochemistry, 12: 100-115.

Arumäe, U., Siigur, J., Neuman, T., and Saarma, M., 1987, Monoclonal antibodies against *Vipera lebetina* venom nerve growth factor cross-react with other snake venom nerve growth factors, Mol. Immunol., 24: 1295-1302.

Arumäe, U., Neuman, T., Sinijärv, R., and Saarma, M. ,1989, Sensitive time-resolved fluoroimmunoassay of nerve growth factor and the disappearance of nerve growth factor from rat pheochromocytoma PC12 cell culture medium, J. Immunol. Meth., 122: 59-65.

Bailey, G.S., Banks, B.E.C., Carstairs, J.R., Edwards, D.C., Pearce, F.L., and Vernon, C.A., 1976, Immunological properties of nerve growth factors, Biochim. Biophys. Acta, 437: 259-263.

Barde, Y.-A., Edgar, D., and Thoenen, H., 1982, Purification of a new neurotrophic factor from mammalian brain, Eur. Mol. Biol. Org. J., 1: 549-553.

Belew M., and Ebendal, T., 1986, Chick embryo nerve growth factor. Fractionation and biological activity, Exp. Cell Res., 167: 550-558.

Bruce, G., and Heinrich, G., 1989, Production and characterization of biologically active recombinant human nerve growth factor. Neurobiol. Aging, 10: 89-94.

Chapman, C.A., Banks, B.E.C., Carstairs, J.R., Pearce, F.L., Vernon, C.A., 1979, The preparation of nerve growth factor from the prostate of the guinea-pig and isolation of immunogenically pure material from the mouse submandibular gland, Fed. Eur. Biochem. Soc. Lett., 105: 341-344.

Davies, A.M., Thoenen, H., and Barde, Y.-A., 1986, The response of chick sensory neurons to brain-derived neurotrophic factor, J. Neurosci., 6: 1897-1904.

Donohue, S.J,, Head,R.J., Stitzel, R.E., 1989, Elevated nerve growth factor levels in young spontaneously hypertensive rats, Hypertension, 14: 421-426.

Ebendal, T., 1979, Stage-dependent stimulation of neurite outgrowth exerted by nerve growth factor and chick heart in cultured embryonic ganglia, Dev. Biol., 72: 276-290.

Ebendal, T., 1989a, NGF in CNS: Experimental data and clinical implications, Progr. Growth Factor Res., 1: 143-159.

Ebendal, T., 1989b, Use of collagen gels to bioassay nerve growth factor activity, in: "Nerve Growth Factors", R.A. Rush, ed., pp. 81-93. John Wiley & Sons, Ltd., Chichester.

Ebendal, T., and Persson, H, 1988, Developmental expression of nerve growth factor, in: "Neural Development and Regeneration", B. Haber, J. R. Perez-Polo, J. de Vellis and A. Gorio, eds., NATO ASI Series, Life Sciences Vol. H22, pp.233-234. Springer, Berlin.

Ebendal, T., Jordell-Kylberg, A., and Söderström, S., 1978, Stimulation by tissue explants on nerve fibre outgrowth in culture, in: "Formshaping Movements in Neurogenesis", C.-O. Jacobson, and T. Ebendal, eds, Zoon 6, pp. 235-243. Almqvist & Wiksell Int., Stockholm.

Ebendal, T., Olson, L., Seiger, Å., and Hedlund, K.-O., 1980, Nerve growth factors in the rat iris, Nature, 286: 25-28.

Ebendal, T., Olson. L., and Seiger, Å., 1983, The level of nerve growth factor (NGF) as a function of innervation. A correlative radio-immunoassay and bioassay study of the rat iris, Exp. Cell. Res., 148: 311-317.

Ebendal, T., Olson, L., Seiger, Å., and Belew, M., 1984, Nerve growth factors in chick and rat tissues. in: "Cellular and Molecular Biology of Neuronal Development", I.B. Black, ed., pp. 231-242. Plenum Publishing Corp., New York.

Ebendal, T., Larhammar, D., and Persson, H., 1986, Structure and expression of the chicken ß nerve growth factor, Eur. Mol. Biol. Org. J., 5: 1483-1487.

Ebendal, T., Hallböök, F., Ibañez, C., Persson, H., Olson, L., and Lärkfors, L., 1990, Activity and immunological properties of recombinant nerve growth factor (ß-NGF), in: "Brain Repair", Wenner-Gren Center International Symposium Series, Vol. 56, A. Björklund, A.J. Aguayo, and D. Ottoson, eds., pp. 57-71, MacMillan Press, London.

Eriksdotter-Nilsson, M., Skirboll, S., Ebendal, T., Hersch, L., Grassi, J, Massoulié, J., and Olson, L., 1989, NGF treatment promotes development of basal forebrain tissue grafts in the anterior chamber of the eye, Exp. Brain Res., 74: 89-98.

Ernfors, P., Hallböök, F., Ebendal, T., Shooter, E.M., Radeke, M.J., Misko, T.P., and Persson, H., 1988, Developmental and regional expression of ß-nerve growth factor receptor mRNA in the chick and rat, Neuron, 1: 983-996.

Ernfors, P., Ebendal, T., Olson, L., Mouton, P., Strömberg, I., and Persson, H., 1989, A cell line producing recombinant nerve growth factor evokes growth responses in intrinsic and grafted central cholinergic neurons, Proc. Natl. Acad. Sci. USA, 86: 4756-4760.

Ernfors, P., Ibáñez, C.F., Ebendal, T., Olson, L., and Persson, H., 1990a, Molecular cloning and neurotrophic activities of a protein with structural similarities to nerve growth factor: Developmental and topographical expression in the brain, Proc. Natl. Acad. Sci. USA, 87: 5454-5458.

Ernfors, P., Lindefors, N., Chan-Palay, V., and Persson, H., 1990b, Cholinergic neurons of nucleus basalis express elevated levels of nerve growth factor receptor mRNA in senile dementia of the Alzheimer type, Dementia, (in press).

Ernfors, P., Wetmore, C., Olson, L., and Persson, H., 1990c, Identification of cells in rat brain and peripheral tissues expressing mRNA for members of the nerve growth factor family, Neuron, 5: 511-526.

Fahnestock, M., and Bell, R.A., 1988, Molecular cloning of a cDNA encoding the nerve growth factor precursor from *Mastomys natalensis*, Gene, 69: 257-264.

Fischer, W., Wictorin, K., Björklund, A., Williams, L.R., Varon, S., and Gage, F.H., 1987, Amelioration of cholinergic neuron atrophy and spatial memory impairment in aged rats by nerve growth factor, Nature, 329: 65-68.

Friedman, W.J., Ernfors, P., and Persson, H., 1990, *In situ* hybridization reveals both transient and persistent expression of HDNF/NT-3 mRNA in the rat brain during postnatal development, J. Neurosci., (submitted).

Furukawa, S., Kamo, I., Furukawa, Y., Akazawa, S., Satoyoshi, E., Itoh, K., and Hayashi, K., 1983, A highly sensitive enzyme immunoassay for mouse ß nerve growth factor, J. Neurochem., 40: 734-744.

Gluzman, Y., 1981, SV40-transformed simian cells support the replication of early SV40 mutants, Cell, 23: 175-182.

Goedert, M., Fine, A., Hunt, S.P., and Ullrich, A., 1986, Nerve growth factor mRNA in peripheral and central rat tissues and in the human central nervous system: Lesion effects in the rat brain and levels in Alzheimer's disease, Mol. Brain Res, 1: 85-92.

Hallböök, F., Ebendal, T., and Persson, H., 1988, Production and characterization of biologically active recombinant beta nerve growth factor, Mol. Cell. Biol., 8: 452-456.

Hallböök, F., Ayer-LeLièvre, C., Ebendal, T., and Persson, H., 1990, Expression of nerve growth factor receptor mRNA during early development of the chicken embryo: Emphasis on the cranial ganglia, Development, 108: 693-704.

Harper, G., and Thoenen, H., 1980, Nerve growth factor: Biological significance, measurement, and distribution, J. Neurochem., 34: 5-16.

Harper, G.P., Glanville, R.W., and Thoenen, H., 1982, The purification of nerve growth factor from bovine seminal plasma, J. Biol. Chem., 257: 8541-8548.

Harper, G.P., Barde, Y.-A., Edgar, D., Ganten, D., Hefti, F., Heumann, R., Naujoks, K.W., Rohrer, H., Turner, J.E., and Thoenen, H., 1983, Biological and immunological properties of the nerve growth factor from bovine seminal plasma: Comparison with the properties of mouse nerve growth factor, Neuroscience, 8: 375-387.

Hedlund, K.-O., and Ebendal, T., 1978, Different ganglia from the chick embryo in studies of neuron development in culture, in: "Formshaping Movements in Neurogenesis", C.-O. Jacobson, and T. Ebendal, eds., Zoon 6, pp. 217-223. Almqvist and Wiksell Int., Stockholm.

Hefti, F., 1986, Nerve growth factor promotes survival of septal cholinergic neurons after fimbrial transections, J. Neurosci., 6: 2155-2162.

Hefti, F., Weiner, W.J., 1986, Nerve growth factor and Alzheimer's disease, Ann. Neurol., 20: 275-281.

Heinrich, G., and Meyer, T.E., 1989, Nerve growth factor is present in human placenta and semen but undetectable in normal and Paget's disease blood: Measurements with an anti-mouse-NGF enzyme immunoassay using a recombinant human NGF reference, Biochem. Biophys. Res. Commun., 155: 482-486.

Hellweg, R., Hock, C., and Hartung, H.-D., 1989, An improved rapid and highly sensitive enzyme immunoassay for nerve growth factor, Technique, 1: 43-48.

Heumann, R., Korsching, S., Scott, J., and Thoenen, H., 1984, Relationship between levels of nerve growth factor (NGF) and its messenger RNA in sympathetic ganglia and peripheral target tissues, Eur. Mol. Biol. Org. J., 3: 3183-3189.

Hofmann, H.-D., and Unsicker, K., 1982, The seminal vesicle of the bull: A new and very rich source of nerve growth factor, Eur. J. Biochem., 128: 421-426.

Hohn, A., Leibrock, J., Bailey, K., and Barde, Y.-A., 1990, Identification and characterization of a novel member of the nerve growth factor/brain-derived neurotrophic factor family, Nature, 344: 339-341.

Ibáñez, C.F., Hallböök, F., Ebendal, T., and Persson, H., 1990a, Structure-function studies of nerve growth factor: functional importance of highly conserved amino acid residues, Eur. Mol. Biol. Org. J., 9: 1477-1483.

Ibáñez, C.F., Hallböök, F., Söderström, S., Ebendal, T., and Persson, H., 1990b, Biological and immunological properties of recombinant human, rat and chicken nerve growth factors: A comparative study. Eur. J. Neurosci., (submitted).

Kaisho, Y., Yoshimura, L., and Nakahama, K., 1990, Cloning and expresssion of a cDNA encoding a novel human neurotrophic factor, Fed. Eur. Biochem. Soc. Lett., 266: 187-191.

Korsching, S., and Thoenen, H., 1983, Nerve growth factor in sympathetic ganglia and corresponding target organs of the rat: Correlation with density of sympathetic innervation, Proc. Natl. Acad. Sci. USA, 80: 3513-3516.

Korsching, S., and Thoenen, H., 1987, Two-site enzyme immunoassay for nerve growth factor, Meth. Enzymol., 147: 167-185.

Korsching, S., Auburger, G., Heumann, R., Scott, J., and Thoenen, H., 1985, Levels of nerve growth factor and its mRNA in the central nervous system of the rat correlate with cholinergic innervation, Eur. Mol. Biol. Org. J., 4: 1389-1393.

Kromer, L. F., 1987, Nerve growth factor treatment after brain injury prevents neuronal death, Science, 235: 214-216.

Lärkfors, L., and Ebendal, T., 1987, Highly sensitive enzyme immunoassays for ß-nerve growth factor, J. Immunol. Meth., 97: 41-47.

Lärkfors, L., Ebendal, T., Whittemore, S.R., Persson, H., Hoffer, B., and Olson, L., 1987, Decreased level of nerve growth factor (NGF) and its messenger RNA in the aged rat brain, Mol. Brain Res., 3: 55-60.

Lärkfors, L., Ebendal, T., Whittemore, S.R., Persson, H., Hoffer, B., and Olson, L., 1988, Developmental appearance of nerve growth factor in the rat brain: Significant deficits in the aged forebrain, in: "Transplantation into the Mammalian CNS", D.M. Gash, and J.R. Sladek, Jr, eds., Progr. Brain Res. 78; pp. 27-31. Elsevier, Amsterdam.

Leibrock, J., Lottspeich, F., Hohn, A., Hofer, M., Hengerer, B., Masiakowski, P., Thoenen, H., Barde, Y.-A., 1989, Molecular cloning and expression of brain-derived neurotrophic factor, Nature, 341: 149-152.

Levi-Montalcini R (1966): The nerve growth factor: Its mode of action on sensory and sympathetic nerve cells. Harvey Lect. 60: 217-259.
Levi-Montalcini, R., 1987, The nerve growth factor 35 years later, Science, 237: 1154-1162.
Levi-Montalcini, R., and Angeletti, P.U., 1968, Nerve growth factor, Physiol. Rev., 48: 534-569.
Luthman, H., and Magnusson, G., 1983, High efficiency polyoma DNA transfection of chloroquine treated cells, Nucl. Acids Res., 11: 1295-1305.
Maisonpierre, P.C., Belluscio, L., Squinto, S., Ip, N.Y., Furht, M.E., Lindsay, R.M., and Yancopoulos, G.D., 1990, Neurotrophin-3: A neurotrophic factor related to NGF and BDNF, Science, 247: 1446-1451.
Meier, R., Becker-André, M., Götz, R., Heumann, R., Shaw, A., and Thoenen, H., 1986, Molecular cloning of bovine and chick nerve growth factor (NGF): Delineation of conserved and unconserved domains and their relationship to the biological activity and antigenicity of NGF, Eur. Mol. Biol. Org. J., 5: 1489-1493.
Mobley, W.C., Schenker, A., and Shooter, E.M., 1976, Characterization and isolation of proteolytically modified nerve growth factor, Biochemistry, 15: 5543-5552.
Nagata, Y., Ando, M., Takahama, K., Iwata, M., Hori, S., and Kato, K., 1987, Retrograde transport of endogenous nerve growth factor in superior cervical ganglion of adult rats, J. Neurochem., 49: 296-302.
Nakata, Y., Nomoto, H., Takahashi, I., Furukawa, Y., Furukawa, S., and Hayashi, K., 1988, Comparison of the biochemical and immunological properties of nerve growth factors from various animals, J. Clin. Biochem. Nutr., 4: 73-86.
Olson, L., Ebendal, T., Eriksdotter-Nilsson, M., Ernfors, P., Friedman, W., Persson, H., Wetmore, C., and Strömberg, I., 1990, Strategies to increase NGF levels and effects thereof on lesioned and grafted brain tissue, in: "Brain Repair", Wenner-Gren Center International Symposium Series, Vol. 56, A. Björklund, A.J. Aguayo, and D. Ottoson, eds., pp. 87-98, MacMillan Press, London.
Perry, E.K., Tomlinson, B.E., Blessed, G., Bergmann, K., Gibson, P.H., and Perry, R.H., 1978, Correlation of cholinergic abnormalities with senile plaques and mental test scores in senile dementia, Br. Med. J., 2: 1457-1459.
Persson, H., Ernfors, P., Friedman, W., Hallböök, F., Ayer-LeLievre, C., Ebendal, T., Olson, L., Henschen, A., Mouton, P., and Strömberg, I., 1990, Expression of ß-nerve growth factor and its receptor in the mammalian central nervous system, in: "Brain Repair", Wenner-Gren Center International Symposium Series, Vol. 56, A. Björklund, A.J. Aguayo, and D. Ottoson, eds., pp. 73-86, MacMillan Press, London.
Pontén, J., 1976, The relationship between in vitro transformation and tumor formation *in vivo*, Biochim. Biophys. Acta, 458: 397-422.
Rodriguez-Tébar, A., Dechant, G., and Barde, Y.-A., 1990, Binding of brain-derived neurotrophic factor to the nerve growth factor receptor, Neuron, 4: 487-492.
Rosenberg, M.B., Friedmann, T., Robertson, R.C., Tuszynski, M., Wolff, J.A., Breakefield, X.O., and Gage, F.H., 1988, Grafting genetically modified cells to the damaged brain: Restorative effects of NGF expression, Science, 242: 1575-1578.
Rosenthal, A., Goeddel, D.V., Nguyen, T., Lewis, M., Shih, A., Laramee, G.R., Nikolics, K., and Winslow, J.W., 1990, Primary structure and biological activity of a novel human neurotrophic factor, Neuron, 4: 767-773.
Schwarz, M.A., Fisher, D., Bradshaw, R.A., and Isackson, P.J., 1989, Isolation and sequence of a cDNA clone of ß-nerve growth factor from the guinea pig prostate gland, J. Neurochem., 52: 1203-1209.
Scott, J., Selby, M., Urdea, M., Quiroga, M., Bell, G.I., and Rutter, W.J., 1983 Isolation and nucleotide sequence of a cDNA encoding the precursor of mouse nerve growth factor, Nature, 302: 538-540.
Selby, M.J., Edwards, R., Sharp, F., and Rutter, W.J. 1987a, Mouse nerve growth factor gene: Structure and expression, Mol. Cell. Biol., 7: 3057-3064.
Selby, M.J., Edwards, R.H., and Rutter, W.J., 1987b, Cobra nerve growth factor: Structure and evolutionary comparison, J. Neurosci. Res., 18: 293-298.
Server, A.C., Herrup, K., Shooter, E.M., Hogue-Angeletti, R.A., Frazier, W.A., and Bradshaw, R.A., 1976, Comparison of the nerve growth factor proteins from cobra venom (*Naja naja*) and mouse submaxillary gland, Biochemistry, 15: 35-39.

Shelton, D.L., and Reichardt, L.F., 1986, Studies on the expression of the ß nerve growth factor (NGF) gene in the central nervous system: Level and regional distribution of NGF mRNA suggest that NGF functions as a trophic factor for several distinct populations of neurons, Proc. Natl. Acad. Sci. USA, 83: 2714-2718.

Söderström, S., Hallböök, F., Ibáñez, C.F., Person, H., and Ebendal, T., 1990, Recombinant human ß-nerve growth factor (NGF): Biological activity and properties in an enzyme immunoassay, J. Neurosci. Res., (in press).

Strömberg, I., Herrera-Marschitz, M., Ungerstedt, U., Ebendal, T., and Olson, L., 1985, Chronic implants of chromaffin tissue into the dopamine-denervated striatum. Effects of NGF on graft survival, fiber growth and rotational behavior, Exp. Brain Res., 60: 335-349.

Strömberg, I., Wetmore, C.J., Ebendal, T., Ernfors, P., Persson, H., and Olson, L., 1990, Rescue of basal forebrain cholinergic neurons after implantation of genetically modified cells producing recombinant NGF, J. Neurosci. Res., 25: 405-411.

Suda, K., Barde, Y.-A., and Thoenen, H., 1978, Nerve growth factor in mouse and rat serum: Correlation between bioassay and radioimmunoassay determinations, Proc. Natl. Acad. Sci. USA, 75: 4042-4046.

Taylor, J.W., Ott, J., and Eckstein, F., 1985, The rapid generation of oligonucleotide-directed mutations at high frequency using phosphorothiolate-modified DNA, Nucl. Acids Res., 13: 8765-8785.

Thoenen, H., and Barde, Y.-A., 1980, Physiology of nerve growth factor, Physiol. Rev., 60: 1284-1335.

Thoenen, H., Bandtlow, C., and Heumann, R., 1987, The physiological function of nerve growth factor in the central nervous system: Comparison with the periphery, Rev. Physiol. Biochem. Pharmacol., 109: 145-178.

Todaro, G.J., and Green, H., 1963, Quantitative studies of the growth of mouse embryo cells in culture and their development into established lines, J. Cell Biol., 17: 299-313.

Ueyama, T., Saito, H., and Yohro, T., 1986, Nerve growth factor and the submandibular gland of the house musk shrew *Suncus murinus*: A morphological and comparative immunological study, Biomed. Res., 7: 379-387.

Ullrich, A., Gray, A., Berman, C., and Dull, T.J., 1983, Human ß-nerve growth factor gene sequence highly homologous to that of mouse, Nature, 303: 821-825.

Weskamp, G., and Otten, U., 1987, An enzyme-linked immunoassay for nerve growth factor (NGF): A tool for studying regulatory mechanisms involved in NGF production in brain and in peripheral tissues, J. Neurochem., 48: 1779-1786.

Westin, G., Gerster, T., Muller, M.M., Schaffner, G., and Schaffner, W., 1988, OVEC, a versatile system to study transcription in mammalian cells and cell-free extracts, Nucl. Acids Res., 15: 6787-6798.

Wetmore, C., Ernfors, P., Persson, H., and Olson, L., 1990, Localization of brain-derived neurotrophic factor mRNA to neurons in the brain by *in situ* hybridization, Exp. Neurol., 109: 141-152.

Whittemore, S.R., Ebendal, T., Lärkfors, L., Olson, L., Seiger, Å., Strömberg, I., and Persson, H., 1986, Developmental and regional expression of ß nerve growth factor messenger RNA and protein in the rat central nervous system, Proc. Natl. Acad. Sci. USA, 83: 817-821.

Whittemore, S.R., Persson, H., Ebendal, T., Lärkfors, L., Larhammar, D., and Ericsson, A., 1988a, Structure and expression of ß-nerve growth factor in the rat central nervous system, in: "Neural Development and Regeneration" B. Haber, A. Gorio, J. de Vellis, and J.R. Regino Perez-Polo, eds., NATO ASI, Life Sciences, Vol. H22, pp. 245-256. Springer, Berlin.

Whittemore, S.R., Friedman, P.L., Larhammar, D., Persson, H., Gonzalez-Carvajal, M., and Holets, V.R., 1988b, Rat ß-nerve growth factor sequence and site of synthesis in the adult hippocampus, J. Neurosci. Res., 20: 403-410.

Wilcock, G.K., Esiri, M.M., Bowen, D.M., and Smith, C.C.T., 1982, Alzheimer's disease. Correlation with cortical choline acetyltransferase activity with the severity of dementia and histological abnormalities, J. Neurol. Sci., 57: 407-417.

Williams, L.R., Varon, S., Peterson, G.F., Wictorin, K., Björklund, A., and Gage, F.H., 1986, Continuos infusion of nerve growth factor prevents basal forebrain neuronal death after fimbria fornix transection, Proc. Natl. Acad. Sci. USA, 83: 9231-9235.

Wion, D., Perret, C., Fréchin, N., Keller, A., Béhar, G., Brachet, P., and Auffray, C., 1986, Molecular cloning of the avian ß-nerve growth factor gene: Transcription in the brain, Fed. Eur. Biochem. Soc. Lett., 203: 82-86.

Yang, Y.-C., Ciarletta, A.B., Temple, P.A., Chung, M.P., Kovacic, S., Witek-Giannotti, J.A.S., Leary, A.C., Kriz, R., Donahue, R.E., Wong, G.G., and Clark, S.C., 1986, Human IL-3 (multi-CSF): Identification by expression cloning of a novel hematopoietic growth factor related to murine IL-3, Cell, 47: 3-10.

SCHWANN CELL PROLIFERATION DURING POSTNATAL DEVELOPMENT, WALLERIAN DEGENERATION AND AXON REGENERATION IN TREMBLER DYSMYELINATING MUTANT

Herbert KOENIG, Anh DO THI, Badia FERZAZ, Annie RESSOUCHES

Université de Bordeaux I, CNRS-URA 1126
Avenue des Facultés 33405 TALENCE CEDEX FRANCE

INTRODUCTION

During mouse ontogenesis, Schwann cells in peripheral nerves migrate along the axons and begin the process of ensheathment (1,2). Subsequently, Schwann cells divide intensely, from the end of gestation (1) till the first days of postnatal development (3,4). The high rate of Schwann cell proliferation decreases rapidly and stops totally during the process of myelination (1,3).

However, Schwann cells retain their capacity to divide in response to nerve injuries, mainly when myelin degradation occurs. Division of Schwann cells was shown in Wallerian degeneration (5,6,7), primary and secondary neuro-intoxications (8,9,10,11), several human hereditary neuropathies like Charcot-Marie-Tooth (12) and murine mutations showing segmental demyelination like Dystonia musculorum (13) and Trembler (14). The proliferation of Schwann cells in injured nerves, may be triggered either by the release of myelin degenerating products (15,16) or by the contact with axolemma (17) or axolemmal fragments (18).

Trembler mouse, a spontaneous mutant showing dominant inheritance (19) is regarded as the animal model system for Dejerine-Sottas (19) or Charcot-Marie-Tooth (14) human hypertrophic neuropathies.
The mutation is characterized by a severe segmental hypomyelination of the peripheral nerve fibers, an increased number of Schwann cells (x10) and "onion-bulb" formations (14,19). In fact, the individual nerve fibers (myelinated and unmyealinated), at the exception of Remak fibers, are surrounded by a multilayered basal lamina sheath (19,20)(and Fig.1).
Perkins et al. (14) have shown that ventral roots and sciatic nerves of 10-day old Trembler mice contain two times more Schwann cell nuclei as normal mice nerves. The difference increased subsequently to reach five-fold in 1-month and ten-fold in 4-month animals, as a result of continuous Schwann cell multiplication (unpublished).

In our presentation we report (i) the kinetic analysis of Trembler Schwann cell abnormal proliferation throughout postnatal development, in relation with myelination and demyelination (ii) the differential proliferative responses of Trembler and control Schwann cells in an experimental system which allows to separate "in vivo" the effects of

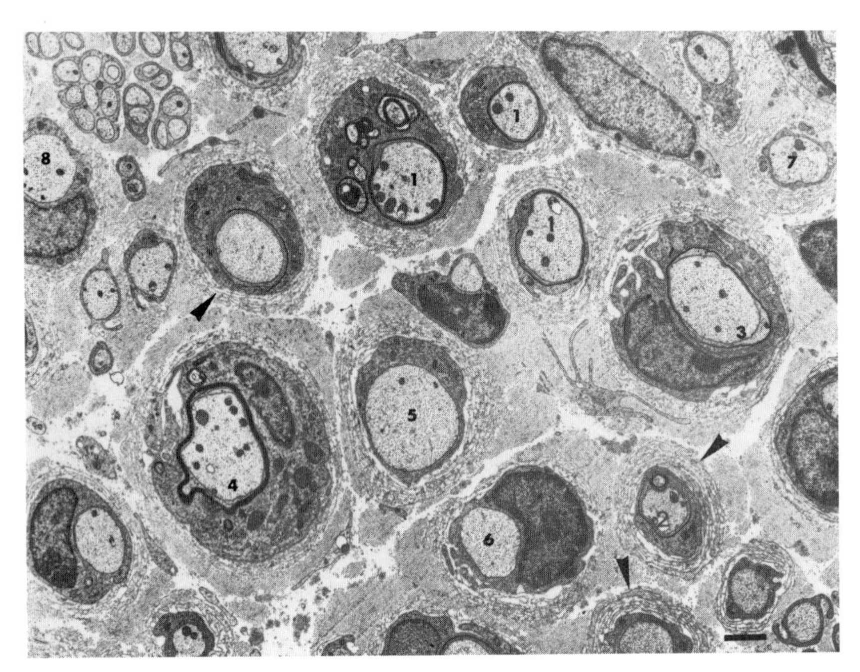

Fig. 1. Cross section of Trembler adult sciatic nerve (2 months). Myelin sheaths are very thin (1,2,3,4). Several axons are totally unmyelinated (5,6,7,8). Seven Schwann cell nuclei are present (N). In normal nerves, only one or none would be observed. Note the multilayered basal lamina (➤). Bar = 1μm.

two potential mitogens, present in myelin breakdown debris and in regenerating axons.

The density of Schwann cell nuclei was obtained by counting their number in cross-sections of nerve EM micrographs. In fact, the nucleus density reflects the density of the Schwann cells. In the nerves of Trembler, the nuclei were counted separately for myelinated and isolated unmyelinated fibers (Fig. 1), the so-called promyelin fibers (19).

SCHWANN CELL PROLIFERATION DURING ONTOGENESIS

In newborn animals, the axons are small (0.65 μm in control and 0.55 in Trembler). They are grouped in bundles inside a common Schwann cells. As in 2-day animals, only few axons were segregated from the bundles, we started our nucleus quantitative analysis in 3-day mice. At this stage of development, most axons were individualized in a 1:1 relationship with Schwann cells and myelinisation had already started. In control nerves, 66-70% of the axons were myelinated with an average 15 lamellae sheath. In Trembler, only 33-36% of the axons had a myelin sheath (13 lamellae).

Fig. 2 shows that in 3 day-old Trembler, the number of Schwann cell nuclei is 33% higher than in control nerves. Assuming that the density of Schwann cell nuclei counted in cross sections reflects the length of the internodes (in control nerves) and of the myelinated and non-myelinated segments in Trembler, our data indicate that the length of the axon covered by each cell is already one-third shorter at this stage of the mutant development.

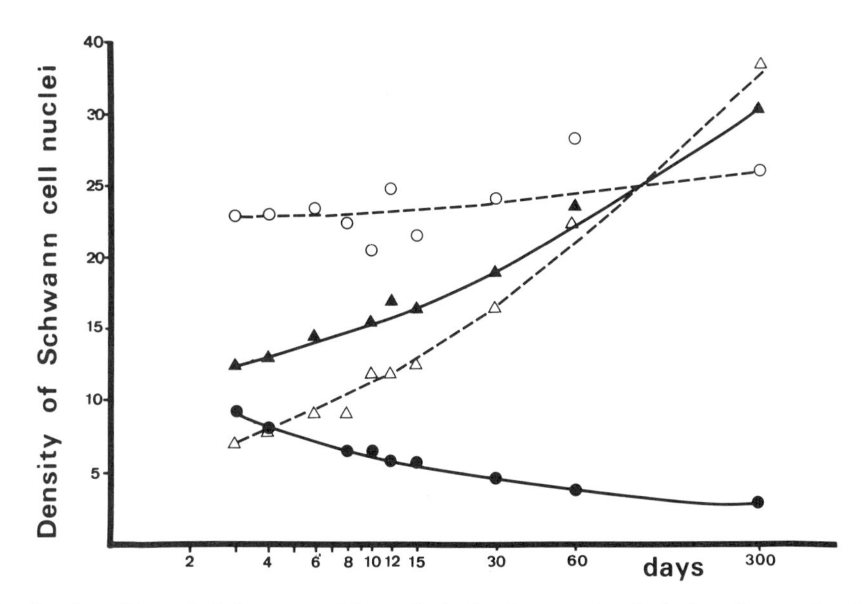

Fig.2. Density of Schwann cell nuclei during postnatal development, from 3 to 300 days. The density of nuclei corresponds to the percentage of axons with a nucleus. Control fibers (●——●); Total Trembler fibers (▲—▲);Trembler myelinated axons (○----○) and unmyelinated axons (△----△) as seen in Fig. 1.

In control nerves, the density of Schwann cell nuclei decreased gradually from 1 for 10 axons at day 3, to 1 for 35 fibers at day 300. As the mitoses in Schwann cells cease within the first week after birth (3), the fall in cell density corresponds to the elongation of the internodes throughout the growth of the mouse. Thus, the internodes would elongate 2 times from day 6- to 60-day old control animals (8.4% to 3.9% nucleated Schwann cells).

Conversely in Trembler nerve, the number of Schwann cells increased regularly (Fig.2), from day 4 (1 nucleus/8 axons) to day 300 (1 nucleus/ 3 axons). Taking into account that the body growth rate of Trembler mice is similar to control, we calculated that mean axonal length covered by each Schwann cell would decrease by a factor 3 from day 6-to day 60- in mutants. In fact, the length of the cells is uneven, particularly in the unmyelinated segments of the axons (Koenig, unpublished).

As a result of the reverse evolution of nucleus density, the ratio of Trembler versus control Schwann cell density, as reflected by their nuclei, increased progressively (Table 1).

Table 1. Ratio of Trembler (Tr) versus Control (C) density of Schwann cell nuclei during ontogenesis (3 to 300 days postnatal).

Age in days	3	4	6	8	10	12	30	60	300
Ratio TR/C	1.3	1.5	1.6	2.3	2.5	2.8	4.0	6.3	10.5

The evolution of the Schwann cell nuclei differed in myelinated and unmyelinated fibers of Trembler nerves (Fig. 2). In 3-day-old animals, we found one nucleus per 4 myelinated axons and only one per 14 unmyelinated axons. At later stages, the nuclear density remained stable in myelinated fibers whereas it increased tremendously in unmyelinated fibers. In 3-to 10-month old mutants, the Schwann cell density was similar in both types of axons (Fig. 2).

As a result of the enhanced Schwann cell proliferation, the axonal length they cover decreased 3 times in the unmyelinated segments of the fibers. In the myelinated parts of the fibers, the internodal length remained almost constant.

Discussion : In the developing peripheral nerve of Trembler mutant, both myelin degradation products (15, 16) and contact with axolemma (17) might actually stimulate Schwann cell division. Naked axons, detached from the bundles, are still present in the nerve of 15 day-old animals. The migrating Schwann cells which later surround these axons are probably stimulated to divide, as a consequence of their contact with the axolemma.

On the other hand, myelin debris are observed all along ontogenesis (Fig. 1). From day 3 to 8 and at day 60, the Schwann cell cytoplasm of approximately 4% of the axons contained such debris. The percentage rose to 7-8% between days 10 and 30. As the majority of the myelin degradation profiles was located in the myelinating Schwann cells, it is likely that the mitogenic molecules are secreted and stimulate the division of the non-myelinating cells. Actually, we have shown that serum of adult mutant mice (Trembler), contained a mitogenic factor for normal and Trembler Schwann cells in culture (Do Thi and Koenig, unpublished) as we previously suggested (34).

SCHWANN CELLS IN AN EXPERIMENTAL DEGENERATION-REGENERATION SYSTEM

In order to discriminate between the stimulating effects of myelin breakdown products and axonal contact on Schwann cell mitoses "in vivo", we used an experimental model system inspired from Politis and al. (21). The surgery procedure was performed in Trembler and normal mice :

1) the tibial nerve was transected and both cut ends were ligated to prevent the entry of the regenerating axons in the distal stump;
2) this stump was allowed to undergo Wallerian degeneration for a period of 2 months. At this time, the myelin breakdown debris were totally removed from the remaining Schwann cells (Fig. 3) and
3) the peroneal nerve was sectioned and its proximal stump was grafted on the distal stump of the tibial nerve, the ligature was removed.

1. Schwann cell proliferation during Wallerian degeneration

Numerous myelin breakdown profiles appear rapidly in the cytoplasm of the Schwann cells in the distal stump of tibial nerve (Table 2). The increase of the myelin debris coincides with the onset of Schwann cell proliferation (Table 2 and Fig. 5). One day after surgery, the density of nuclei was identical to the control levels (Fig. 5), both in normal (590 nuclei/mm2) and Trembler (3800 nuclei/mm2) nerves. The increase of Schwann cells differed in the two strains (Fig. 5).

In normal mice, myelin debris were most abundant between day 11 and day 20. They entirely disappeared by day 60 (Fig. 3) as shown by immunofluorescence using an antibody directed against myelin protein Po. In the first phase of the degenerative process, Schwann cells multiply actively inside the basal lamina of each nerve fiber to form longitudinal columns, the so-called bands of Bungner.

In normal nerve, the density of Schwann cells, as reflected by nucleus counts, was enhanced by a factor of 5 within 2 weeks and remained relatively constant thereafter (Fig. 5).

In Trembler, degenerative and proliferative processes were accelerated, as compared to normal nerve (Fig. 5 and Table 2). The peak of myelin debris was observed at 5-6 days. Anti-Po fluorescence failed to reveal any remaining debris as early as 8-10 days after surgery. Dividing Schwann cells are scattered among the remnants of the multi-layered basal lamina. They shown thin filopode-like digitations surrounded by their own individual basal lamina (Fig. 3, inset).

In the mutant nerve, the number of Schwann cells rose rapidly; the peak was obtained on day 5 after surgery. However, the density of nuclei was increased by 20% only (Fig. 5). In contrast with normal nerves, the fall in nucleus density was substantial, it reached values even slightly below the control level.

Table 2. Evolution of myelin debris in Trembler and normal tibial nerve Wallerian degeneration (days after section).

	Onset of myelin debris	Maximum of debris observed	Total removal of debris
Trembler	2 days	4-5 days	8-10 days
Normal	3-4 days	11-20 days	60 days

Discussion : In both normal and Trembler nerves, the density of Schwann cell nuclei is closely correlated with degenerating myelin

1) the maximum cells counted (Fig. 5) coincides with the peak of myelin debris (Table 2)

2) the limited increase in Trembler is probably due to the low amount of myelin present in the hypomyelinated nerve fibers

3) the slight decrease of the nucleus density, after the peak, in the normal nerve (- 15%) may be related to residual myelin debris for 2 months (Table 2). Similarly, in the mutant, the decline of nuclei to a level even below control is likely the consequence of the absence of debris. Actually we have calculated that the total amount of myelin in 2-month-old Trembler represents only 5% of the myelin in normal mice nerve (Koenig and Do Thi, unpublished). Thus, the thin myelin sheaths were rapidly degraded so that the Schwann cells could not supply "growth factors" in sufficient quantities to insure a high rate and a long-lasting proliferation. Conversely, in normal nerves the large amount of myelin to be degraded and the persistance of debris for a long period would explain the tremendous increase of Schwann cell density and its relative stability.

Similar results were obtained in cell cultures. Salzer and Bunge (16) showed that the Schwann cells which participate in the digestion of the myelin debris during wallerian degeneration were the proliferating

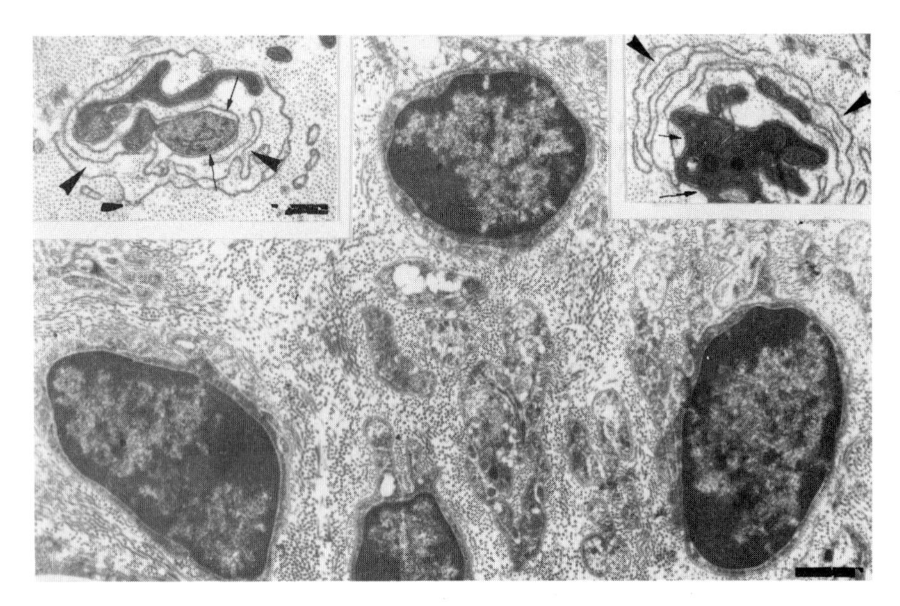

Fig.3. Wallerian degeneration : distal stump or normal mouse tibial nerve, 2 months after transection and ligation. Four Schwann cell nuclei are scattered among abundant collagen fibers. Inset: Trembler degenerated nerve, thin Schwann cell digitations, fringed by their own basal lamina (⟶) are located inside the remains of "old" multilayered basal lamina (➤) Bar = 1 µm.

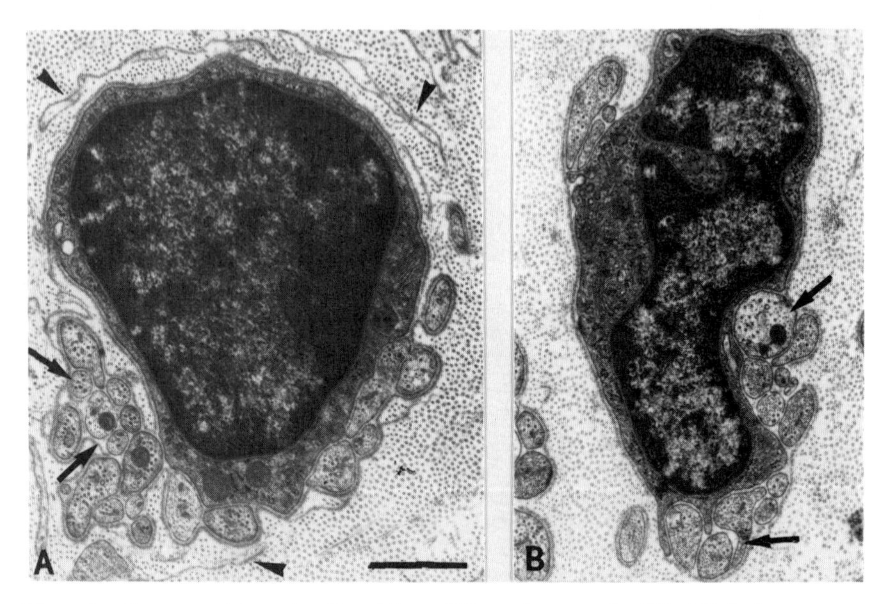

Fig. 4. Regeneration in Trembler (A) and normal (B) tibial nerve, 6 days after peroneal nerve grafting. Trembler bundle of axons (A) contains more axonal profiles (➞) than normal bundle (B). The remaining "old" basal lamina are visible in both nerve types (➤) Bar = 1 µm.

cells. Nevertheless the degeneration results cannot exclude that the low proliferation rate of Trembler Schwann cells is related to modification their genetic programm.

2. Schwann cell proliferation during axonal regeneration

Two days after grafting, the first axons which regenerate from the peroneal nerve penetrate inside the distal stump of the tibial nerve. They grow along the Schwann cells of the Bungner's bands. The proliferation of the Schwann cells was immediately stimulated and reached rapidly a peak (Fig. 5). The maximal level of nucleus density (measured at 4 mm distal from the graft), was identical in Trembler and normal nerves. However, the peak occured earlier in the mutant nerve (Fig. 5). In both strains, the maximum nucleus density coincides with the presence of a high number of regenerated axons. The rapid decrease of Schwann cell nuclei, after the peaks, was synchronous with the onset of myelination. The first myelin lamellae were observed at day 5 and day 8 in Trembler and normal nerves, respectively.

Discussion : In degeneration-regeneration experiments it is difficult to determine if regenerating axons trigger Schwann cell mitoses in addition to myelin debris. The experimental system we used confirmed that a second wave of Schwann cell proliferation occured in mouse, as shown in cat (24) after the peroneal axons entered into the distal stump of the tibial nerve.

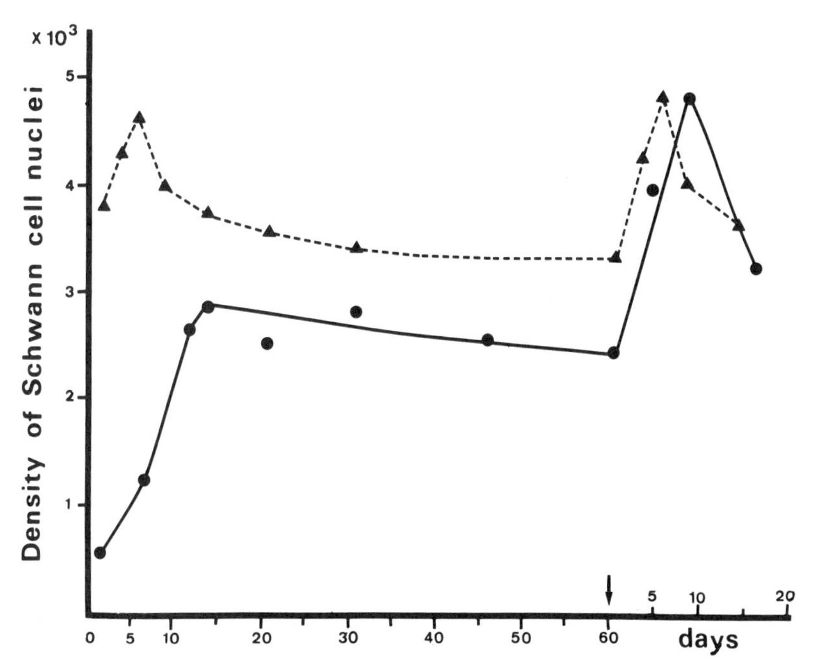

Fig. 5. Density of Schwann cell nuclei (per mm2 of nerve). Nerves were examined during Wallerian degeneration at 1,3, 5, 8, 13,30 and 60 days after surgery in Trembler (▲----▲) and after 1,6,11,13,20,30,45,60 days in normal mice (●——●) Regeneration was examined at 3,5,8,15 days after grafting the peroneal nerve in Trembler and 4,8,17 days in normal mice. Values of day 1 correspond to control. The arrow indicates the time of peroneal nerve grafting.

Similarly to Pellegrino and Spencer (24), we found that in axon-free regions (at 14 mm of the graft), the number of Schwann cells was not increased at the peak periods detected at 4 mm in the tibial nerve.

In addition, our kinetic analysis showed that :

1) the divisions of normal and Trembler mouse Schwann cells were stimulated rapidly after the grafting

2) the peaks of nucleus density paralleled the number of regenerated axons actually present at the level examined

3) the density of Schwann cell nuclei increased faster in the mutant nerve (Fig. 5). The delay observed in normal nerves was the result of the slower rate of axonal regeneration (1.4 mm per day) as compared to Trembler (3.5 mm per day) as we have previously shown (20)

4) the Schwann cell mitoses were blocked when the first myelin sheaths were formed, as Peters and Muir (1) and ourselves noted during ontogenesis. In both situations, the mitogenic factor(s) seem to be dependent on myelination signals

5) the fact that the maximal density of nuclei was identical in normal and Trembler after axonal stimulation (Fig. 5) indicates that the mutation does not affect the proliferative capacities of Schwann cells. We obtained similar results "in vitro" (Do Thi and Koenig, to be published). Trembler Schwann cells divide less actively in culture than normal sciatic nerve cells (23). However, their rate of division were identical under the influence of exogenous factors such as myelin and nerve extracts, Trembler serum and bFGF. It is likely that in our "in vivo" and "in vitro" experimental conditions both Schwann cells (normal and mutant) have reached the limits of their mitotic potential (14).

DISCUSSION AND CONCLUDING REMARKS

Our results indicate that the proliferative abnormality of Trembler Schwann cells is not genetically programmed. Axonal growth and regeneration, as well as the consecutive Schwann cell divisions are dependent upon epigenetic factors, at present unknown. In Trembler mutant, the sequence of events occuring during postnatal development and regeneration was not modified as compared to normal nerve. After formation of axon bundles, the individual axons segregate, then Schwann cells divide, before myelination starts. However, the time course of these processes was accelerated in Trembler nerve.

In the experimental model system of degeneration-regeneration we used, the proliferation of Schwann cells was stimulated both by myelin degradation products followed by contact with regenerating axons. The molecular nature of the mitogen(s) is currently unknown.

Two types of plasminogen activator (PA) are present in the peripheral nervous system (25,26). One may speculate that the PA system could be a valuable candidate to trigger Schwann cell proliferation in the different conditions we have studied. In fact, both neurons and glia contain and release PA (25,26,27,28). Urokinase-type (UK) is released by neurons (29) particularly at the growing tips of axons (25). UK stimulates Schwann cell divisions (29). The tissue plasminogen activator-type (tPA) is secreted predominantly by dividing Schwann cells (28) and is not mitogenic for Schwann cells (29). UK inhibition increases neuritic outgrowth in cell cultures (30).

Another possible source of plasminogen activator in the degenerative and regenerative processes are the macrophages invading the peripheral nerve. Avian and mammalian stimulated macrophages secrete PA (31).

In our experiments, the maximal densities of regenerated axons and macrophages precede the peak of Schwann cell nuclei; all the peaks appear at earlier time intervals in Trembler lesioned nerves (unpublished).

Thus, we may speculate that :

1) during the growth of axons (in late foetal and early postnatal development and in nerve regeneration) the UK-type of PA released by the growth cones and adjacent regions, would bind to UK-receptors present at the surface of "neuronal" membrane (30,32). The contacting Schwann cells would be stimulated by the neuronal bound UK-type of plasminogen activator and divide (30). During normal mice ontogenesis and nerve regeneration, the release of UK would cease when axonal growth stops. As a consequence, Schwann cell proliferation would be arrested and myelination would start.

In Trembler nerve, the continuous multiplication of Schwann cells in adult animals (14,20) could result from a maintained release of UK along the demyelinated segments of the axons. Actually, we have shown that the serum of Trembler adult mice contains a neurite-growth promoting factor (34) and is capable to stimulate Schwann cell proliferation in cell cultures (23). The relationship between the plasminogen activator system and the extracellular matrix components involved in neuritic outgrowth promoted by Trembler serum (34) remains to be determined. Similarly, the increased rate of axonal regeneration in Trembler lesioned nerve (20) could also be associated with UK and its inhibitors.

2) in the course of Wallerian degeneration, Schwann cell divisions may also be triggered by plasminogen activators. In rat transected sciatic nerve, PA activity was markedly increased (35). As in experimental allergic encephalomyelitis, demyelination is only observed where macrophages are present, it was suggested that PA is involved in the degradation of myelin sheaths (36). The proliferation of cultured Schwann cells is stimulated by phagocytosis of myelin membrane (37). Myelin phagocytosis and the subsequent Schwann cell divisions are only possible in the presence of macrophages (38). Bigbee et al. (37) suggest that a mitogenic factor, sensitive to inhibition of lysosomes, could be contained within myelin. We would propose that the degenerating axons release urokinase which would stimulate the division of the surrounding Schwann cells. This mitogen could induce the mitoses directly or indirectly after release from the macrophages as suggested by Bigbee et al. (37).

One more clue in favor of its probable involvement, is the fact that PA contributes to the degradation of extracellular matrix components (40). Actually, the number of basal lamina layers surrounding Trembler Schwann cells is decreased in degenerating (Fig. 3, inset) and regenerating nerves (Fig. 4) as in control mutant nerve (Fig. 1).

Finally, one question remains to be answered : the role of the tPA contained in the Schwann cells (29) and of its neuronal binding sites (30). It is possible that tPA contributes to the degradation and removal of the myelin sheaths in all pathological situations in which demyelination occurs. In fact, Schwann cells participate, with the macrophages, to the degradation of myelin in the course of Wallerian degeneration (39).

To sustain our hypothesis, it will be necessary to compare the UK, tPA and inhibitor content of Trembler and normal nerves and serum at the critical periods we have determined during the postnatal development and the degeneration-regeneration cycle.

ACKNOWLEDGEMENTS

We would like to thank Professor René Couteaux for helpful discussions and advice . This work was supported by INSERM grant n° 85.6.014 and AFM grants.

REFERENCES

1. A. Peters and A.R. Muir, The relationship between axons and Schwann cells during development of peripheral nerves in the rat, Q.J. Exp. Physiol., 64:117 (1959).
2. C.C. Speidel, In vivo studies of myelinated nerve fibers, Int. Rev. Cytol., 16:173 (1964).
3. A.K. Asbury, Schwann cell proliferation in developing mouse sciatic nerve, J. Cell Biol., 34:736 (1967).
4. L.C. Terry, G.M. Bray and A.J. Aguayo, Schwann cell multiplication in developing rat unmyelinated nerves. A radioautographic study, Brain Res., 69:144 (1974).
5. G. Thomas, Quantitative histology of Wallerian degeneration; nuclear populations in two nerves of different fibre spectrum, J. Anat., 82:135 (1948).
6. M.Abercrombie and J. Santler, An analysis of growth in nuclear population during Wallerian degeneration, J. Cell Comp. Physiol., 50:429 (1957).
7. W.G. Bradley and A.K. Asbury, Duration of synthesis phase in neurilemma cells in mouse sciatic nerve during degeneration, Exp. Neurol., 26:275 (1970).
8. P.S. Spencer, R.G. Pellegrino, S. M. Ross, M.J. Politis and M.I. Sabri, The regulation of Schwann cell function in degenerative disorders of the nervous system, in : Molecular Pathology of Nerve and Muscle, A.D. Kidman, Ed., Humana Press, Clifton 3-19 (1983).
9. J.B. Cavanagh and M.F. Gysbers, Dyingback above nerve ligature produced by acrylamide, Acta Neuropathol., 51:169 (1980).
10. P.S. Spencer and H.H. Schaumburg, Pathobiolgy of neurotoxic axonal degeneration, in : Physiology and Pathobiology of Axons, S.G. Waxman, Ed., Raven Press, New York, pp 265-282 (1978).
11. J.W. Griffin and D.L. Price, Schwann and glial responses in ß, ß' iminodiproprionitrile intoxication. I. Schwann cell and oligodendrocyte in growths. J. Neurocytol., 10:995 (1981).
12. P.J. Dyck, Experimental hypertrophic neuropathy : pathogenesis of onion-bulb formations, Arch. Neurol., 21:73 (1969).
13. T.H. Moss, Segmental demyelination in the peripheral nerves of mice affected by a hereditary neuropathy (Dystonia musculorum), Acta Neuropathol., 53:51 (1981).
14. C.S. Perkins, A.J. Aguayo and G.M. Bray, Schwann cell multiplication in Trembler mice, Neuropathol. Applied Neurobiol., 7:115 (1981).
15. S.M. Hall and N. Gregson, The effects of mitomycin C on the process of regeneration in the mammalian peripheral nervous system, Neuropathol. and Applied Neurobiol., 3:65 (1977).
16. J.L. Salzer and R.P. Bunge, Studies of Schwann cell proliferation. I. An analysis in tissue culture of proliferation during development, Wallerian degeneration, and direct injury, J. Cell Biol., 84:739 (1980).
17. J.L. Salzer, R.P. Bunge and L. Glaser, Studies of Schwann cell proliferation. III. Evidence for surface localization of the neurite mitogen, J. Cell Biol., 84:767 (1980).

18. G. Sobue and D. Pleasure, Adhesion of axolemma fragments to Schwann cells : a signal and target specific process closely linked to axolemmal induction of Schwann cell mitosis, J. Neurosci., 5:379 (1985).
19. M. M. Ayers and R. McD. Anderson, Onion bulb neuropathy in the Trembler mouse : A model of hypertrophic interstitial neuropathy (Dejerine-Sottas) in man, Acta Neuropathol., 25:54 (1973).
20. B. Ferzaz, H. Koenig and A. Ressouches, Axonal regeneration in Trembler mouse, a Schwann cell mutant, C.R. Acad. Sci. Paris, 309:377 (1989).
21. M.J. Politis, N. Sternberger, K. Ederle and P.S. Spencer, Studies on the control of myelinogenesis. IV. Neuronal induction of Schwann cell myelin-specific protein synthesis during nerve fiber regeneration, J. Neurosci., 2:1252 (1982).
22. P.M. Wood and R.P. Bunge, Evidence that sensory axons are mitogenic for Schwann cells, Nature, 256:662 (1975).
23. H.L. Koenig, N.A. Do Thi and B. Ferzaz, Proliferation of Trembler mouse Schwann cells in culture and during Wallerian degeneration, Soc. Neurosci., 14:483 (1988).
24. R.G. Pellegrino and P.S. Spencer,Schwann cell mitosis in response to regenerating peripheral axons in vivo, Brain Res., 341:16 (1985).
25. A. Krystosek and N.W. Seeds, Peripheral neurons and Schwann cells secrete plasminogen activator, J. Cell Biol., 98:773 (1984).
26. R. N. Pittman, Release of plasminogen activator and a calcium-dependent metalloprotease from cultured sympathetic and sensory neurons, Devl Biol., 110:91 (1985).
27. A. Alvarez-Buylla and J.E. Valinsky, Production of P.A. in cultures of superior cervical ganglia and isolated Schwann cells, Proc. Natl Acad. Sci., USA, 82:3519 (1985).
28. N. Kalderon, Schwann cell proliferation and localized proteolysis : Expression of P.A. activity predominates in the proliferating cell populations, Proc. Natl. Acad. Sci. USA, 81:7216 (1984).
29. A. Baron-Van Evercooren, P. Leprince, B. Rogister, P.P. Lefebvre, P. Delree, I. Selak and G. Moonen, Plasminogen activators in developing P.N.S. cellular origin and mitogenic effect, Devl Brain Res., 36:101 (1987).
30. R.N. Pittman, J. K. Ivins and H.M. Buettner, Neuronal plasminogen activators : Cell surface binding sites and involvement in neurite outgrowth, J. Neurosci., 9:4269 (1989).
31. J.C. Unkeless, S. Gordon and E. Reich, Secretion of plasminogen activator by stimulated macrophages, J. Exp. Med., 139:834 (1974).
32. J. Pöllänen, K. Hedman, L.S. Nielsen, K. Danø and A. Vaheri, Ultra-structural localization of plasma membrane-associated urokinase-type plasminogen activator at focal contacts, J. Cell Biol., 106:87 (1988).
33. U. Hedner, Studies on an inhibitor of plasminogen activation in human serum, Thromb. Diath. Haemorrh., 30:414 (1973).
34. J. Koenig, D. Hantaz, S. de la Porte, N.A. Do Thi, J.M. Bourre, F. La Chapelle and H.L. Koenig, "In vitro" evidence for a neurite growth-promoting activity in Trembler mouse serum, Int. J. Devl Neurosci., 7:281 (1989).
35. A. Bignami, G. Cella and N.H. Chi, Plasminogen activators in rat neural tissues during development and in Wallerian degeneration, Acta Neuropathol., 58:224 (1982).
36. W.T. Norton, C.F. Brosnan, W. Cammer and E. Goldmuntz, Some aspects of mechanisms of inflammatory demyelination, in : NATO ASI Series H43 : Cellular and molecular biology of myelination, G. Jeserich et al., Eds, Springer-Verlag, pp. 101-113 (1990).

37. J.W. Bigbee, J.E. Yoshino and G.H. De Vries, Morphological and proliferative responses of cultured Schwann cells following rapid phagocytosis of a myelin-enriched fraction, J. Neurocytol., 16: 487 (1987).
38. W. Beuche and R.L. Friede, The role of non-resident cells in Wallerian degeneration, J. Neurocytol., 13:767 (1984).
39. G. Stoll, J.W. Griffin, C.Y. Li and B.D. Trapp, Wallerian degeneration in the peripheral nervous system : participation of both Schwann cells and macrophages in myelin degradation, J. Neurocytol., 18:671 (1989).
40. R.N. Pittman and H.M. Buettner, Degradation of extracellular matrix by neuronal proteases, Devl. Neurosci., 11:361 (1989).

The Trembler mice used in this work were provided by Dr.JM Bourre (INSERM U26,Paris) and bred in our laboratory.The strain was first obtained in Dr.N.Baumann's lab.(Inserm,U134,Paris) by F.Lachapelle. The recessive vestigial tail introduced in B6D2/TR strain allowed to recognize normal mice with short tails and Trembler,with long tails.

BASIC FGF AND ITS ACTIONS ON NEURONS: A GROUP ACCOUNT WITH SPECIAL EMPHASIS ON THE PARKINSONIAN BRAIN

Dörte Otto, Claudia Grothe,
Reiner Westermann and Klaus Unsicker

Department of Anatomy and Cell Biology,
University of Marburg, Robert-Koch-Str. 6
D-3550 Marburg, Germany

INTRODUCTION

The fibroblast growth factors (FGFs) represent a family of peptide growth factors with relatively high amino acid sequence homologies ranging between 39 and 55 % (Table 1; see Baird and Böhlen, 1990, for a review). The first two FGFs characterized, acidic and basic FGF (aFGF and bFGF, respectively), occur in most mesenchymal and neuroectodermal tissues. Five other members of the FGF family were identified as proteins encoded by tumor oncogenes. They possess bFGF-like biological activities in terms of their mitogenic effects. Acidic and basic FGF including some N-terminal truncated forms exert similar biological functions, aFGF being 10-100 times less potent than bFGF.

Table 1. The FGF family

NAME	SEQUENCE HOMOLGY	SOURCE	bFGF-LIKE ACTIVITY
bFGF aFGF	 55%	mesenchymal and neuroectodermal tissues	+
INT-2	42%	site of integration of MMTV	+
hst, K-FGF	39%	human stomach tumor Kaposi`s sarcoma	+
FGF-5	50%	tumor oncogene	+
KGF	39%	embryonic human fibroblasts	+
FGF-6	35%	mouse cosmid library	+

Table 2 summarizes the in vitro and in vivo effects of bFGF: general in vitro effects comprise mitogenic activities for mesenchymal cells, e.g. fibroblasts and endothelial cells, chemotactic effects and regulation of protein synthesis. Accordingly, bFGF has been shown in vivo to induce neovascularization and to affect wound healing and several developmental events.

Table 2. Synopsis of in vitro and in vivo effects of bFGF

in vitro effects: general
- mitogenic for mesenchymal cells
- chemotactic
- induction/suppression of cell specific protein synthesis

in vitro effects: nervous system
- neurotrophic: promotion of survival and neurite outgrowth
  functions related to transmitter metabolism
- glial cell proliferation and differentiation

in vivo effects: general
- neovascularization, wound healing
- embryonic development

in vivo effects: nervous system
- ontogenetic neuron death
- PNS-lesion: sciatic nerve transection
- CNS-lesion: retino-tectal and septo-hippocampal

During the past few years, evidence has accumulated to suggest that bFGF is also a trophic factor for neurons (Morrison et al. 1986; Unsicker et al. 1987; Walicke et al. 1986).It has a lesser degree of selectivity and affects a larger number of neuron populations than does nerve growth factor (NGF; see Levi-Montalcini 1987, for a review), whose limited specificity may eventually turn out to be exceptional. Like many peptide hormones, as e.g. EGF, PDGF and TGFß, FGF exerts a variety of related but different effects on its target cells. Thus, bFGF influences neuronal survival and differentiation during development and modulates functions related to transmitter metabolism. Its effects on glial cells comprise cell proliferation and differentiation (see Westermann et al. 1990a).

Our group has focused on elucidating the physiological role of bFGF in the nervous system. Furthermore, to evaluate the therapeutical potential of bFGF we have pharmacologically applied the protein in several lesion paradigms.

BASIC FGF AND ONTOGENETIC NEURON DEATH

Dreyer and coworkers (1989) were able to show that bFGF influences ontogenetic neuron death in the chick embryo (Fig. 1).

Between embryonic day (E) 8 and 14 the initial neuron population of chick ciliary ganglia is reduced to about 50 % of its inital population size. The application of 3 μg of bFGF on E 8, 9, 11, 13 onto the chorionallantoic membrane maintained a significantly larger proportion of neurons that would naturally die, as demonstrated by controls that were injected with the non-trophic protein cytochrome c. Although a direct action of bFGF on neurons cannot be deduced based on these cell counts, it may be conceived that bFGF is able to mimic an endogenous factor supporting chick ciliary neurons or trigger the production and availability of a yet unidentified physiological agent.

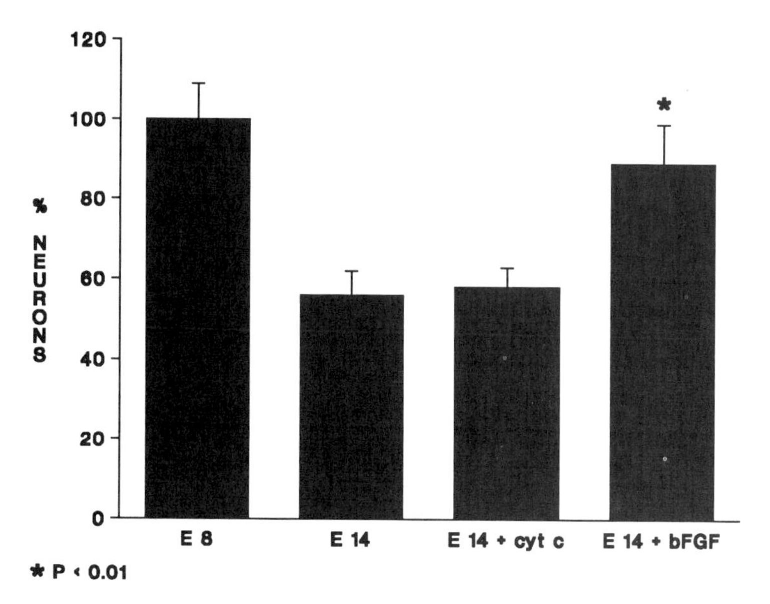

Figure 1. Basic FGF rescues chick ciliary ganglion neurons from ontogenetic neuron death in vivo. Neurons present at embryonic day (E) 8 were set as 100 % At E 14 their number had declined by about 50 %. Basic FGF, but not cytochrome c, maintained a significant proportion of neurons that otherwise would have died.

BASIC FGF AND AXOTOMY-INDUCED NEURON DEATH

Other evidences for a trophic role for bFGF have been put forward by studies on the axotomy-induced neuron death examplified in adult rats by one central and one peripheral nervous system lesion model, i.e. the fimbria-fornix and the sciatic nerve transection, respectively (Fig 2).

Target deprivation and the concomitant loss of a trophic input required for the maintenance of the neuronal cell body are supposedly at least one reason for axotomy-induced neuron death. Our results support this concept in so far as we could show that the loss of Nissl-stained neurons in axotomized, saline-treated rats was significantly reduced by the application of a single dose of bFGF to the lesion site (Otto et al. 1987, 1989). In both the fimbria-fornix and the sciatic nerve lesion the rescue effect brought about by bFGF is not as great as that seen with NGF. This observation suggests that NGF is the more specific factor, although in the CNS both of them occur in a target region of septal neurons, i.e. the hippocampal formation.

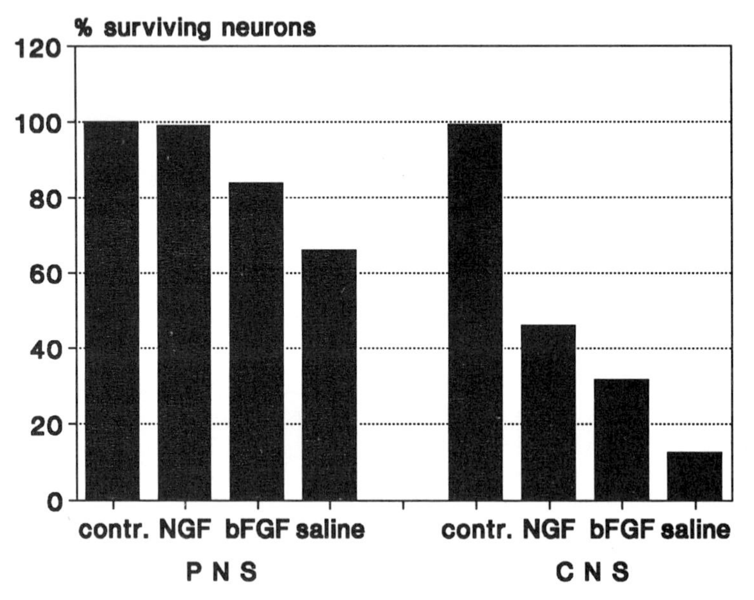

Figure 2. Axotomy-induced neuron death and its prevention by NGF or bFGF in the medial septal nucleus (CNS) after a fimbria-fornix lesion and in dorsal root ganglia (PNS) after sciatic nerve transection. For details see Otto et al., 1987, 1989).

BASIC FGF AND TARGET ORGAN-DEPENDENT NEURON MAINTENANCE

The discovery of an intraneuronal localization of bFGF-like (IR) immunoreactivity in conjunction with the findings that neuronal cells store and secrete neurotrophic agents promoted a series of studies on the neuron-like adrenal chromaffin cells that aimed at identifying those molecules. Our data suggested that one of these factors is bFGF: bFGF-like IR can be detected in rat and bovine adrenal chromaffin cells in situ and in culture. The subcellular distribution seems to be restricted to the secretory granules. Second, an 18 kD bFGF can be partially purified from bovine adrenal medulla and chromaffin vesicles, where it coexists with

immunoreactive forms of higher molecular weight (Grothe and Unsicker 1990; Westermann et al. 1990b). Third, the neurotrophic activity from isolated chromaffin granules can be blocked by anti-bFGF antibodies (Blottner et al. 1989). The presence of bFGF or a related molecule with neurotrophic activity in the adrenal medulla, which can be addressed as a modified sympathetic ganglion, pointed out the possibility that these factors may act in a retrograde neurotrophic fashion on the presynaptic spinal cord neurons innervating the chromaffin cells.
Accordingly, the ablation of the target should result in the death of ipsilateral spinal cord neurons at the corresponding levels. Moreover, subsequent substitution of the target by a trophic factor support should maintain these neurons. The medulla of rats was selectively destroyed and replaced by a gel foam soaked with either cytochrome c or bFGF. Within four weeks, 25 % of the Nissl-stained presynaptic neurons located in the intermediolateral column of the spinal cord between thoracic levels Th 7-10 had disappeared, when the gel foam had contained cytochrome c. The substitution of the medulla with 2 μg of bFGF completely prevented the neuron losses (Blottner et al. 1989).

Although it remains to be shown whether bFGF is directly acting on preganglionic spinal cord neurons and transported by retrograde axonal flow we found that interruption of this pathway by cutting the splanchnic nerve abolished the rescue effects of bFGF suggesting a retrograde action of FGF or other agents induced by the local administration of bFGF (Blottner et al. 1990).

BASIC FGF AND MORBUS PARKINSON

Neurotrophic factors from adrenal chromaffin cells may also be responsible for the functional recovery following transplantation of adrenal tissue into Parkinsonian brains. Previous data including findings by Ferrari et al. (1989) that bFGF is a neurotrophic protein for dopaminergic midbrain neurons in vitro suggested that bFGF might be involved in the beneficial effect mediated by adrenal medulla grafts. Grafting adrenal medulla directly into the brain parenchyma is an approach already in clinical use to treat the symptoms of patients with Parkinson's disease (PD). Experiments in animal models of PD have shown that the survival of adrenal chromaffin cells implanted into the brain is subjected to great variations. Even though, grafted rats with a 6-hydroxy-dopamine (6-OHDA) lesion show reduced rotational behaviour suggesting an improved function of their nigrostriatal dopaminergic system. In mice and monkeys treated with the neurotoxin 1-methyl-4-phenyl-1,2,3,6-tetrahydropyridine (MPTP) chromaffin grafts enhance the regeneration of the dopaminergic host system. This is true even for MPTP treated mice, which have received freeze-thawed adult adrenal medulla (Bohn, personal communication). There is as yet no conclusive explanation for these observations, but the implication of mechanisms as e.g.liberation of trophic factors and inflammation related events are being discussed.

In a recent study (Otto and Unsicker 1990) we tried to mimic the effects of adrenal medulla grafts in MPTP lesioned mice on the host dopaminergic system by administering bFGF in a single dose via a gel foam implant. Young male C57BL mice received three intraperitoneal injections of 30 mg/kg MPTP at 24 h intervals. The first injection was given at the day of or 8 days prior to surgery. This protocol destroys striatal dopaminergic fibers within one week,leaves the majority of nigral cell bodies unaffected and thus maintains a regenerative capacity. Pieces of gel foam served as implants into the right striatum and were soaked either with 4 μg cytochrome c as a non-trophic control protein or with 4 μg bovine or human recombinant bFGF. Two weeks after implantation the striatal dopaminergic sytem was examined for morphological and neurochemical changes. MPTP treatment reduced striatal dopamine levels to about 20 % of control values. Basic FGF diminished this reduction significantly to about 55 %. Moreover, the application of another neurotrophic protein, ciliary neurotrophic factor (CNTF), had no effect corroborating the specificity of the bFGF effect. In unlesioned animals, application of bFGF did not affect dopamine levels.

The dense network of very fine tyrosine-hydroxylase (TH) immunoreactive fibers in untreated mice virtually disappeared in MPTP/cytochrome c treated animals. On the bFGF- treated side TH-immunoreactive fibers were again detectable after two weeks. The fiber network was especially dense close to the implant within a zone of 150-200 μm and decreased further away from it. Delayed application of bFGF one week after initiation of the MPTP lesion lead to a comparable TH-fiber pattern suggesting that bFGF, in fact, is capable to partially restore a dopaminergic innervation of the denervated striatum. Surprisingly, TH-immunoreactive fibers were only detectable on the ipsilateral side in spite of a bilateral increase of dopamine contents. This fact pointed at side differences in the dopamine metabolism. Determination of TH-activities revealed a bilateral depletion of 80 % in MPTP/cytochrome c mice as compared to controls. Basic FGF treated animals did not reveal a significant change in TH-activity on either side. We then used TH-immununblots of striatal homogenates for detecting eventual differences in the amount of TH-protein. This semiquantitative method revealed a reduction of TH-protein in MPTP/cyt c-treated mice as compared to controls. Basic FGF failed to enhance TH-protein on the side contralateral to the gel foam. Yet, on the ipsilateral side the amount of TH-protein was apparently increased, a result being in line with the immunohistochemical data. Levels of dihydroxyphenylacetic acid (DOPAC), a metabolite of dopamine, were determined to obtain an estimate of dopamine turnover. In MPTP/cytochrome c-treated mice, DOPAC was not detectable on either side. Application of bFGF increased the ipsilateral DOPAC content to about 25 % of control levels. However, in the contralateral striatum levels were found to be twice as high, suggesting an elevated dopamine turnover. It is not clear as yet, at which level the underlying compensatory mechanisms for the regulation of dopamine metabolism occur.

Supposedly, our data result from a combination of events including sprouting of fibers, cell survival and regulatory processes in dopamine metabolism. Moreover, they demonstrate that the application of a sole protein can partially mimic the effects of adrenal medulla grafts in this lesion paradigm. Certainly, our data do not permit the conclusion that FGF is the agent directly responsible for these effects.

Indirect effects may be mediated by glial cells and/or by immigrated inflammatory cells. Both are known to secrete agents and/or mediators upon appropriate stimulation, which might be involved in recovery-related events (Guilian and Lachman 1985; Nieto-Sampedro et al. 1983). Based on immunocytochemical data,we observed a marked gliosis throughout both striata, the highest density being found close to the implant. Staining pattern as well as apparent number of immunoreactive astrocytes turned out to be independent from the previous treatment. The same is true for the peroxidase-positive cells that invaded the gel foam and which we are currently trying to identify. Our suggestion that bFGF may trigger the production of endogenous trophic agents was supported by in vivo experiments. We tested the eluates from gel foam soaked with cytochrome c or bFGF that had been implanted for two weeks for their neurotrophic activity on chick ciliary ganglion neurons. The neurotrophic activities elicited by bFGF-containing or by cytochrome c-containing gel foam eluates could not be significantly inhibited by the addition of bFGF-antibodies to the medium, which were able to block the pure bFGF-activity (40 ng/ml). It is conceivable that we are dealing here with a multitude of neurotrophic agents. However, we suggest that the composition of the eluates from the gel foams treated with cytochrome c or treated with bFGF has to be different since only the bFGF treatment was capable to induce recovery in the lesioned animals.

CONCLUSIONS AND PERSPECTIVES

Data accumulated in our and other laboratories are consistent with the notion that bFGF and other members of the FGF family may be relevant for differentiation and regenerative events in the nervous system. Issues to be addressed in the future include regulation of these proteins and their mRNAs, distribution, function and regulation of their receptors and actions in the context of other growth factors. Results from such studies will eventually permit to decide whether these proteins or agents that interfere with their expression are applicable in human neurodegenerative disorders.

ACKNOWLEDGEMENTS

Work of our groups described in this article was supported by grants from the German Research Foundation (Un 34/13-1) and Federal Ministry of Science and Technology (BMFT).

REFERENCES

Baird, S., and Böhlen P., 1990, Fibroblast growth factors, in: Peptide growth factors and their receptors I, M.B. Sporn and A.B. Roberts, Hdb. Exper. Pharmacol. Vol. 95/I, pp. 369-403

Blottner, D., Westermann, R., Grothe C., Böhlen, P., and Unsicker, K.,1989, Basic fibroblast growth factor in the adrenal gland: possible trophic role for preganglionic neurons in vivo, Eur. J. Neur. 1:471-478

Blottner, D., and Unsicker, K., 1990, Maintencance of intermediolateral spinal cord neurons by fibroblast growth factor administered to the medullectomized rat adrenal gland: dependence on intact adrenal innervation and cellular organization of implants, Eur. J. Neurosci. 2: 378-382

Bohn, M.C., Cupit L., Marciano F., Gash DM., 1987, Adrenal medulla grafts enhance recovery of striatal dopaminergic fibers, Science 237: 913-916

Dreyer, D., Lagrange ,A., Grothe C., and Unsicker, K., 1989, Basic fibroblast growth factor prevents ontogenetic neuron death in vivo, Neurosci. Lett. 99: 35-38

Ferrari, G., Minozzi, M-C., Toffiano, G., Leon, A., Skaper, SD., 1989, Basic fibroblast growth factor promotes the survival and development of mesencephalic neurons in culture, Dev. Biol. 133: 140-147

Grothe, C., Unsicker, K., 1990, Immunocytochemical mapping of basic fibroblast growth factor in the developing and adult rat adrenal gland, Histochem. 94: 141-147

Giulian, D., Lachmann L.B., 1985, Interleukin-1-stimulation of astroglial proliferation after brain injury,Science 228: 497-499

Morrison, R.S., Sharma, A., De Vellis, J., Bradshaw, R.A., 1986, Basic fibroblast growth factor supports the survival of cerebral cortical neurons in primary culture, Proc. Natl. Acad. Sci. USA 83; 7537-7541

Nieto-Sampedro M., Manthorpe, M., Barbin, G., Varon, S., Cotman, CW., 1983, Injury-induced neurontrophic activity in adult rat brain: correlation with survival of delayed implants in the wound cavity, J. Neurosci. 3: 2219-2229

Otto, D., Unsicker, K., Grothe C., 1987, Pharmacological effects of nerve growth factor and fibroblast growth factor applied to the transectioned sciatic nerve on neuron death in adult rat dorsal root ganglia, Neurosci. Lett. 83: 56-160

Otto, D., Frotscher M., Unsicker K., 1989, Basic fibroblast growth factor and nerve growth factor administered in gel foam rescue medial septal neurons after fimbria fornix transection, J. Neurosci. Res. 22: 83-91

Otto, D., Unsicker, K., 1990, Basic FGF reverses chemical and morphological deficits in the nigrostriatal system of MPTP-treated mice, J. Neurosci. 10(6): 1912-1921

Unsicker, K., Reichert-Preibsch, H., Schmidt, R., Pettmann, B., Labourdette, G., Sensenbrenner, M., 1987b, astroglial and fibroblast growth factors have neurontrophic functions for cultured peripheral and central nervous system neurons, Proc. Natl. Acad. Sci. USA 83: 817-821

Walicke, P., Cowan, WM., Ueno, N., Baird, A., Guillemin, R., 1986, Fibroblast growth factor promotes survival of dissociated hippocampal neurons and enhances neurite extension, Proc. Natl. Aca. Sci. USA 83: 3012-3016

Westermann, R., Grothe, C., Unsicker K., 1990a, Basic Fibroblast growth factor (bFGF), a multifunctional growth factor for neuroectodermal cells, J. Cell Sci. (in press)

Westermann, R., Johannsen M., Unsicker, K., Grothe, C.,1990b, Basic fibroblast growth factor (bFGF) immunoreactivity is present in chromaffin granules, J. Neurochem.

MOLECULAR AND MORPHOLOGICAL CORRELATES FOLLOWING NEURONAL DEAFFERENTATION: A CORTICO-STRIATAL MODEL

G.M. Pasinetti, H.W. Cheng, J.F.* Reinhard, C.E. Finch and T.H. McNeill

Andrus Gerontology Center, University of Southern California Los Angeles, CA 90089; *The Wellcome Research Laboratories, Research Triangle Park, NC 27709.

The ability of neurons to remodel the extent and configuration of their axons and dendrites plays an important role in maintaining function in the central nervous system in normal aging (Cotman and Anderson, 1983; Coleman and Flood, 1987). Conversely, the lack of an appropriate compensatory response of surviving cells to phenomena in the aged brain such as spontaneous neuron loss, deafferentation, or neurotransmitter deficits, is hypothesized to represent a common pathophysiological process in age-related neurodegenerative disorders (Coleman and Flood, 1986). Although the mechanisms governing synaptic remodelling in the adult brain are unknown, we hypothesize that it involves altered genomic expression in surviving neurons of afferent projection systems, whose terminals are induced to sprout and reinnervate deafferentated tissue (Cotman and Nieto-Sampedro, 1984). Moreover, since astrocytes participate in the process of removing degenerating axons and dendrites following a deafferentation lesion (Gage et al., 1988), alterations in the genomic response of these cells could be a critical factor leading to incomplete or delayed reorganization of new synaptic circuits (Scheff et al., 1989).

While changes in the cytoarchitecture of surviving neurons occur after experimental lesions (Pasinetti et al., 1989; Cotman and Nieto Sampedro, 1984) or during aging (Coleman and Flood, 1986; McNeill and Koek, 1990), few studies have addressed alterations in gene expression of surviving neurons or astrocytes which may underlie the morphological events leading to synaptic remodelling and functional recovery (Steward et al., 1990; Pourier et al., 1990, May et al., 1990; Geddes et al., 1990).

The common goal of this study has been to identify potential molecular markers that may be used to study and characterize the cellular mechanisms which are involved in the phenomenon of synaptic remodelling required to maintain normal functional in the aged brain. These data also provide insight into the cellular and molecular events that underlie the failed compensatory response of surviving neurons thought to occur during neurodegenerative disorders such as Parkinson's disease (McNeill et al., 1988).

Since synaptic remodelling is most evident after injury to the nervous system, in this study we used an established cortico-striatal rat lesion model (Cheng et al., 1988) to identify changes in the prevalence as well as temporal occurrence of mRNA sequences following neuronal deafferentation. Experimental manipulation of the cortico-striatal pathway is particularly well suited for this type of study because previous retrograde tracing studies have shown that the striatum (ST) is interconnected with the cortex and midbrain through a distinct set of topographically-defined neurochemical pathways (Fig. 1). These include: 1) long projection neurons of the striato-nigral system which originate from separate population of medium spiny neurons (MSI) of the ST that contain the neurotransmitter GABA or substance P, some of which may also contain enkephalin (Chang et al., 1981); 2) dopaminergic perikarya from the midbrain which originate from the pars compacta of the substantia nigra and project topographically to the ST to form predominately axospinous synapses on MSI neurons (Freund et al., 1984); and 3) glutamatergic neurons from both the ipsilateral and contralateral cortex which synapses on the dendritic spines of MSI striatal neurons (Freund et al., 1984)

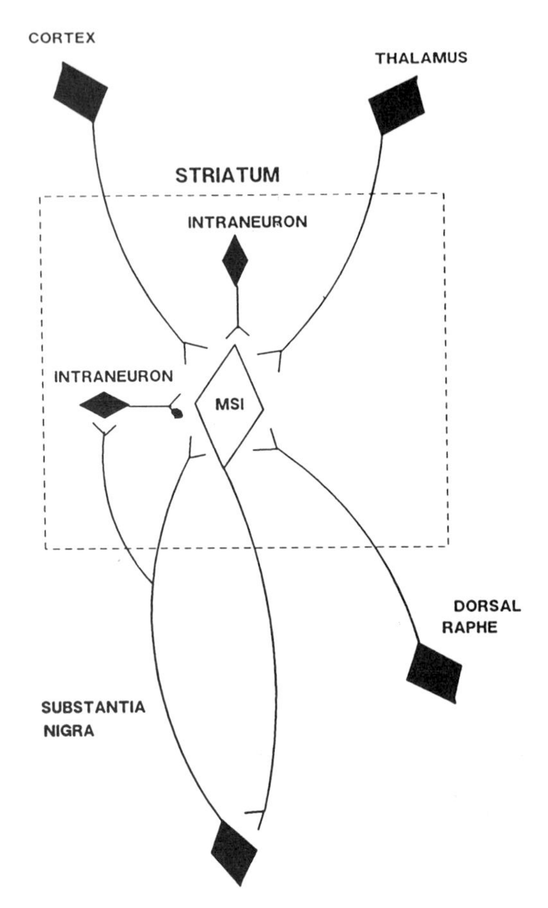

Fig. 1 Schematic diagram of the principal afferent projection systems that synapse on Medium Spiny I (MSI) neurons of the ST. The major fiber pathways include: glutamatergic fibers from the cortex and thalamus; dopaminergic neurons from the nigra; and serotoninergic neurons from the dorsal raphe.

Previous studies showed that the loss of afferent input from the ipsilateral cortico-striatal pathway leads to an initial loss of dendritic spine density on striatal MSI neurons which returns to control levels by 20 days postlesion (Cheng et al., 1988). This recovery of spine density can be blocked by a subsequent lesion of the contralateral cortex (Cheng et al., in prep). As a follow up to this morphological investigation , we examined the temporal sequence of molecular events that are thought to be involved in synaptic remodelling of the ST caused by experimental brain lesions (cortico-striatal lesions by pre-frontal cortex aspiration). In particular, we quantified the mRNA prevalence of growth associated protein (GAP-43), sulfated glycoprotein 2 (SGP-2) and the astrocytic specific glial fibrillary acidic protein (GFAP) using blot hybridization techniques. In addition, we defined changes in mRNA content in specific regions and cells of the ST and contralateral cortex using in situ hybridization. Details of reported findings are not given because journals require that data in submitted manuscripts not be simultaneously published.

GAP-43 is an axonally transported protein that is especially prominent in growth cones. It is expressed exclusively in neurons and is believed integral to axonal outgrowth and new synapse formation (Kosik et al., 1988). GFAP is a member of the class of intermediate filament proteins (Capetanaki et al., 1984) which has been used as a marker for reactive astroglia in many studies (Isacson et al., 1987). Indirect pharmacological evidences suggest that reactive astrocytes play an important role in synaptic remodelling by removing the degenerative axon and dendritic profiles from the lesion site. In addition, glucocorticoid treatment in rats decreases GFAP mRNA in the hippocampus (Nichols et al.,1990) and slows reactive synaptogenesis normally induced by deafferentation (Scheff et al., 1990). SGP-2 is the major secretory product of Sertoli cells in the testis and is thought to be a lipid-transport molecule (Collard and Griswold, 1987). Although the function of SGP-2 is presently unclear, recent findings indicate its involvement in degenerative processes (Buttyan et al., 1989, Duguid et al., 1989). In particular, the prevalence of SGP-2 mRNA is increased in the hippocampus in Alzheimer's disease (May et al., 1990) and Pick's disease (Duguid et al., 1989).

Our preliminary data suggest that the temporal sequence of reinnervation of the ipsilateral ST, deafferented by unilateral cortical lesion, correlated with an increase in mRNA prevalence for both GFAP and SGP-2 in the ipsilateral ST at 3 days postlesion and reached a maximum at 10 days postlesion. In addition the increase in striatal GFAP protein content (assayed by immunoassay) paralleled the increase in GFAP mRNA. By 27 days postlesion SGP-2 mRNA prevalence was still elevated over the levels found in the intact ST, whereas GFAP mRNA rapidly declined toward control levels. On the contrary, by 27 days the striatal GFAP protein content was still elevated suggesting a slower turnover rate of GFAP protein or an increased efficiency GFAP mRNA translation (Fig. 2). Moreover, these data indicate that the dynamic response of striatal GFAP mRNA precede the GFAP protein changes in response to deafferentation.

Changes in mRNA prevalence for GFAP were correlated with an increase in the density of immunoreactive GFAP positive astrocytic fibers and radioactive label for GFAP mRNA by in situ hybridization in the dorsal part of the ipsilateral ST. The increase in mRNA prevalence for SGP-2 by both northern blot analysis and in situ hybridization was correlated with the appearance of large clusters of SGP-2 immunoreactive punctate varicosities adjacent to as well as over neurons of the deafferented ST at 10 days post-lesion. Changes in the prevalence of mRNA for both markers paralleled the time course required for the homotypic reinnervation of the lesioned ST by axonal fibers from the contralateral cortex
that has been described previously in our morphological studies. In the contralateral cortex, mRNA prevalence for GFAP was increased at 3 days postlesion and rapidly declined to control values by 27 days postlesion In contrast, GAP-43 mRNA prevalence was increased only slightly at 10 and 27 days postlesion and was not significantly elevated above the values found in intact controls.

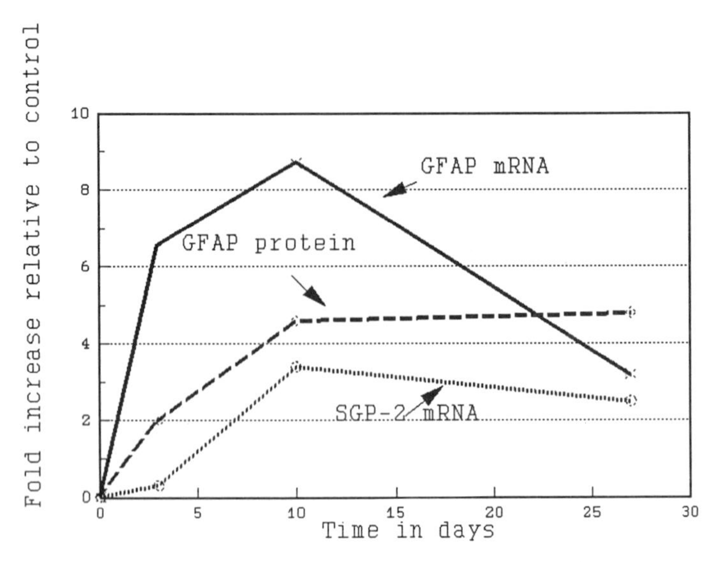

Fig. 2 Striatal GFAP and SGP-2 mRNA and GFAP protein changes following prefrontal cortex deafferentation

Data from our studies are consistent with the notion that reactive astrocytes play and important role in remodeling the synaptic connectivity of the ST following cortical deafferentation and are involved in removing the degenerative axon and dendritic profiles from the lesion site (Gage et al., 1988). Whether they also secrete potential trophic factors that signal neurons to sprout collaterals and direct axonal outgrowth following injury to the CNS is unknown (Gage et al., 1988, Needels et al., 1986). In addition, the increase of SGP-2 at the lesion site indicate similar events to what has been previously described in the prostate following orchidectomy (Buttyan et al., 1989), and suggest that the presence of SGP-2 may provides an early indicator of an ongoing degenerative process (May et al., 1990). Whether the gene is expressed in order to protect cells from neurotoxic insults that could lead to cell death or whether it is actively expressed in neurons as part of a cytocidal death program is unclear. However, since previous ultrastructural studies suggest that striatal cell death is not a primary feature of the unilateral cortical lesion we suggest that SGP-2 may serve a beneficial function analogous to a heat-shock response to enable deafferented neurons to survive the neurotoxic insult of deafferentation. This hypothesis is based, in part, on previous studies that the human analogue of SGP-2 (SP-40,40) inhibits complement-dependent cytolysis by blocking the formation of the terminal complement attack complex (C5b-9) on cell membranes (Tsuruta et al., 1990). In addition, previous studies have found that rat SGP-2, a major secretory product of Sertoli cells in the testis, is homologous with human complement cytolysis inhibitor (CLI) and is thought to carry out a "protective function" by preventing complement-mediated events in the reproductive tract (Jenne and Tschopp, 1989). We hypothesize that the "protective" role proposed for SGP-2 in the hippocampus during Alzheimer's disease and in response to kainic acid lesions in rat (May et al., 1990), extends to the ST during reactive synaptogenesis. Speculatively SGP-2 is synthesized and secreted at the lesion site in order to protect striatal neurons from phagocytic attack while reactive astrocytes remove surrounding degenerative axon and dendritic spine profiles from the lesion site. However, whether the gene is expressed by both neuronal and/or non-neuronal cells at the lesion site is currently unknown and we are currently pursuing this question using electron microscopy and a combination of in situ hybridization and immunocytochemistry.

Our data also suggest that the prevalence of GAP-43 mRNA does not significantly increase in the cell soma of neurons of the contralateral cortex that are induced to sprout and innervate vacant postsynaptic sites in the deafferented ST (Cheng et al., 1988). While the activation of GAP-43 related growth mechanisms is required for axonal elongation, GAP-43 may not play a central role in the type of paraterminal sprouting required to form new synapses on dendritic spines of target neurons within a few microns of the original degenerating axon. This hypothesis is consistent with studies suggesting that, in the CNS, injury of a neuron's axon far away from the cell body does not enhance GAP-43 expression in the cell soma (Kalil and Skene, 1986, Reh et al., 1987, Skene, 1989) whereas the transection of an axon very close to the parent cell body can induce GAP-43 gene expression in the cell body of the axotomized neuron in order to induce axonal sprouting (Lozano et al., 1987). However, further studies will need to be completed before a final conclusion can be reached.

In conclusion, our study provide information which are pertinent to our understanding of differential gene expression in surviving neurons and/or astrocytes associated with the morphological events leading to synaptic remodelling in the ST of adult and aged brain. Moreover, we hypothesize that the failure of these compensatory mechanisms at the molecular level may underlie the degeneration of the dendritic mantle of striatal target neurons in Parkinson's disease (PD) since the pathogenesis of PD involves multiple neurotransmitter pathways including the cortico-striatal and nigro-striatal systems (McNeill et al., 1988). In addition, these data may provide insight into the sequence of morphological and molecular events that characterize the hierarchical response of surviving afferent fibers to reinnervate the deafferented ST based on either a homotypic or heterotypic pattern of reinnervation.

Acknowledgements: The authors would like to thank Mr. Lloyd Price for his assistance in the manuscript preparation, Dr. M.D. Griswold for the antiserum and cDNA to SGP-2, Dr. R. Neve for GAP-43 cDNA. This research was supported by the National Institute on Aging to THM (AG 05445 and 00300) and CEF (AG 07909) and by grants from the National Parkinson Foundation and the United Parkinson Foundation.

References

Buttyan R., Olsson C.A., Pintar J., Chang C., Bandyk M., NG P.Y., and Sawczuk I.S., 1989, Induction of the TRPM-2 gene in cells undergoing programmed cell death. Mol. and Cell. Biol., 9:3473.

Capetanaki Y.G., Ngai J., and Lazarides E., 1984, Regulation of the expression of genes coding for the intermediate filament subunits vimentin, desmin, and glial fibrillary acidic protein, in: "Molecular Biology of the Cytoskeleton," Cold Spring Harbor Press.

Chang H.T., Wilson C.J., and Kitai S.T., 1981, Single neostriatal efferent axons in the globus pallidus: a light and electron microscopy study, Science, 213:915.

Cheng H.W., Anavi Y., Goshgarian H., McNeill T.H., and Rafols J.A., 1988, Loss and recovery of striatal dendritic spines following lesions in the cerebral cortex of adult and aged mice, Soc. Neurosci. Abst., 14:1219.

Coleman P.D., and Flood D.G., 1986, Dendritic proliferation in the aging brain as a compensatory repair mechanisms, Prog. Brain Res., 70:227.

Coleman P.D., and Flood D.G., 1987, Neurons numbers and dendritic extent in normal aging and Alzheimer disease, Neurobiol. Aging, 8:521.

Collard M.W., and Griswald M.D., 1987, Biosynthesis and molecular cloning of sulfated glycoprotein 2 secretedby rat sertoli cells, Biochem.. 26:3297.

Cotman C.W., and Nieto-Sampedro M., 1984, Cell biology of Synaptic plasticity. Science, 225:1287.

Cotman C.W., and Anderson K.J., 1983, Synaptic plasticity and functional stabilization in the hippocampal formation: possible role in Alzheimer disease, in: "Advances in Neurology," S.G. Waxman, ed., Vol. 47 Functional recovery in Neurological Disease, Raven Press, New York, NY.

Duguid J.R., Bohmont C.W., Liu N., and Tourtellotte W., 1989, Changes in brain gene expression shared by scrapie and Alzheimer disease, Proc. Natl. Acad. Sci (USA), 86:7260.

Freund T.F., Powell J.F., and Smith A.D., 1984, Tyrosine hydroxylase immunoreactivity boutons in synaptic contact with identified striatonigral neurons with particular reference to dendritic spines, Neuroscience, 13:1189.

Gage F.H., Olejniczak P., and Armstrong D., 1988, Astrocytes are important for sprouting in the septohippocampal circuit, Exp. Neurol., 102:2.

Geddes S.W., Wong J., Choi B.H., Kim R. C., Cotman C.W., and Miller F.D., 1990, Increased expression of embrionyc growth associated mRNA in Alzheimer disease, Neurosci. Lett., (in press).

Isacson O., Fisher W., Wictorin K., Dawbarn D., and Bjorklund A., 1987, Astroglial response in the excitotoxically lesioned neostriatum and its projections areas in the rat, Neuroscience, 20:1043.

Kalil K., and Skene J.H.P., 1986, Elevated synthesis of an axonally transported protein correlates with axon outgrowth in normal and injured pyramidal tracts, J. Neurosci., 6:2563.

Kosik K.S., D'Orecchio Lisa., Bruns G.A., Benowitz L.I., MacDonald P., Cox D.R., and Neve R., 1988, Human GAP-43: its deduced aminoacid sequence and chromosomal localization in mouse and human, Neuron, 1:127.

Jenne D.E., and Tschopp J., 1989, Molecular structure and functional characterization of a human complement cytolysis inhibitor found in blood and seminal plasma: identity to sulfated glycoprotein 2., a constituent of rat testis fluid, Proc. Natl. Acad. Sci. (USA), 86:7123.

Lozano A.M., Doster S.K., Aguayo A.J., and Willard M.B., 1987, Immunoreactivity to GAP-43 in axotomized and regenerating retinal ganglion cells of adult rats, Abstr. Soc. Neurosci., 13:1389.

May P.C., Lampert-Etchelles M., Johnson S.A., Poirier J., Master J., and Finch C.E., 1990, Dynamics of gene expression for hippocampal glycoprotein elevated in alzheimer's disease and in response to experimental lesion in rat, Neuron, (in press).

McNeill T.H., Brown S.A., Rafols J.A., and Shoulsson I., 1989, Atrophy of medium spiny I striatal dendrites in advanced Parkinson's disease, Brain Res., 455:158.

McNeill T.H. and Koeck L.L., 1990, Differential effects of advancing age on neurotransmitter, cell loss in the substantia nigra and striatum of the C57BL/6N mouse, Brain Res., (in press).

Needles D.L., Nieto-Sampedro M., and Cotman C.W., 1986, Induction of a neurite factor in rat brain following injury or deafferentation, Neuroscience, 18:517.

Nichols N.R., Osterburg H.H., Masters J.N., Millar S.L., and Finch S.L., 1990, Messenger RNA for glial fibrillary acidic protein is decreased in rat brain following acute and chronic corticosterone, Mol. Brain Res., 7:1.

Pasinetti G.M., Lerner S.P., Johnson S.A., Morgan D.G., Telford N.A., and Finch C.E.F., 1989, Chronic lesions differentially decrease tyrosine hydroxylase messenger RNA in dopaminergic neurons of substantia nigra, Mol. Brain Res., 5:203.

Poirier J., May P.C., Osterburg H.H., Geddes J., Cotman C., and Finch C.E., 1990, Selective alterations of RNA in rat hippocampus after enthorhinal cortex lesioning, Proc. Natl. Acad. Sci. (USA), 87:303.

Reh T.A., Redshaw J.D., and Bisby M.A., 1987, Axons of the pyramidal tract do not increase their transport of growth-associated proteins after axotomy, Mol. Brain Res., 2:1.
Scheff W.S., and Dkosky S.T., 1989, Glucocorticoid suppression of lesion-induced synaptogenesis: effect of temporal manipulation of steroid treatment, Exp. Neurol., 105:260.
Skene, J.H.P., 1989. Axonal growth-associated proteins, Ann. Rev. Neurosci., 12:127.
Steward O., Torre E.R., Philips L.L., and Trimmer P.A., 1990, The process of reinnervation in the dente gyrus of adult rats: time course of increases in mRNA for glial fibrillary acidic protein, J. Neurosci., 10:2373.
Tsuruta J., Wong K., Fritz B., and Griswold B., 1990, Structural analysis of sulphated glycoprotein 2 from aminoacid sequence, Relationship to clusterin and serum protein 40, 40. Biochem. J., 268:571.

MONOSIALOGANGLIOSIDE GM1 AND MODULATION OF NEURONAL PLASTICITY IN CNS REPAIR PROCESSES

Stephen D. Skaper, Silvio Mazzari, Guido Vantini
Laura Facci, Gino Toffano and Alberta Leon

Fidia Research Laboratories
Via Ponte della Fabbrica, 3/A
35031 Abano Terme, Italy

INTRODUCTION

Today we understand the brain as a dynamic, not static, organ. Central nervous system (CNS) neurons are endowed with the capacity to react to chemical signals presented from their microenvironment with morpho-functional modifications, a process termed plasticity. This phenomenon has provided the foundation for studies directed at elucidating the pathophysiological correlates of neuronal life and death, with the ultimate objective of developing strategies to improve neurological outcome following various types of CNS insults, in particular cerebrovascular insufficiency (stroke), head and spinal trauma, and neurodegenerative diseases.

Identifying the molecular mechanisms underlying stroke and CNS trauma and the development of therapeutic modalities for their treatment represent important goals of current neurobiological research. Agents that reduce, in the acute phase the onset of damage and subsequent neuronal loss in the penumbra zone, and that promote reparative processes in surviving neurons may allow for a functional neurological recovery. In this chapter, we will review current ideas concerning the pathophysiology of cerebral damage, focusing on the role of neurotoxic substances in secondary neuronal death and the action of neuronotrophic factors during reparative events. The effects and possible mechanisms of action of monosialoganglioside GM1 treatment after cerebral ischemia and trauma are also discussed, with respect to the ability of GM1 to both limit excitotoxin-induced injury and potentiate the effects of neuronotrophic factors.

ENDOGENOUS GANGLIOSIDES IN THE NERVOUS SYSTEM

Gangliosides comprise a family of glycosphingolipids (Svennerholm, 1963) which are normal components of virtually all vertebrate cell membranes. The greatest concentration of gangliosides in mammals is found in CNS grey matter, especially in the region of synaptic terminals (Ledeen, 1983). These molecules occur in the outer leaflet of the membrane lipid bilayer with their hydrophobic ceramide moiety, while the saccharide portion (containing one or more sialic acid groups) is oriented towards the extracellular space. With their par-

Plasticity and Regeneration of the Nervous System
Edited by P.S. Timiras *et al.*, Plenum Press, New York, 1991

ticular localization and plasma membrane orientation, gangliosides may participate in the recognition of receptor-triggered signals; transduction of these signals across the cell membrane (second messengers); activation of membrane enzymes and permeability of ion channels; and cell-cell and cell-substratum adhesion processes (Dal Toso et al., 1988; Skaper et al., 1989a).

There is good evidence that gangliosides are involved in a variety of functions related to the maturation and/or repair of neuronal tissue. For example, changes in the level of GM1 occur during CNS development (Willinger and Schachner, 1980; Krakun et al., 1986). The suggestion that GM1 plays a role in neurite growth is supported by the fact that exogenous GM1 promotes neurite outgrowth in vitro (Leon et al., 1982; Skaper et al., 1985) and facilitates postnatal CNS development (Karpiak et al., 1984. Mahadik and Karpiak, 1986). A possible relationship between levels of endogenous GM1 and formation of "mega-neurites" has been proposed (Purpura and Baker, 1977).

EXOGENOUS GM1 EFFECTS IN VITRO

Neurotoxic Agents: Relationship to Secondary Neuronal Injury In Vivo

Groups of neurons subjected to an ischemic/anoxic event or hypoglycemia can die during and/or immediately following the insult (primary death) or at later times. The latter case comprises secondary neuronal loss in proximity to the area of injury (penumbra/perilesion zone) or, at times, in even more distant areas. Within a limited window these neurons, while metabolically compromised, are still potentially functional and, therefore, represent a target for pharmacologic intervention.

Secondary neuronal death in perilesion areas correlates with the synaptic accumulation of excitatory amino acid (EAA) transmitters (glutamate and aspartate) in large amounts (Benveniste et al., 1984; Auer, 1986). In some pathological states, these EAAs can reach toxic levels, resulting in persistent overactivation of their post-synaptic receptors. Protracted increases of intracellular free Ca^{2+} follow, with induction of Ca^{2+}-dependent intracellular processes (e.g. proteases, phospholipases, protein kinase C) culminating in cell death (Rothman and Olney, 1986; Choi, 1988).

The recognition of this phenomenon of excitotoxicity has resulted in an intensive search for molecules with the capacity to attenuate or block the neurotoxic effects of EAAs at the receptor level. This pharmacologic approach can present practical problems, as EAAs are responsible for critical brain physiological functions, e.g. memory and learning. In principle, such drugs should not disrupt normal EAA physiology (Morris, 1989) but should only act to limit or prevent the biochemical cascade triggered by overstimulation of EAA receptors. Glutamate cytotoxicity, at least in cultured cerebellar granule cells, does not appear to depend on the opening of voltage-sensitive Ca^{2+} channels, since drugs which antagonize these channels are not neuroprotective (Manev et al., 1989; our unpublished observations).

When examined for a possible action in modulating neuronal responsiveness to the cytotoxicity of exogenous EAAs in vitro, GM1 was found to reduce such death for cerebellar granule cells (Favaron et al., 1988; Skaper et al., 1990) and cortical neurons (Favaron et al., 1988); similar results have been reported using retinal neurons (Facci et al., 1990a). Figure 1 illustrates the ability of administered GM1 to protect cultured hippocampal pyramidal neurons from glutamate-induced injury. The neuroprotective effects of GM1 have been demonstrated under conditions where neuronal injury results from endogenous glutamate receptor activation, as well (Skaper et al., 1989b; Facci et al., 1990b).

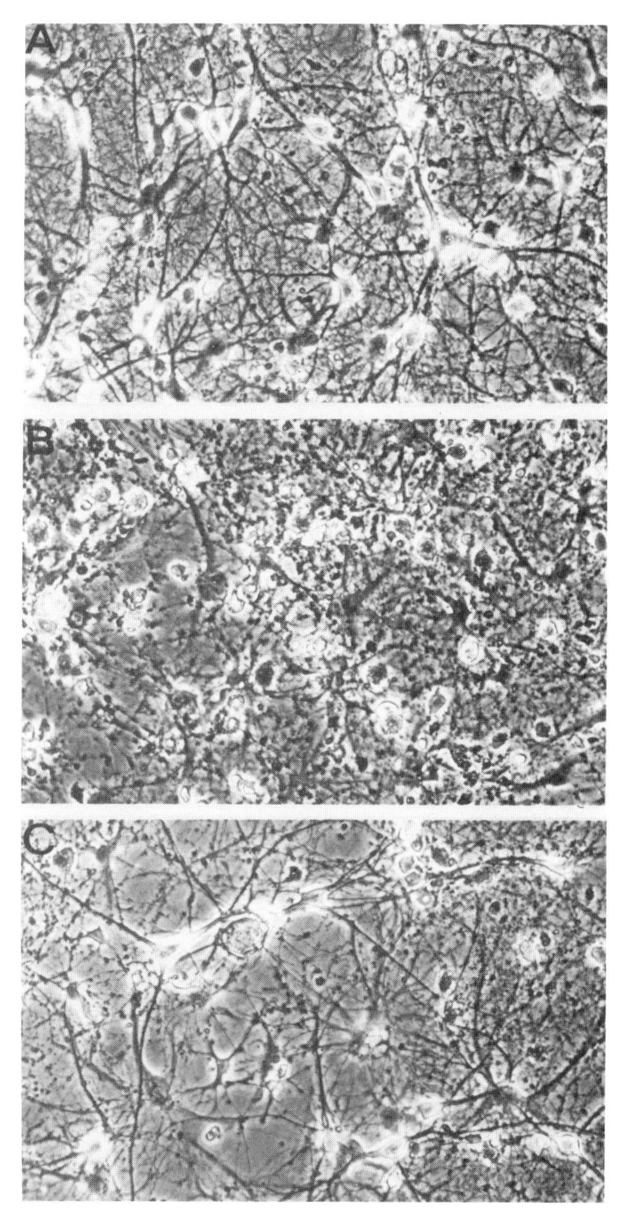

Fig. 1

Ganglioside GM1 prevents glutamate-induced death of hippocampal pyramidal neurons. Cells prepared from embryonic day 18 rat fetuses were maintained in vitro 15 days, followed by GM1 treatment (200 μM, 2 hr) prior to a 20 min pulse of 100 μM glutamate in Locke's solution. Cultures were photographed 24 hr later. (A) control; (B) glutamate; (C) GM1 + glutamate.

It is important to note that GM1 limits EAA neurotoxicity without disturbing normal cellular functions regulated by these transmitters (Favaron et al., 1988). At the molecular level, GM1 reduces the cytosol-to-membrane translocation and activation of protein kinase C (Vaccarino et al., 1987), a probable consequence of Ca^{2+} influx through glutamate receptor-linked ion channels. The neuroprotective effect of GM1 may be due, in part, to the observed reduction in $^{45}Ca^{2+}$ accumulation (Manev et al., 1989) and improved recovery of cytoplasmic levels of free Ca^{2+} (Milani et al., 1990) in the post-glutamate period.

Neuronotrophic Factors: Relation to Reparative Phenomena In Vivo

The adult mammalian CNS possesses a certain degree of self-repair capability in response to perturbations from its surrounding environment. This plasticity can manifest itself as the ability of nerve cells to modify, at various times following injury, their metabolism, excitability and gene expression.

The expression of neuronal plasticity is subject to regulation by, among other molecules, neuronotrophic factors present in the CNS. Neuronotrophic factors comprise protein molecules which influence the survival and development of nerve cells. The classical example of such macromolecules is nerve growth factor (NGF), although several other factors have now been well characterized. Removal of target tissue or interruption of synaptic contact deprives a neuron of adequate trophic supplies, leading to atrophy and death (Levi-Montalcini, 1987; Barde, 1989). Drug strategies aimed at increasing trophic factor efficacy in the brain have the potential for not only promoting neuronal survival at acute times but also for facilitating reparative processes at longer times.

The steps which translate extracellular trophic signals into intracellular events very likely occur at the plasma membrane level. In this context exogenous GM1, which functionally incorporates into neuronal membranes (Leon et al., 1981; Facci et al., 1984), can potentiate neuronal responsiveness to trophic factors in vitro. Ganglioside GM1, added to the culture medium potentiates the neuritogenic effects of NGF in not only sensory and sympathetic neurons (Ferrari et al., 1983; Katoh-Semba et al., 1984; Doherty et al., 1985; Skaper and Varon, 1985; Skaper et al., 1985; Spoerri, 1986), but also in cholinergic CNS neurons (Cuello et al., 1989). GM1 potentiates the neuronotrophic actions of other defined molecules (Skaper et al., 1985) and in different neuronal cell types (Skaper et al., 1985; Leon et al., 1988).

PHARMACOLOGICAL EFFECTS OF GM1 IN VIVO

An extensive literature is available describing the effects of administered GM1 in various models of brain injury. While beyond the scope of this chapter, the following section will briefly summarize the key findings for GM1 applied to traumatic, toxic or ischemic lesions. Recent data which implicate anti-neurotoxic and pro-neuronotrophic actions of GM1 in vivo - as already demonstrated in vitro - will also be discussed.

Using a unilateral nigrostriatal hemitransection model, Toffano et al., (1983) provided the first demonstration that systemically administered GM1 could improve the outcome after traumatic CNS lesioning. A number of laboratories have shown, with this model, that GM1 prevents the loss of nigral dopaminergic neurons, stimulates sprouting by the surviving striatal dopaminergic neurons, reduces the associated metabolic imbalance and circulation between the intact and lesioned sides and produces an improvement of behavioral parameters (Agnati et al., 1983; Sabel et al., 1984). Ganglioside GM1 treatment also stimulates recovery of other dopaminergic neurons, as well as serotoninergic and cholinergic neurons. The biochemical and behavioral deficits which accompany administration of 1-methyl-4-phenyl-1,2,3,6-tetrahydropyridine, a drug inducing neurodegenerative changes in mammalian brain and clinical symptoms in humans resembling Parkinson's disease, are reduced by treatment with GM1 (Hadjiconstantinou et al., 1986; Weihmuller et al., 1988). The reader is referred to more detailed accounts of these studies elsewhere (c.f. Skaper et al., 1989a).

The cytotoxic effects in vitro of glutamate and related EAA receptor agonists have a counterpart in vivo. Direct injection into the rodent brain of N-methyl-D-aspartic acid (NMDA) (Lipartiti et al., 1989), quinolinic acid (Lombardi et al;, 1989) or ibotenic acid (Mahadik et al., 1988; Casamenti et al., 1989) produces widespread neuronal cell loss. Systemically applied GM1 attenuates this excitoxin-induced tissue injury (Mahadik et al., 1988; Casamenti et al., 1989; Lipartiti et al., 1989; Lombardi et al., 1989). In addition, hypoxia-induced neurotransmitter deficits in neonatal rats are partially corrected by exogenous GM1 (Hadjiconstantinou et al., 1990).

Neuroprotective effects of systemically administered GM1 have been evaluated in animal models of global and focal cerebral ischemia. In some cases, the inner ester derivative of GM1 (siagoside or AGF2) was used. This latter ganglioside, like GM1, attenuates glutamate-induced neuronal injury in vitro (Skaper et al., 1989a). Table 1 summarizes these studies. Reduction of brain injury is observed morphologically, biochemically and functionally, in species ranging from rodents to primates. Such parameters include mortality, Na^+,K^+-ATPase activity, hippocampal neuronal loss, brain edema, Na^+ and Ca^{2+} balance, cyclo- and lypoxygenase metabolites, and behavioral deficits. Pharmacokinetic studies have shown that GM1, given systemically, reaches the brain intact (Bellato et al., 1989; Ghidoni et al., 1989).

TABLE 1

Experimental models of cerebral ischemia: reduction of secondary brain damage by monosialoganglioside

Model	Source
Unilateral permanent CCAo	Karpiak et al. (1987)
Unilateral permanent MCAo + CCAo	Karpiak et al. (1990)
Unilateral transitory MCAo	Komatsumoto et al. (1988)
Transitory global ischemia	Cahn et al. (1989)
Bilateral transitory CCAo + hypoxia	Petroni et al. (1989)
4-vessel occlusion	Seren et al. (1990)

CCAo = common carotid artery occlusion
MCAo = middle cerebral artery occlusion

RELATIONSHIP BETWEEN GANGLIOSIDE EFFECTS IN VIVO AND IN VITRO

A critical feature of the observed in vitro GM1 effects is a verification that the same mechanisms operate in vivo. Exogenous GM1 potentiates the protective and reparative effects of NGF administered intraventricularly following lesions to central cholinergic neuronal systems (Casamenti et al., 1989; Cuello et al., 1989; Di Patre et al., 1989). Ganglioside GM1 alone has been reported to prevent retrograde degeneration of septal cholinergic neurons produced by devascularizing lesions (Cuello et al., 1986), perhaps a consequence of potentiating endogenous NGF or other neuronotrophic factors. Administered GM1 also potentiates exogenous NGF action on the biochemical recovery of vinblastine-induced sympathectomy in peripheral organs (Vantini et al., 1988). These results indicate that the molecular basis of the GM1 effects in facilitating functional recovery after acute brain damage most probably reflect a potentiating action of one or more endogenous trophic factors in the CNS.

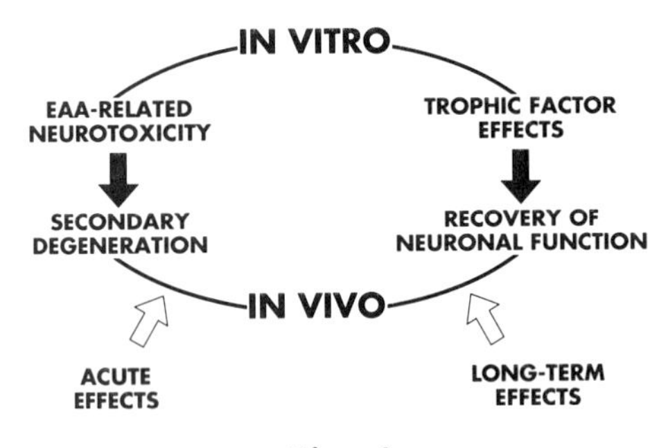

Fig. 2

A schematic representation of GM1 effects following acute brain injury.

As discussed earlier, GM1 reduction of excitatory amino acid cytotoxicity in vitro can be replicated in vivo, where systemic administration of GM1 diminshes neuronal damage induced by several excitotoxins. Thus, the mechanism(s) by which the ganglioside limits secondary neuronal death associated with ischemia or brain trauma would be expected to parallel those operating in vitro.

CONCLUDING REMARKS

A crucial role for synaptically mediated excitatory activity is clearly established in the pathogenesis of ischemic brain damage. While NMDA antagonists may be effective in protecting against tissue injury accompanying such pathologies, they also carry adverse colateral consequences, especially those acting at the phencyclidine site. Antagonists of NMDA receptors may also impact on synaptic plasticity, with effects on learning and memory (Collingridge and Bliss, 1987; Morris, 1989); neuropathological changes have also reported (Olney et al., 1989). More realistic pharmacologic strategies should focus on molecules which exert their cerebroprotective effects at a point downstream of the NMDA receptor/ion channel complex, without affecting normal receptor physiology. This review has provided good evidence to support ganglioside GM1 in this latter context.

Systemic administration of GM1 is effective in reducing acute nerve cell damage and in improving medium and long-term functional recovery following brain injury. Our working hypothesis is that GM1 protective effects in the acute injury phase result from attenuation of excitotoxicity, while functional recovery in the longer-term could reflect GM1 potentiation of neuronotrophic molecules. These ideas are presented schematically in Fig. 2. Interestingly, preliminary data indicate that trophic factors may also be effective in reducing hypoxic-ischemic brain damage (Shigeno and Mima, 1988).

The pharmacotherapeutic potential of GM1 in man is suggested by clinical studies demonstrating improved neurological outcome in acute stroke patients (Argentino et al., 1989). Further clinical trials are currently underway to better define the therapeutic utility of the ganglioside in human CNS pathologies, including stroke and head trauma.

REFERENCES

Agnati, L.F., Fuxe, K., Calzà, K., Benfenati, F., Cavicchioli, L., Toffano, G., and Goldstein, M., 1983, Gangliosides increase the survival of lesioned nigral dopamine neurons and favour the recovery of dopaminergic synaptic function of rats by collateral sprouting, Acta Physiol. Scand., 119:347.

Argentino, C., Sacchetti, M.L., Toni, D., Savoini, G., D'Arcangelo, E., Erminio, F., Federico, F., Ferro Milone, F., Gallai, V., Gambi, D., Mamoli, A., Ottonello, G.A., Ponari, O., Rebucci, G., Senin, U., and Fieschi, C., 1989, GM1 ganglioside therapy in acute ischemic stroke, Stroke, 20:1143.

Auer, R.N., 1986, Progress review: hypoglycemic brain damage, Stroke, 17:699.

Barde, Y.A., 1989, Trophic factors and neuronal survival, Neuron, 2: 1525.

Bellato, P., Milan, F., and Toffano, G., 1989, Disposition of exogenous tritium-labelled GM1 monosialoganglioside in the rat, Clin. Trials J., 26:1.

Benveniste, H., Drejer, J., Schousboe, A., and Diemer, N.H., 1984, Elevation of the extracellular concentrations of glutamate and aspartate in rat hippocampus during transient cerebral ischemia monitored by intracerebral microdialysis, J. Neurochem., 43:1369.

Cahn, R., Borzeix, M.G., Aldinio, C., Toffano, G., and Cahn, J., 1989, Influence of monosialoganglioside inner ester on neurologic recovery after global cerebral ischemia in monkeys, Stroke, 20:652.

Casamenti, F., DiPatre, P.L., Milan, F., Petrelli, L., and Pepeu, G., 1989, Effects of nerve growth factor and GM1 ganglioside on the number and size of cholinergic neurons in rats with unilateral lesion of the nucleus basalis, Neurosci. Lett., 103:87.

Choi, D.W., 1988, Glutamate toxicity and diseases of the nervous system, Neuron, 1:623.

Collingridge, G.L., and Bliss, T.V., 1987, NMDA receptors - their role in long-term potentiation, Trends Neurosci., 10:288.

Cuello, A.C., Garofalo, L., Kenigsberg, R.L., and Maysinger, D., 1989, Gangliosides potentiate in vivo and in vitro effects of nerve growth factor on central cholinergic neurons, Proc. Natl. Acad. Sci. USA, 86:2056.

Cuello, A.C., Stephens, P.H., Tagari, P.C., Sofroniew, M.V., and Pearson, R.C.A., 1986, Retrograde changes in the nucleus basalis of the rat, caused by cortical damage, are prevented by exogenous ganglioside GM1, Brain Res., 376:373.

Dal Toso, R., Skaper S.D., Ferrari, G., Vantini G., Toffano, G., and Leon, A., 1988, Ganglioside involvement in membrane-mediated transfer of trophic information. Relationship to GM1 effects following CNS injury, in: "Pharmacological Approaches to the Treatment of Spinal Cord Injury," D.G. Stein and B.A. Sabel, eds., Plenum Press, New York.

DiPatre, P.L., Casamenti, F., Cenni, A., and Pepeu, G., 1989, Interaction between nerve growth factor and GM1 monosialoganglioside in preventing cortical choline acetyltransferase and high affinity choline uptake decrease after lesion of the nucleus basalis, Brain Res., 480:219.

Doherty, P., Dickson, J.G., Flanigan, T.P., and Walsh, F.S., 1985, Ganglioside GM1 does not initiate, but enhances neurite regeneration of nerve growth factor-dependent sensory neurons, J. Neurochem., 44:1259.

Facci, L., Leon, A., Toffano, G., Sonnino, S., Ghidoni, R., and Tettamanti, G., 1984, Promotion of neuritogenesis in mouse neuroblastoma cells by exogenous gangliosides. Relationship between the effect and the cell association of ganglioside GM1, J. Neurochem., 42:299.

Facci, L., Leon, A., and Skaper, S.D., 1990a, Excitatory amino acid neurotoxicity in cultured retinal neurons: involvement of N-methyl-D-aspartate (NMDA) and non-NMDA receptors and effect of ganglioside GM1, J. Neurosci. Res., in press.

Facci, L., Leon, A., and Skaper, S.D., 1990b, Hypoglycemic neurotoxicity in vitro: involvement of excitatory amino acid receptors and attenuation by monosialoganglioside GM1, Neuroscience, in press.

Favaron, M., Manev, H., Alho, H., Bertolino, M., Ferret, B., Guidotti, A., and Costa, E., 1988, Gangliosides prevent glutamate and kainate neurotoxicity in primary neuronal cultures of neonatal rat cerebellum and cortex, Proc. Natl. Acad. Sci. USA, 85:7351.

Ferrari, G., Fabris, M., and Gorio, A., 1983, Gangliosides enhance neurite outgrowth in PC12 cells, Dev. Brain Res., 8:215.

Ghidoni, R., Fiorilli, A., Trinchera, M., Venerando, B., Chigorno, V., and Tettamanti, G., 1989, Uptake, cell penetration and metabolic processing of exogenously administered GM1 ganglioside in rat brain, Neurochem. Internatl., 15:455.

Hadjiconstantinou, M., Rosetti, Z.L., Paxton, R.C., and Neff, N.H., 1986, Administration of GM1 ganglioside restores the dopamine content in striatum after chronic treatment with MPTP, Neuropharmacology, 25:1075.

Hadjiconstantinou, M., Yates, A.J., and Neff, N.H., 1990, Hypoxia-induced neurotransmitter deficits in neonatal rats are partially corrected by exogenous GM1 ganglioside, J. Neurochem., 55:864.

Karpiak, S.E., Li, Y.S., and Mahadik, S.P., 1987, Gangliosides (GM1 and AGF2) reduce mortality due to ischemia: protection of membrane function, Stroke, 18:184.

Karpiak, S.E., Vilim, F., and Mahadik, S.P., 1984, Gangliosides accel-

erate rat neonatal learning and levels of cortical acetylcholinesterases, Dev. Neurosci., 6:127.
Karpiak, S.E., Mahadik, S.P., and Wakade, C.G., 1990, Ganglioside reduction of ischemic injury, CRC Crit. Rev. Neurobiol., 5:221.
Komatsumoto, S., Greenberg, J.H., Hickey, W.F., and Reivich, M., 1988, Effect of the ganglioside GM1 on neurologic function, electroencephalogram amplitude, and histology in chronic middle cerebral artery occlusion in cats, Stroke, 19:1027.
Katoh-Semba, R., Skaper, S.D., and Varon, S., 1984, Interaction of GM1 ganglioside with PC12 pheochromocytoma cells: serum-and NGF-dependent effects on neuritic growth (and proliferation), J. Neurosci. Res., 12:299.
Krakun, I., Rösner, H., and Cosovic, C., 1986, Topographical distribution of the gangliosides in the developing and adult human brain, in: "Gangliosides and Neuronal Plasticity," G. Tettamanti, R.W., Ledeen, K. Sandhoff, Y. Nagai, and G. Toffano, eds., Fidia Research Sereies, Vol. 6, Liviana Press, Padova.
Ledeen, R.W., 1983, Gangliosides, in: "Handbook of Neurochemistry," A. Lajtha, ed., Plenum Press, New York.
Leon, A., Facci, L., Toffano, G., Sonnino, S., and Tettamanti, G., 1981, Activation of (Na^+,K^+)ATPase by nanomolar concentrations of GM1, J. Neurochem., 37:350.
Leon, A., Facci, L., Benvegnù, D., and Toffano, G., 1982, Morphological and biochemical effects of gangliosides in neuroblastoma cells, Dev. Neurosci., 5:108.
Leon, A., Benvegnù, D., Dal Toso, R., Presti, D., Facci, L., Giori, O., and Toffano, G., 1984, Dorsal root ganglia and nerve growth factor: a model for understanding the mechanism of GM1 effects on neuronal repair, J. Neurosci. Res., 12:277.
Leon, A., Dal Toso, R., Presti, D., Benvegnù, D., Facci, L., Kirschner, G., Tettamanti, G., and Toffano, G., 1988, Development and survival of neurons in dissociated fetal mesencephalic serum-free cell cultures: II. Modulatory effects of gangliosides, J. Neurosci., 8:746.
Levi-Montalcini, R., 1987, The nerve growth factor thirty five years later, Science, 237:1154.
Lipartiti, M., Mazzari, S., Lazzaro, A., Zanoni, R., Seren, M.S., and Leon, A., 1989, Monosialogangliosides reduce NMDA neurotoxicity in neonatal rats, Soc. Neurosci. Abstr., 15:764.
Lombardi, G., Zanoni, R., and Moroni, F., 1989, Systemic treatments with GM1 ganglioside reduce quinolinic acid-induced striatal lesions in the rat, Eur. J. Pharmacol., 174:123.
Mahadik, S.P., and Karpiak, S.E., 1986, GM1 ganglioside enhances neonatal cortical development, Neurotoxicology, 7:161.
Mahadik, S.P., Vilim, F., Korenvosky, A., and Karpiak, S.E., 1988, GM1 ganglioside protects nucleus basalis from excitotoxin damage: reduced cortical cholinergic losses and animal mortality, J. Neurosci. Res., 20:479.
Manev, H., Favaron, M., Guidotti, A., and Costa, E., 1989, Delayed increase of Ca^{2+} influx elicited by glutamate: role in neuronal death, Mol. Pharmacol., 36:106.
Milani, D., Guidolin, D., Facci, L., Pozzan, T., Buso, M., Leon, A., and Skaper, S.D., 1990, Excitatory amino acid-induced alterations of cytoplasmic free Ca^{2+} in individual cerebellar granule neurons: role in neurotoxicity, J. Neurosci. Res., in press.
Morris, R.G.M., 1989, Synaptic plasticity and learning: selective impairment of learning in rats and blockage of long-term potentiation in vivo by the N-methyl-D-aspartate receptor antagonist AP5, J. Neurosci., 9:3040.
Olney, J.W., Labruyere, J., and Price, M.T., 1989, Pathological changes induced in cerebrocortical neurons by phencyclidine and related drugs, Science, 244:1360.

Petroni, A., Bertazzo, A., Sarti, S., and Galli, C., 1989, Accumulation of arachidonic acid cyclo- and lipoxygenase products in rat brain during ischemia and reperfusion: effects of treatment with GM1-lactone, J. Neurochem., 53:747.

Purpura, D.P., and Baker, H.J., 1977, Neurite induction in mature cortical neurones in feline GM1 ganglioside storage disease, Nature, 266:553.

Rothman, S.M., and Olney, J.W., 1986, Glutamate and the pathophysiology of hypoxic/ischemic brain damage, Ann. Neurol., 19:105.

Sabel, B.A., Slavin, M.D., and Stein, D.G., 1984, GM1 ganglioside treatment facilitates behavioral recovery from bilateral brain damage, Science, 225:340.

Seren, M.S., Rubini, R., Lazzaro, A., Zanoni, R., Fiori, M.G., and Leon, A., 1990, Protective effects of the inner ester derivative of monosialoganglioside following transitory forebrain ischemia in rats, Stroke, in press.

Shigeno, T., and Mima, T., 1988, Prevention of hippocampal cell death after cerebral ischemia by intraventricular administration of nerve growth factor, in: "Proceedings of the International Symposium on Alzheimer's Disease", H. Saininen, ed., University of Kuopio Press, Kuopio (abstract).

Skaper, S.D., and Varon, S., 1985, Ganglioside GM1 overcomes serum inhibition of neuritic outgrowth, Internatl. J. Dev. Neurosci., 3:187.

Skaper, S.D., Katoh-Semba, R., and Varon, S., 1985, GM1 ganglioside accelerates neurite outgrowth from primary peripheral and central neurons under selected culture conditions, Dev. Brain Res., 23:19.

Skaper, S.D., Leon, A., and Toffano, G., 1989a, Ganglioside function in the development and repair of the nervous system: from basic science to clinical application, Mol. Neurobiol., 3:173.

Skaper, S.D., Facci, L., Milani, D., and Leon, A., 1989b, Monosialoganglioside GM1 protects against anoxia-induced neuronal death in vitro, Exp. Neurol., 106:297.

Skaper, S.D., Facci, L., and Leon, A., 1990, Gangliosides attenuate the delayed neurotoxicity of aspartic acid in vitro, Neurosci. Lett., in press.

Spoerri, P.E., 1986, Facilitated establishment of contacts and synapses in neuronal cultures: ganglioside-mediated neurite sprouting and outgrowth, in: "Gangliosides and Neuronal Plasticity," G. Tettamanti, R.W. Ledeen, K. Sandhoff, Y. Nagai, and G. Toffano, eds., Fidia Research Series, Vol. 6, Liviana Press, Padova.

Svennerholm, L. , 1963, Chromatographic separation of human brain gangliosides, J. Neurochem., 10:613.

Toffano, G., Savoini, G., Moroni, F., Lombardi, G., Calzà, L., and Agnati, L.F., 1983, GM1 ganglioside stimulates regeneration of dopaminergic neurons in the central nervous system, Brain Res., 261:163.

Vaccarino, F., Guidotti, A., and Costa, E., 1987, Ganglioside inhibition of glutamate-mediated protein kinase C translocation in primary cultures of cerebellar neurons, Proc. Natl. Acad. Sci. USA, 84:8707.

Vantini, G., Fusco, M., Bigon, E., and Leon, A., 1988, GM1 ganglioside potentiates the effect of nerve growth factor in preventing vinblastine-induced sympathectomy in newborn rats, Brain Res., 488:252.

Weihmuller, F.B., Hadjicontantinou, M., Bruno, J.P., and Neff, N.H., 1988, Administration of GM1 ganglioside eliminates neuroleptic-induced sensorimotor deficits in MPTP-treated mice, Neurosci. Lett., 92:207.

Willinger, M., and Schachner, M., 1980, GM1 ganglioside as a marker for neuronal differentiation in mouse cerebellum, Dev. Biol., 74:101.

NERVE GROWTH FACTOR IN CNS REPAIR AND REGENERATION

Silvio Varon, Theo Hagg, Marston Manthorpe

Department of Biology, School of Medicine, M-001
University of California, San Diego
La Jolla, CA 92093

INTRODUCTION

Increasing recognition is given by neuroscientists to the fact that considerable plasticity is retained even in the adult mammalian central nervous system (CNS), i.e. that maintenance and function of adult CNS neurons continue to depend on and be regulated by agents presented to them by their humoral, extracellular matrix and cellular micro-environments. One major class of macro-molecular regulatory agents is that of special proteins controlling the performance of selected target neurons named neuronotrophic factors (NTFs), of which Nerve Growth Factor (NGF) is the first discovered and best characterized representative (Levi-Montalcini, 1987). NGF and other newly recognized NTFs have been traditionally investigated in the context of lesions in the peripheral nervous system (PNS) and in the course of their early development, both in vivo and in vitro. A recently articulated CNS neuronotrophic hypothesis (Varon et al., 1984) has proposed that i) adult CNS neurons also require support from their own NTFs, and ii) relative deficits of endogenous NTF may underlie neuronal damages in experimental and pathologic situations, and iii) exogenous NTF administration may reduce or prevent neuronal damage in such situations.

RESPONSES TO NGF BY ADULT RAT CNS CHOLINERGIC NEURONS

Much evidence for this hypothesis has been gathered from the investigations of certain NGF-responsive cholinergic neurons (for reviews see: Whittemore and Seiger, 1987; Varon et al., 1989; Hefti et al., 1989; Hagg et al., 1991). Among them are basal forebrain neurons in the medial septum/diagonal band (which project to the hippocampal formation) and in the nucleus basalis (projecting to the cerebral cortex), as well as interneurons in the neostriatum (synapsing on other striatal neurons). The projective cholinergic neurons find NGF in their innervation territories, take it back to their cell bodies by retrograde axonal transport, display NGF receptors on soma and processes, and can respond to NGF with increases in their transmitter-synthesizing enzyme choline acetyltransferase (ChAT). Recent in vivo work on the adult rat has demonstrated that these CNS cholinergic neurons depend on NGF when acutely or chronically damaged, as well as in their normal state.

Acute damage

Partial or complete transection of the fimbria-fornix tract which carries axonal projections from medial septum (MS) -- as well as diagonal band -- cholinergic neurons to the hippocampal formation (HF) deprives the axotomized neurons of their HF-supplied endogenous NGF and leads to the disappearance of MS neuronal cell bodies recognizable by their acetylcholinesterase (AChE) or their ChAT content (Gage et al., 1986, 1988). Intraventricular administration of NGF prevents part or all of the numerical loss of AChE or ChAT-positive MS neurons (Hefti, 1986; Williams et al., 1986; Kromer, 1987). The axotomy-induced loss and its prevention by exogenous NGF do not reflect an actual death but rather a reduction of their cholinergic marker content beyond detectable levels, since an already incurred loss can be reversed by NGF treatments that are delayed by weeks or even months (Hagg et al., 1988, 1989a). Reduction and restoration of the ChAT marker are paralleled by corresponding changes in the neuronal content of immunostainable, lower affinity NGF receptor (NGFR) as well as in the neuronal somal size (Hagg et al., 1989a). These findings indicate that i) NGF exerts a truly trophic action on these MS neurons, involving several neuronal properties, and ii) opportunities to correct neuronal damage persist for a very long time.

Chronic damage

The NGF-sensitive basal forebrain cholinergic neurons play a crucial role in the cognitive functions of the brain. Cognitive deficits are accompanied by cholinergic deficits in the acute experimental lesions, in brain aging and in Alzheimer's type dementia. Aged rats with an overt impairment in spatial memory have provided an opportunity to test the potential benefits of NGF (Fischer et al., 1987). Four weeks of continuous intraventricular NGF infusion improved their memory-related response in a water maze task to practically the level of a normal, unimpaired rat. Concurrent with this improvement was a significant reversal from a reduced somal size of striatal and nucleus basalis cholinergic neurons. Other experiments with acute cholinergic lesions had also revealed a positive effect of NGF administration (Will and Hefti, 1985).

Normal animal

According to the CNS trophic hypothesis, normal CNS cholinergic neurons are serviced by the available endogenous NGF. One may speculate that endogenous NGF levels are normally adequate to elicit a maximal performance from these neurons. Alternatively, upward changes in the available NGF may provide yet another level of extrinsic neuronal regulation. Evidence for the latter case is accumulating from various experiments. In the delayed NGF treatment of axotomized MS neurons, the NGF-mediated reversal if injury-elicited somal atrophy proceeds above the normal size level to actual hypertrophy of some of these cholinergic neurons (Hagg et al., 1989a). Neostriatal cholinergic neurons with no experimental injury respond to intraventricular NGF with a considerable increase in somal size, as well as an enhanced ChAT immunoreactivity (Hagg et al., 1989b). NGF treatments in intact rats cause MS neurons to increase both their mRNA for NGF receptor (in the septum)(Cavicchioli et al., 1989) and their ChAT content (in their hippocampal innervation territory)(Fusco et al., 1989).

CNS AXONAL REGENERATION IN THE ADULT RAT AND THE ROLE OF NTFs

Adult mammalian CNS axons are apparently unable to regenerate over long distances within the CNS tissue itself. This failure, initially attributed to an inability of the adult CNS neurons themselves, is now recognized as largely due to an adverse or deficient quality of the adult CNS environment. When brain or spinal cord neurons are presented with PNS tissue, i.e. a segment of peripheral nerve, they can regrow axons for several centimeters much as PNS axons are known to do (Aguayo, 1985). The question, then, is what is responsible for the apparent "resistance" of adult CNS tissue to be re-innervated. One possibility is the presence of axonal growth inhibitors, possibly related to the presence of CNS myelin or the myelin-producing CNS oligodendroglia (Schwab, 1990). An extension of the CNS neuronotrophic hypothesis proposed an alternate though not necessarily conflicting view, namely that CNS tissue resistance reflects an inadequate level of local NTFs which would be required for vigorous axonal growth.

The septo-hippocampal regeneration model

An opportunity to test such a speculation was again provided by the adult septo-hippocampal cholinergic model (Hagg et al., 1990a), the only one for which we currently know the appropriate trophic factor, NGF. As Figure 1 illustrates in a sagittal view of the model, a complete and bilateral fimbria-fornix transection creates a cavity which separates the

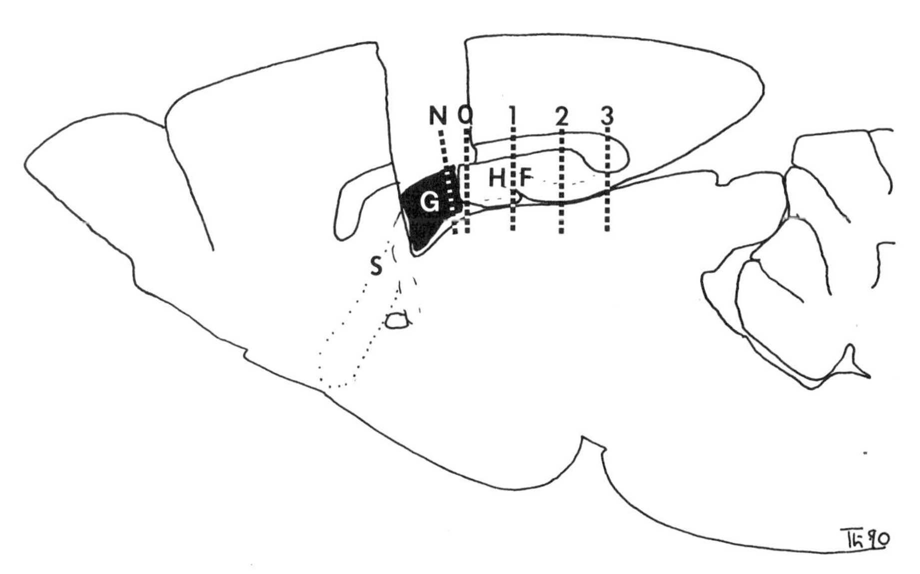

Figure 1. Schematic sagittal representation of the rat brain and septo-hippocampal model used for the evaluation of cholinergic fiber regeneration. Peripheral (sciatic) nerve grafts (G) are placed in a lesion cavity between the septum (S) and the hippocampal formation (HF). In sagittal sections counts are made of the number of AChE-positive fibers intersecting imaginary lines through the hippocampal end of the nerve graft (N), and at 0.1 (0), 1, 2 and 3 mm from the rostral tip of the hippocampal formation.

two septi from their corresponding hippocampal formations. The cavity on either or both sides can be bridged with a short segment of peripheral (sciatic) nerve the two cut faces of which are closely apposed to the septal and hippocampal sides of the cavity, respectively. At various post-implantation times, animals are sacrificed and sagittal sections are stained for AChE to visualize cholinergic axons re-growing from septum to HF. Axonal regeneration progress can be quantified by counting in sagittal sections the number of AChE-positive fibers in the nerve bridge and in the hippocampus at various distances from the septum, specifically: i) near the end of the nerve bridge (line N), to evaluate the number of fibers that crossed the bridge and made themselves potentially available to invade the HF; ii) 0.1 mm into the HF (line 0), to ascertain how many fibers actually enter the HF; iii) 1,2 and 3 mm inside the HF, to measure the penetration of hippocampal tissue by the regenerating axons.

A first study (Hagg et al., 1990a) was aimed at determining the time course of axonal regeneration across a nerve bridge (using gelfoam as a "control" bridge material, in the absence of exogenous NGF infusions. In control (gelfoam-bridged) rats, practically no cholinergic fibers were seen after 1 month within the gelfoam and none was left in the denervated HF. A remarkably different regeneration performance was seen in the nerve-bridged animals, as shown in Figure 2. In the nerve bridge, cholinergic fibers began to appear after 1 week, reached a maximal number by the end of 1 month and maintained it over the total examined period of 6 months. At the entrance of the HF and 1 mm inside it, regenerating axons began appearing after 2 weeks, were numerous by 1 month (although considerably fewer than those available at the bridge exit at that time) and continued to increase in numbers to approximate the number of bridge-available fibers by 6 months. Deeper into the HF, however, the number of

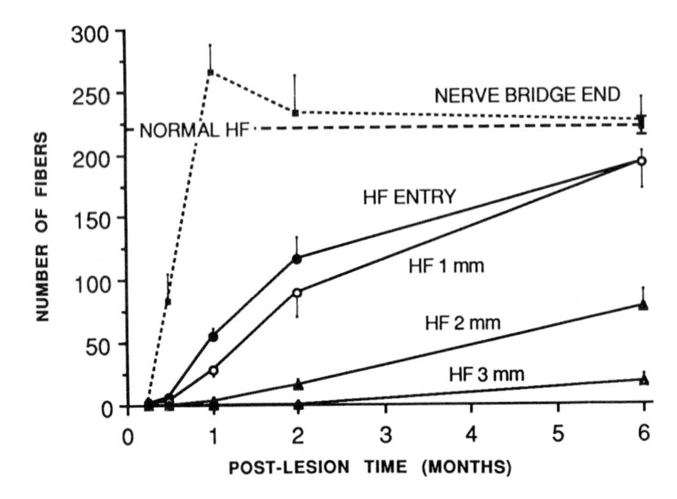

Figure 2. Quantification of regenerated cholinergic fibers in the septo-hippocampal nerve graft and hippocampal formation. Presented are the maximal number of AChE-positive fiber intersections counted at different postlesion times in single sagittal sections at the nerve bridge end, and in the hippocampal formation (HF) at 0.1 (HF entry), 1, 2 and 3 mm from its rostral tip. The broken horizontal line indicates corresponding numbers in normal unoperated animals. Bars = SEM.

regenerating AChE-positive axons increased much more slowly with time, to reach by 6 months only about one third and one tenth of the ingrown fibers at the 2 and 3 mm lines, respectively. These findings suggested that i) the HF "resistance" to penetration was not homogeneous, but rather varied in space and/or time, and that ii) it could, therefore, be

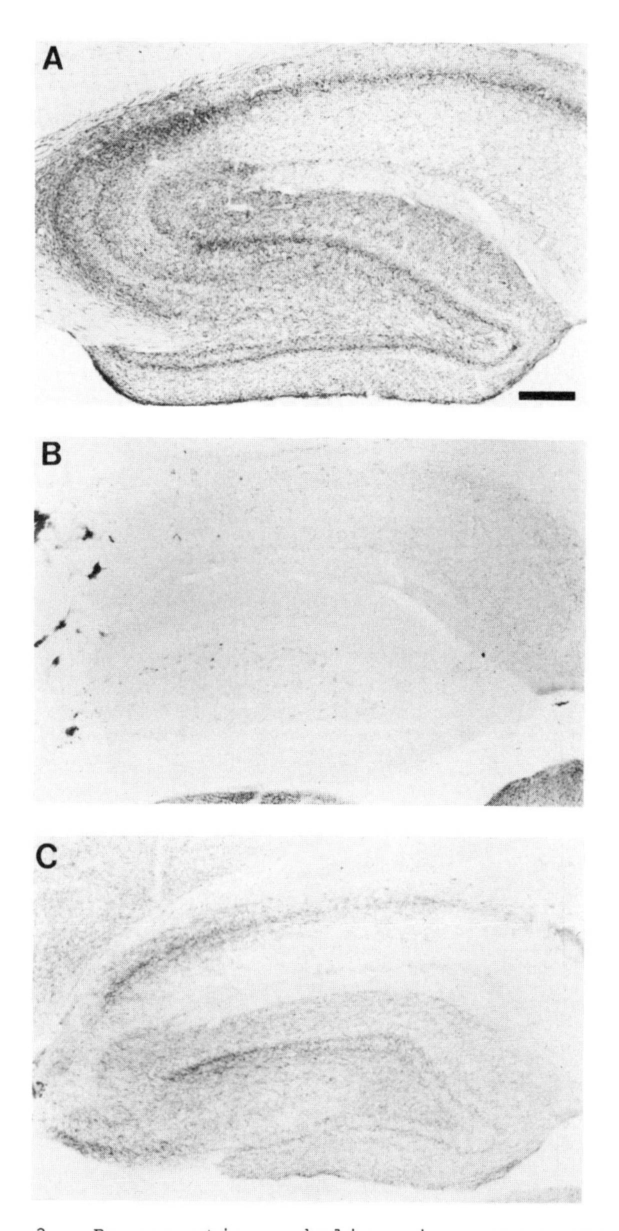

Figure 3. Regenerating cholinergic axons establish approximately normal densities and distributions in the hippocampal formation. Compared to the hippocampal formation of normal animals (A), no AChE-positive fibers are revealed after of one month in control gelfoam implanted animals (B). However, six months after nerve graft implantation (C) AChE-positive fibers had re-innervated the dorsal hippocampal formation and formed a fiber plexus with a distribution and density similar to normal ones. Bar = 250 µm.

susceptible to experimental manipulations. Interestingly, the fiber density and distribution patterns achieved by regenerating cholinergic axons in the most rostral portion of the HF by 6 months (Figure 3) appeared very similar to those of normal, unoperated animals -- encouraging the perception that local patterning cues may be retained in the adult brain tissue even after denervation.

NGF roles in the regeneration model

Preliminary examinations of this septo-hippocampal model after 1 month of unilateral intraventricular infusion (at the level of the septum) of NGF versus only vehicle have yielded two important observations. One was that the presence of the nerve bridge did not help by itself to prevent the disappearance of axotomized MS neurons. This suggests that the axons regenerating into the bridge were originating from the remaining (less or not damaged) septal cholinergic neurons. The second observation was the intraventricular NGF infusion, beside offering the expected protection to the axotomized neurons, also reduced substantially the number of cholinergic axons regenerating into the nerve bridge. This reduction contrasted with a concurrent marked appearance of a ChAT-positive fiber plexus in the dorsolateral septum near the NGF-infused ventricular wall -- a phenomenon repeatedly observed in previous work (Williams et al., 1986; Gage et al., 1988; Hagg et al., 1989b). The combined observations suggested that exogenous NGF may not only encourage axonal growth but also direct it toward the location at which it is presented. One can thus speculate that cholinergic axonal regeneration may require NGF to be available in different locations and, possibly, in appropriate temporal sequences. Specifically, we postulated that regeneration requires NGF in the septum (where the cell bodies lie), in the bridge (where the axons are expected to course, and in the hippocampus (which the axons are supposed to be re-innervated). The latter two locations have been further investigated.

NGF and the nerve bridge

The nerve bridge may be viewed as comprising two components: i) a mechano-chemical scaffold providing a neurite-promoting, axially oriented matrix (Varon and Williams, 1986), and ii) living cells (presumably mainly Schwann cells) representing a local source of humoral factors such as NGF (Varon and Manthorpe, 1982; Muir et al., 1989). It is possible to subject pieces of sciatic nerve to a serie of treatments that would destroy its cells as well as its axons, remove the cellular debris and leave in place axially oriented, basal lamina lined tubular spaces into which axons could regrow. Such an "acellular" nerve preparation was tested for its bridge competence without or with exogenous NGF supplementation (Hagg et al., 1990b). As seen in Figure 4A, the acellular bridge not supplemented with NGF failed to promote its invasion by septal cholinergic axons. In contrast, pre-incubation of the material with NGF before its implantation as a bridge led to vigorous cholinergic growth into it (Figure 4B), quantitatively similar to that observable with a fresh, cellular graft of sciatic nerve. Thus, i) elimination of the living cells did impair the nerve material competence as a CNS regeneration bridge, ii) replacement of the cells with one of their products, NGF, was sufficient to restore the bridge competence, and iii) the acellular preparation remains a suitable "scaffold" when supplemented with the appropriate neuronotrophic factor. This last point opens up future opportunities to identify the relevant constituents of such a

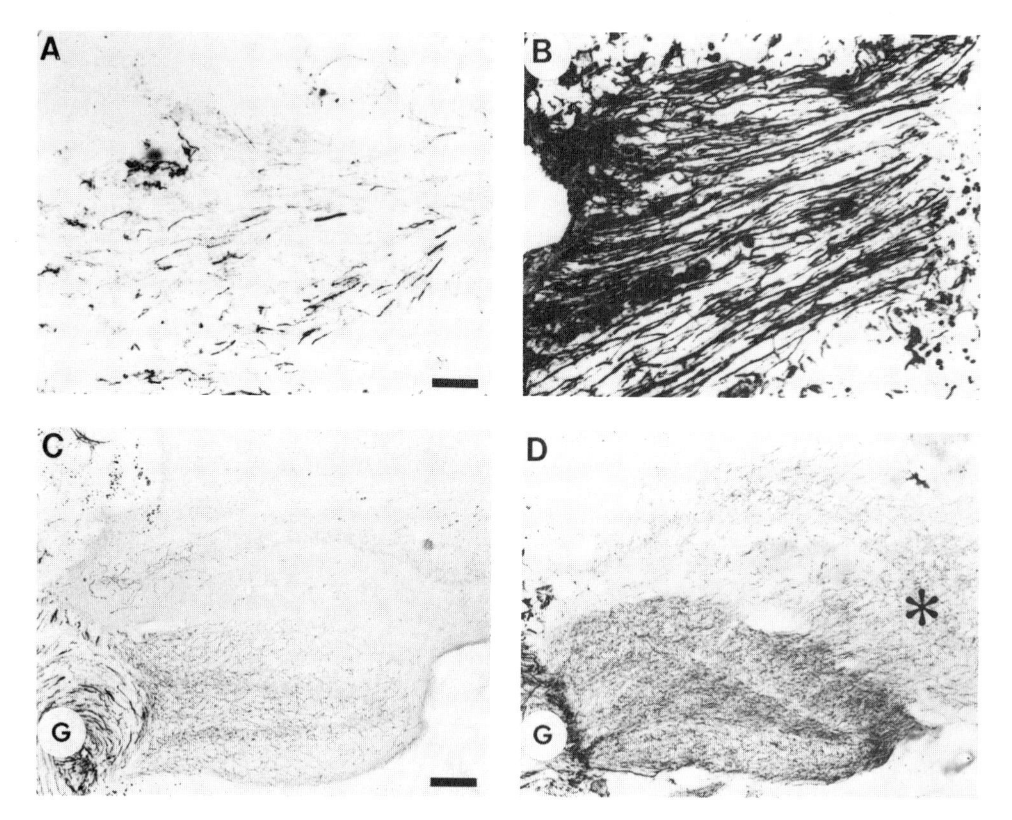

Figure 4. Nerve growth factor promotes cholinergic axonal regeneration into nerve grafts and hippocampal formation. A) One month after implantation, acellular grafts contained only a few AChE-positive fibers. B) In marked contrast, NGF-pre-soaked acellular grafts contained fiber growth comparable to that seen in fresh cellular nerves. Bar for A) and B) = 100 μm. C) One month after implantation of a sciatic nerve graft (G) the dorsal hippocampal formation contains a modest amount of AChE-positive fibers. D) Nerve growth factor infusion for one month into the dorsal hippocampal formation, in addition to a nerve graft (G), considerably increased hippocampal re-innervation. Asterisk = infusion site. Bar for C) and D) = 250 μm.

scaffold and, eventually, to generate equivalent surrogate bridge materials for CNS (and PNS) axonal regeneration.

NGF and the hippocampal tissue

The speculation that CNS tissue resistance to cholinergic reinnervation could be due to insufficient local levels of endogenous NGF was tested next (Hagg et al., 1990c). Animals supplied with fresh (cellular) nerve bridge were subjected to continuous intra-hippocampal NGF infusions for 1 month before being analyzed. Figure 4(C,D) shows AChE-stained sagittal sections of the HF of vehicle-infused and NGF-infused rats, respectively. The massive increases of intra-hippocampal fibers elicited by the local administration of NGF was nearly equal at this 1 month time point to that observable without NGF after a 6 month period (cf Fig. 3, C). The NGF-mediated increase represented genuine axonal regrowth and not just an NGF-induced rise in AChE content (by which the fibers are visualized), since a 3-day local infusion of NGF starting 1 month after implantation did not cause increased visualization of pre-grown fibers. The increased HF reinnervation was not accompanied by an increased number of fibers brought across by the nerve bridge, but rather reflected a more aggressive advance of the fibers entering the HF into its deeper regions.

CONCLUSIONS AND PROJECTIONS

The observations reviewed here make it clear that NGF plays a critical role for cholinergic neurons in the adult mammalian CNS, which concerns i) maintenance, size and functional properties of these neurons in the normal state, ii) protection against and reversal of damages incurred in acute and chronic situations, and iii) axonal regeneration within the appropriate adult CNS innervation territory. It is plausible to expect that such NGF benefits could eventually be extended to human pathologic situations such as Alzheimer's dementia. One important advance in such a direction would be the acquisition of better administration modalities, and several efforts to that aim are underway in various laboratories.

The data obtained thus far are limited to NGF and cholinergic neurons, the only currently identified combination of trophic factor and target neuron. To attempt generalizing the validity of the CNS neuronotrophic hypothesis it will be necessary to extend the present findings to other model systems by, for example, i) demonstrating the usefulness of NGF for CNS target neurons other than the cholinergic ones, and ii) investigating neuronotrophic factors other than NGF for similar competences on their respective target neurons. A successful outcome of such undertakings holds considerable promise for the treatment of a number of human pathologic conditions, from Parkinson's and motor neuron (ALS) diseases to optic nerve and spinal cord injuries. The field is young and rich with potential rewards.

ACKNOWLEDGEMENTS:
Work described here was partially supported by NINDS grants NS-16349 and -25011, and NSF grant BNS-88-08285

REFERENCES

Aguayo, A.J., 1985, Axonal regeneration from injured neurons in the adult mammalian central nervous system, in : "Synaptic Plasticity," C.W. Cotman, ed., Guillford Press, New York, pp. 457-484.

Cavicchioli, L., Flanigan, T., Vantini, G., Fusco, M., Polato, P., Toffano, G., Walsh, F., and Leon, A., 1989, NGF amplifies the expression of NGF receptor messenger RNA in mammalian forebrain cholinergic neurons, Eur. J. Neurosci., 1: 258-262.

Fischer, W., Wictorin, K., Björklund, A., Williams, L.R., Varon, S., and Gage, F.H., 1987, Intracerebral infusion of nerve growth factor ameliorates cholinergic neuron atrophy and spatial memory impairments in aged rats, Nature, 329: 65-68.

Fusco, M., Oderfeld-Nowak, B., Vantini, G., Schiavo, N., Gradowska, M., Zaremba, M. and Leon, A., 1989, Nerve growth factor affects uninjured, adult rat septo-hippocampal cholinergic neurons, Neurosci. , 33: 47-52.

Gage, F.H., Wictorin, K., Fischer, W., Williams, L.R., Varon, S., and Björklund, A., 1986, Retrograde cell changes in medial septum and diagonal band following fimbria-fornix transection : quantitative temporal analysis, Neurosci. 19: 241-255.

Gage, F.H., Armstrong, D.M., Williams, L.R., and Varon, S., 1988, Morphological responses of axotomized septal neurons to nerve growth factor. J. Comp. Neurol. 269: 147-155.

Hagg, T., Vahlsing, H.L., Manthorpe, M., and Varon, S., 1988, Delayed treatment with nerve growth factor reverses the apparent loss of cholinergic medial septum neurons after brain damage, Exp. Neurol., 101: 303-312.

Hagg, T., Fass-Holmes, B., Vahlsing, H.L., Manthorpe, M., Conner, J.M., and Varon, S., 1989a, Nerve growth factor (NGF) reverses axotomy-induced decreases in choline acetyltransferase, NGF-receptor and size of medial septum cholinergic neurons, Brain Res., 505 : 29-38.

Hagg, T., Hagg, F., Vahlsing, H.L., Manthorpe, M., and Varon, S., 1989b, Nerve growth factor effects on cholinergic neurons of neostriatum and nucleus accumbens in the adult rat, Neurosci., 30 : 95-103.

Hagg, T., Vahlsing, H.L., Manthorpe, M., and Varon, S., 1990a, Septo-hippocampal cholinergic axonal regeneration through peripheral nerve bridges : quantification and temporal development, Exp. Neurol., 109: 153-163.

Hagg, T., Gulati, A.K., Behzadian, M.A., Vahlsing, H.L., Varon, S., and Manthorpe, M., 1990b, Nerve growth factor promotes CNS axonal regeneration into acellular peripheral nerve grafts, Exp. Neurol., (submitted).

Hagg, T., Vahlsing, H.L., Manthorpe, M., and Varon, S., 1990c, Nerve growth factor infusion into the denervated adult rat hippocampal formation promotes its cholinergic reinnervation, J. Neurosci., 10 : 3087-3092.

Hagg, T., Manthorpe, M., Vahlsing, H.L., and Varon, S., 1991, Nerve growth factor roles for cholinergic axonal regeneration in the adult mammalian central nervous system, Comm. Dev. Neurobiol., (in press).

Hefti, F., 1986, Nerve growth factor promotes survival of septal cholinergic neurons after fimbrial transections, J. Neurosci., 6: 2155-2162.

Hefti, F., Hartikka, J. and Knusel, B., 1989, Function of neurotrophic factors in the adult and aging brain and their treatment of neurodegenerative diseases, Neurobiol. Aging, 10: 515-533.

Levi-Montalcini, R., 1987, Nerve growth factor 35 years later, Science, 237: 1154-1162.

Muir, D., C. Gennrich, S. Varon, and M. Manthorpe, 1989, Rat sciatic nerve Schwann cell cultures. Responses to mitogens and production of trophic and neurite-promoting factors, Neurochem. Res., 14: 1003-1012.

Schwab, M.E., 1990, Myelin-associated inhibitors of neurite growth, Exp. Neurol., 109 : 2-5.

Varon, S., and Manthorpe, M., 1982, Schwann cells: an in vitro perspective, Adv. Cell Neurobiol., 3: 35-95.

Varon, S., Manthorpe, M., and Williams, L.R., 1984, Neuronotrophic and neurite-promoting factors and their clinical potentials, Dev. Neurosci., 6: 73-100.

Varon, S., and Williams, L.R., 1986, Peripheral nerve regeneration in a silicon model chamber: cellular and molecular aspects. Peripheral Nerve Repair and Regeneration, 1: 9-25.

Varon, S., Hagg, T., and Manthorpe, M., 1989, Neuronal growth factors, in: "Neural Regeneration and Transplantation," F.J. Seil, ed., Frontiers Clinical Neurosciences, Vol. 6, Alan Liss, New York, pp. 101-121.

Whittemore, S.R., and Seiger, Å., 1987, The expression, localization and functional significance of ß-nerve growth factor in the central nervous system, Brain Res. Rev., 12: 439-464.

Will, B., and Hefti, F., 1985, Behavioral and neurochemical effects of chronic intraventricular injections of nerve growth factor in adult rats with fimbria lesions, Behav. Brain Res., 17: 17-24.

Williams, L.R., S. Varon, G.M. Peterson, K. Wictorin, W. Fisher, A. Björklund and F.H. Gage, 1986, Continuous infusion of Nerve Growth Factor prevents forebrain neuronal death after fimbria-fornix transection, Proc. Natl. Acad. Sci. USA, 83: 9231-9235.

ORDERED DISORDER IN THE AGED BRAIN

Luciano Angelucci, Sebastiano Alemà, Laura Ferraris[a], Orlando Ghirardi[a], Assunta Imperato, Maria Teresa Ramacci[a], Maria Grazia Scrocco[a], and Mario Vertechy[a]

Farmacologia 2a, Facoltà di Medicina, Rome University "La Sapienza" and [a] Institute for Research on Senescence, Sigma Tau, Pomezia, Italy

INTRODUCTION

A perusal of experimental work on the neurobiology of brain aging shows that several investigations have been carried out on the theoretical bias that the aging process is quantitative in its nature, thus consisting in the progressive generalized weakening of biochemical (metabolic and specific) activities in the neuron, and leading to the impoverishment of brain functions which are typical of the species. In view of the perennity of the neuron, the factors underlying the aging process have been singled out in the deterioration of the nuclear function, especially with regard to protein synthesis, in reduced resistance to and increased formation of toxic wastes, mostly oxygen radicals, and consequent irreversible damage to enzymes, plasma membrane components, etc. Environmental impact on aging is believed to be exerted through these factors. The homogeneous character of these changes has recently been questioned, as in contrast to the ascertained prevailing selectivity of the changes for brain regions and cell types.[1] Furthermore, the above interpretative frame of the nature and causation of the brain aging process disagrees with several experimental findings, notably the following. There is some indication that cell aging can be a genomically triggered active process: senescence can be induced in immortal cells by hybridization with human fibroblasts, provided the former ones retain human chromosome 1,[2] or in human endometrial cancer cells microcell-transferred with this chromosome.[3] Age-dependent changes in RNA production are restricted to some brain regions and neuronal cell types,[1] as is reduction in protein synthesis.[4] It should be noted that the latter differentially affects the various neuroanatomical components of a neurophysiological system, e.g. the limbic, or the nigro-striatal one. Age-dependent reduction in glucose utilization affects restricted brain structures,[5] which are not necessarily those affected by reduction of protein synthesis.[4] The selectivity of some age-dependent changes may be evident even at the level of a single

Plasticity and Regeneration of the Nervous System
Edited by P.S. Timiras *et al.*, Plenum Press, New York, 1991

cell type. For instance, slow axonal transport of cytoskeletal elements is retarded during aging, while the fast axonal transport of membranous vesicles remains unchanged.[6] The dichotomy in age-dependent changes in axonal transport might explain why choline-acetyltransferase and acetylcholine-esterase (membrane associated proteins, fast transport) activities are not substantially reduced in the aged brain, while glycolytic activity, structural components of the neuron such as growth cone, dendritic spines, synaptic terminals, and functional components such as recycling of synaptic vesicles, receptor mediated endocytosis, formation and remodelling of synaptic connections (cytoplasmic proteins, slow transport) are affected. Finally, age-dependent genomic alterations are possible, so as to lead to the emergence of a heterozygous phenotype from a homozygous one, probably due to somatic intrachromosal gene conversion: it is the case of the Brattleboro rat which with aging progressively acquires the ability to synthesize a normal vasopressin precursor.[7]

The free-radical hypothesis of aging[8] cannot be generalized in the form of a life-long oxidative stress to the various animal species because important exceptions can be encountered, at least in the case of relatively low rates of oxygen metabolism,[9] although even in the rat, a species with a high metabolic rate, no concordant results have been reported.

With regard to the aging of the brain cholinergic system as the basis of age-dependent cognitive impairment, dysfunction of this system in the aged rat is not due to substantial quantitative reduction in choline-acetyltransferase activity[10] but to the impairment of mechanisms coupling acetylcholine storage and functional release.[11]

All the above considerations support the interpretation of the brain aging process as a non-homogeneous one with regard to the various brain regions, the different components of fundamental cellular activities, and its causative mechanisms. In the following we make some experimental contributions to this interpretation.

EXPERIMENTAL FINDINGS

a) Peroxidative activity in aged rat brain

Assuming malonyl dialdehyde (MDA) formation to be an index of the ability of brain tissue to produce and inactivate superoxide, we have found differences between the young (5 month age) and the old (24 month age) male Fischer 344 rat, as shown in Fig. 1. Lipid peroxidation in the old rat was significantly lower in the frontal cortex, striatum and hippocampus, both in basal and stimulated condition, in the hypothalamus in basal condition, in the substantia nigra in the stimulated condition. The pineal gland was the only tissue in which a tendency towards an increased lipid peroxidation in old animals was evident (not shown here). From this study it was apparent that basal lipid peroxidation has a widely discrete distribution in the brain;

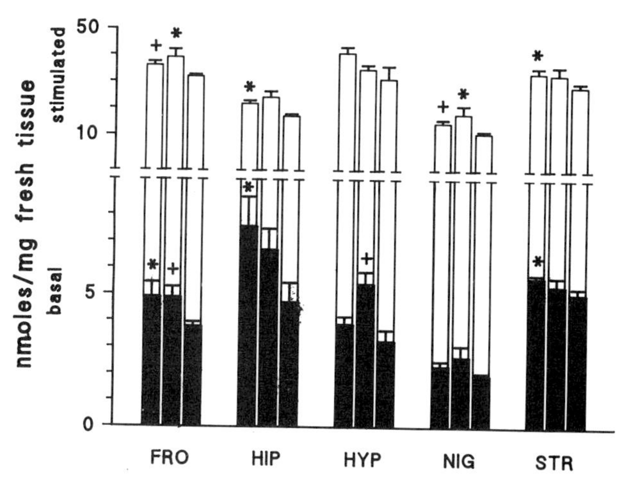

Fig. 1. Basal (filled column) and ADP/Fe^{++} stimulated (plain column) MDA nmoles/mg protein in various brain areas (frontal cortex, striatum, hippocampus, hypothalamus and s. nigra homogenates) of 5 month (n=14, plain columns), 14 month (n=14, hatched columns) and 24 month old (n=19, filled columns) male Fischer 344 rats. * and + : 5 and 1 per cent significant difference versus 24 month old, respectively.

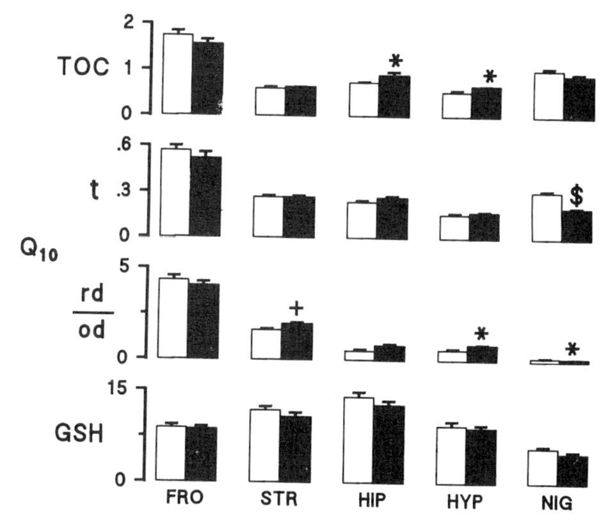

Fig. 2. Antioxidants (tocopherol, ubiquinone: total and reduced/ oxided ratio, and glutathione) nmoles/mg protein in young (5-6 month age, n=8, plain columns) and old (28-30 month age, n=7-8, filled columns) male Fischer 344 rats. * , + , and $: 5, 1 and 0.1 per cent significant difference versus young, respectively.

paradoxically the substantia nigra, i.e. the structure credited with a faster rate of aging, had a much lower MDA production than the hippocampus (2.29 against 7.55 nmoles/mg protein), notwithstanding, as shown in Fig. 2, the lowest content in the former, among the various brain structures, in reduced glutathione and the lowest reduced/oxidation ubiquinone ratio, which are the main scavengers of superoxide. Fig. 2 also shows that no reduction in antioxidant concentrations was found in the brain of the aged rat, apart the substantia nigra in which a decrease in total Q_{10} and in the reduced/ oxidized ratio was found, just the opposite of the change in this ratio in the striatum and hypothalamus. It is quite remarkable that scavenger tocopherol level was increased in the hippocampus and the hypothalamus, and that glutathione, credited with a major role in anti-radical protection, showed no reduction in the brain of the aged rat. It is again evident from this study that antioxidants have a non-uniform distribution in the various brain areas, and that possible age-dependent changes in their levels affect certain areas preferentially. With regard to age-dependent changes in antioxidant enzymes, Fig. 3 shows that only catalase was found to have been reduced in the hippocampus and striatum, whereas superoxide-dismutase and glutathione-peroxidase were unchanged in all areas investigated. Taking into account that cytochrome-oxidase also was unchanged, while glutathione-reductase was increased in all the areas investigated, and that catalase carries out one thousandth of all antioxidant enzyme activity in the brain, one can conclude that the anti-peroxidative balance is not impaired in an age-dependent way, as also indicated by the unchanged Mn^{++} -superoxide - dismutase / cytochrome-oxidase ratio. Even from the findings of this study it is evident that age-dependent changes in antioxidant enzymes, when actual as in the case of catalase, do not uniformly affect the various brain areas. All in all, the findings we have obtained in the Fischer rat indicate that peroxidative damage might not be an important agent in the brain aging process.

b) Mitochondrial dehydrogenase activities in the aged brain

With histoenzymatic techniques we have found, as shown in Fig. 4, that the enzymes involved in the mitochondrial utilization of energetic substrates are reduced in the aged rat's cortex neuropil, hippocampus pyramidal neurons and caudatus perikaria, although to a very different degree and especially in the latter two cases. Glycerolphosphate-dehydrogenase is the most affected. This finding indicates that the primary change characterizing the aged brain might be a deficiency in mitochondrial energy metabolism. This hypothesis is supported by the reduced transcription of mitochondrial DNA in the senescent brain.[12]

c) Loss of cholinergic receptors and function of the cholinergic synapse in the hippocampus of the aged brain

With measurement of binding parameters on membrane preparations we have found, as shown in Fig. 5, that the number of muscarinic binding

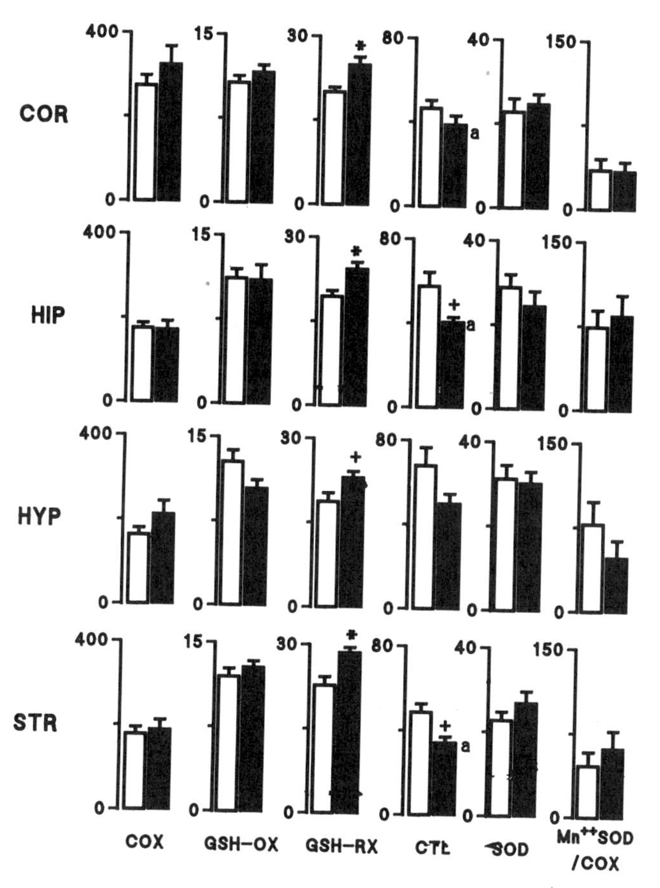

Fig. 3. Antioxidant enzymes activities (cytochrome-oxidase: moles of cytochrome C oxidized x min^{-1} x mg $protein^{-1}$; superoxide-dismutase: units x mg $protein^{-1}$; catalase: moles H_2O_2 consumed x min^{-1}x mg $protein^{-1}$; glutathione-peroxidase and reductase: moles NADPH oxidized x min^{-1} x mg $protein^{-1}$) in various brain areas (cortex, hippocampus, hypothalamus, striatum) of 4 month old (n=6, plain columns) and 26 month old (n=6, filled columns) male Sprague-Dawley rats. * and + : 5 and 1 per cent significant difference versus young, respectively. [a]: values at 15 month age; no difference was present at 27 month age; for all other enzyme activities value at 15 month age was not different versus 4 month age.

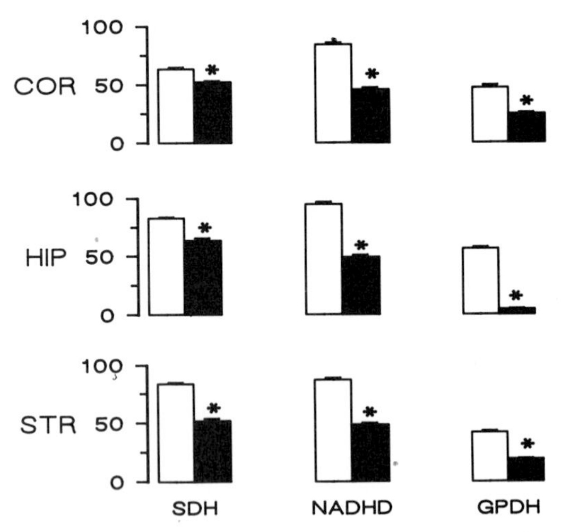

Fig. 4. Mitochondrial dehydrogenase activities in cortex neuropil, hippocampus pyramidal cells and caudatus perikaria of 3 month (n=5, plain column) and 22 month old (n=8, filled column) female Wistar rats. Histoenzymatic evaluation, color intensity in arbitrary units. * : 0.1 per cent significant difference versus young.

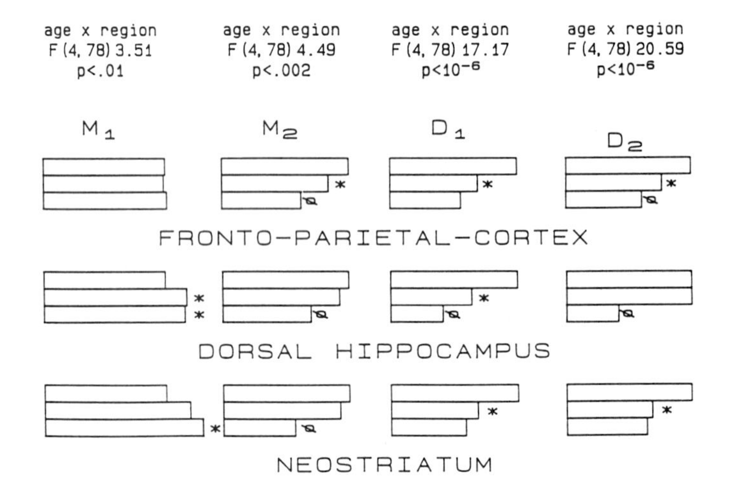

Fig. 5. Number of binding sites (B_{max} : fmoles/mg tissue) for M_1 (0.1-5 nM ^{3}H-QNB plus 100 µM carbachol) and M_2 (same radio ligand plus 300 nM pirenzepine) muscarinic receptors on crude membrane preparations from different brain areas in the male Sprague-Dawley rat at the ages of 4 (upper bar), 12 (middle bar) and 22 months (lower bar). Values normalized on 4 months; n=9-10. * and ∅ : significantly different from 4 or 12 months, respectively. Also shown, the number of binding sites (B_{max} : fmoles/mg tissue) for D_1 (0.05-10 nM ^{3}H-SCH 23390) and D_2 (1-40 nM ^{3}H-spiroperidol plus 40 nM ketanserine) dopaminergic receptors.

sites changes in an age-dependent way, but in a different direction according to the subtype of receptor: M_1 is increased in the neostriatum and dorsal hippocampus, but not in fronto-parietal cortex, while M_2 is reduced in all investigated areas. It should be noted that, according to the different brain areas changes are evident both at a quite early stage of aging (12 months), and in a later stage (22 months). It is to remark that changes in dopaminergic binding sites in the same areas are in a single direction, reduction, and appear at quite an early stage of aging (12 months) when, consequently, a disturbance in the balance between dopaminergic and cholinergic innervation could be induced. To confirm that these changes are relatively disordered with regard to spatial distribution and time of appearance in the brain, the comparative autoradiography densitometry evaluation of the two subtypes of muscarinic receptors revealed, as shown in Fig. 6, that the increase in M_1 receptors occurs only in two regions of the Ammon's horn, while the decrease in M_2 receptors occurs in all regions, but in a stage of aging (12 months) not disclosed by the study of binding parameters on membrane preparations. So, even in the case of age-dependent changes in muscarinic (and dopamine) receptors, we are dealing with a non uniform phenomenon (losses and increases touching discrete regions) by means of which variously assorted quantitative alterations lead to a novel qualitative aspect.

On the basis of what was found with regard to age-dependent changes in muscarinic receptor subtypes, one should expect a functional change in the cholinergic synapse in the hippocampus of the aged rat. In fact, as shown in Fig. 7, we have found that both basal (in agreement with other authors[13]) and K^{++}stimulated release (not shown) of ACh in the hippocampus (and striatum) are reduced in awake, freely moving old rats. However, we unexpectedly found that ACh concentration in the hippocampus (and striatum) is not significantly reduced in the aged rat, at least not to such an extent as to explain the reduction in release. We hypothesized that this uncoupling between concentration and release might be due to the above described age-dependent changes in muscarinic receptors, and investigated the functional responsiveness of the cholinergic synapse probing the receptor subtypes with specific blockers. As shown in Fig. 8, atropine a very potent aspecific blocker of cholinergic receptors produced a huge increase in the release of ACh in the hippocampus, which, however was much smaller in the old rat than in the young animal. This result indicated that in the cholinergic synapse, some mechanisms (receptor, -s) regulating the release of the transmitter are not fully operative in the aged brain, and was consonant with the previous finding of a loss of M_2 subtype muscarinic receptor in the hippocampus of the aged rat. We now know that the M_2 subtype, to which a pre-synaptic location and a negative control of ACh release have been assigned, is not a single one but two subtypes must be distinguished in it: M_2 proper and M_3 (m_3), a distinction which we could not appraise in the binding studies having used non-specific blocker,

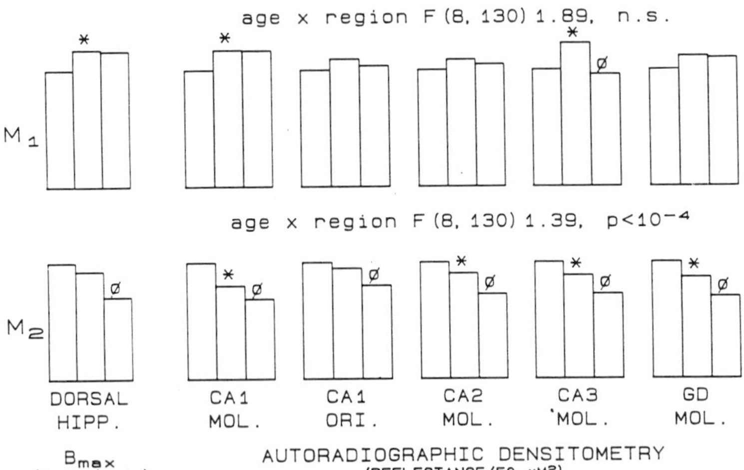

Fig. 6. Number of binding sites (B_{max} : fmoles/mg tissue) for M_1 and M_2 receptors in the dorsal hippocampus (as in figure 5) and autoradiography densitometry of binding for the same receptors (^{3}H-QNB 1 nM, alone or in presence of 100 μM carbachol or 300 nM pirenzepine) on slices of various brain areas in the male Sprague-Dawley rat at the age of 4 (left column), 12 (middle column) and 22 months (right column). Values normalized on 4 months; n=9-10. * and ∅ : significantly different from 4 or 12 months, respectively.

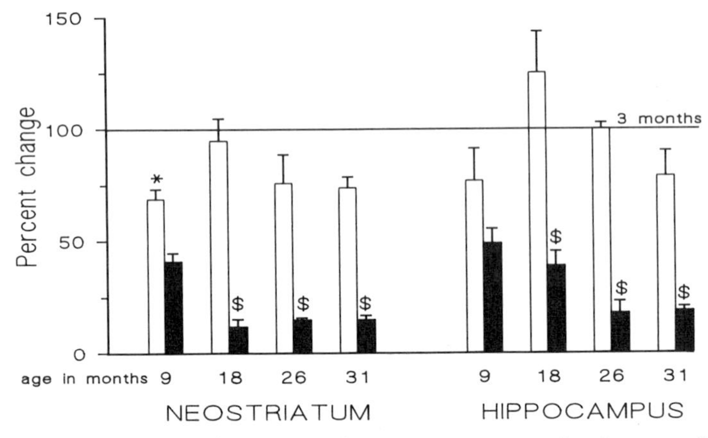

Fig. 7. Concentration (plain columns, average absolute value in 4 month old rats: 760 in the striatum and 250 pmoles per mg protein in the hippocampus) and "in vivo" release (filled columns, transcerebral microdialytic procedure in the awake, freely moving animal; average absolute value in 4 month old rats: 625 in the striatum and 515 fmoles/min in the hippocampus) of ACh (cholinesterease coupled electro-chemical detection HPLC assay) in and from the hippocampus and striatum in male Fischer 344 rats at various ages (n=4-5). Values in per cent changes versus 3 month age. $: 0.1 per cent significant difference versus 3 months.

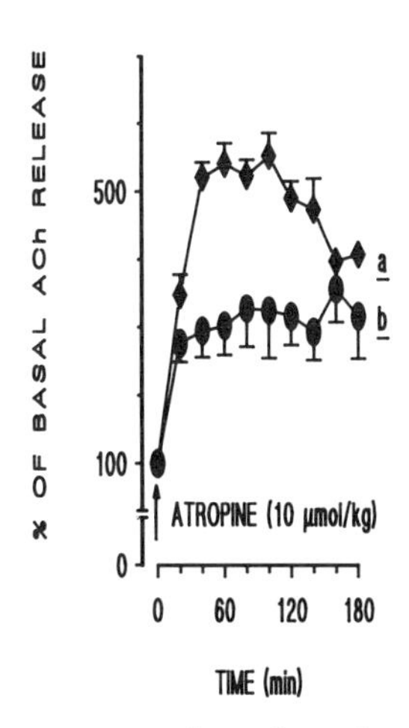

Fig. 8. Per cent change versus basal value in "in vivo" release of ACh from the hippocampus of young (3 month, n=5, diamonds) and old (24 month age, n=3, ovals) awake, freely moving male Fischer 344 rats following i.p. administration of atropine sulphate. Procedures as in Fig. 7. Between under curve areas a and b : 1 per cent significant difference.

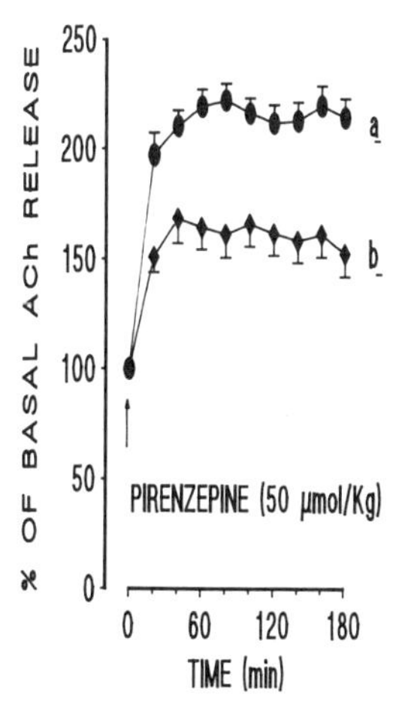

Fig. 9. Per cent change versus basal value in "in vivo" release of ACh from the hippocampus of young (3 month, n=10, diamonds) and old (24 month age, n=15, ovals) awake, freely moving male Fischer 344 rats following i.p. administration of pirenzepine bichlorhydrate. Procedures as in Fig. 7. Between under curve areas a and b : 1 per cent significant difference.

^3H-QNB, in presence of pirenzepine. Of the M_3receptor has been recently shown the primary (versus the M_2) role in negatively regulating the functional release of ACh in the striatum "in vivo."[14] Therefore, the difference between young and old rats in the ability of the cholinergic synapse to respond to the atropine challenge, namely the smaller magnitude of the increase in ACh release, might be due to the blocking limited to the M_1 receptor (reputed as post-synaptically located) and the consequent positive feedback adjustment of pre-synaptic activity from the post-synaptic neuron. In the young animal, the increase in release of ACh due to this mechanism should be smaller than in the old animal, which has an increased number of M_1 (post-synaptic) receptors, but would be accompanied by a component due to the atropine blocking of M_3 (and/or M_2) pre-synaptic receptors negatively modulating the release of ACh. This component would be absent in the old animal, due to the age-dependent loss of these receptors, so that the net increase is smaller than in the young rat. If this were true, one would expect that a challenge of the cholinergic synapse with pirenzepine in a concentration such as to act as a selective blocker of M_1 post-synaptic receptor, would produce in the young animal an increase in ACh release smaller than in the old rat (with an increase in M_1 and impoverishment in M_3 and/or M_2 receptors). Fig. 9 shows that this was our finding. To confirm that what we had assumed, that the difference in the cholinergic synapse between young and old rats could be due to the anatomical (and functional) loss of pre-synaptic muscarinic receptors modulating the release of ACh, we challenged the synapse with selective

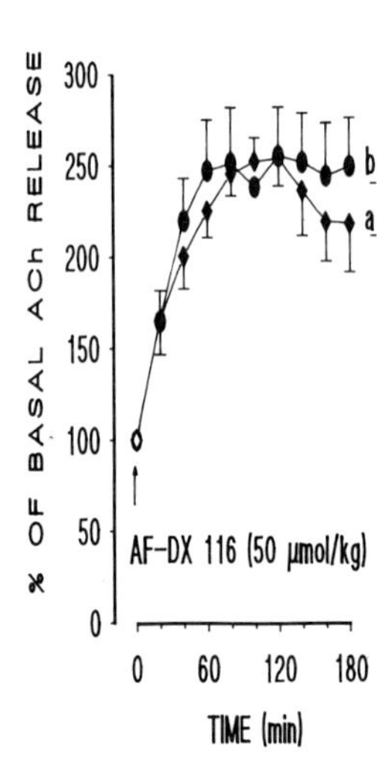

Fig. 10. Per cent change versus basal value in "in vivo" release of ACh from the hippocampus of young (3 month, n=5, diamonds) and old (24 month age, n=4, ovals) awake, freely moving male Fischer 344 rats following i.p. administration of AF-DX 116. Procedures as in Fig. 7. Between under curve areas a and b : no statistically significant difference.

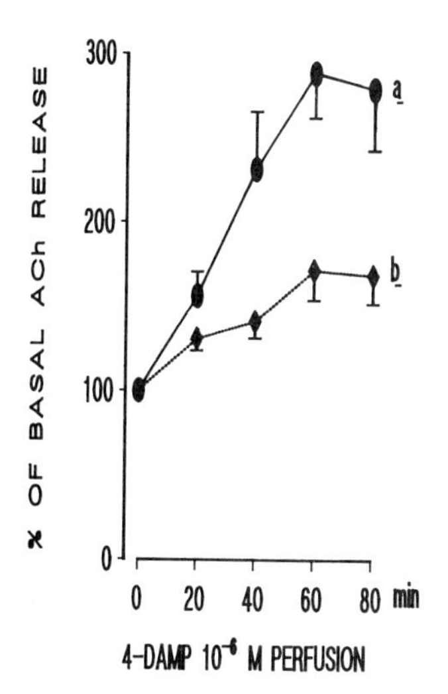

Fig. 11. Per cent change versus basal value in "in vivo" release of ACh from the hippocampus of young (3 month, n=7, ovals) and old (24 month age, n=7, diamonds) awake, freely moving male Fischer 344 rats following administration of 4-DAMP with the dialytic perfusing fluid. Procedures as in Fig. 7. Between under curve areas a and b : 0.1 per cent significant difference.

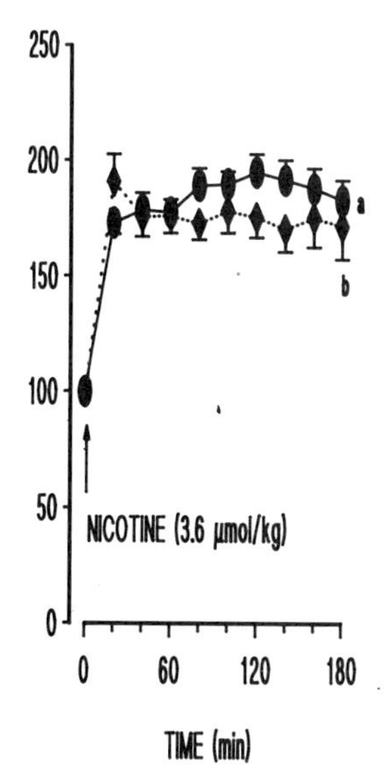

Fig. 12. Per cent change versus basal value in "in vivo" release of ACh from the hippocampus of young (3 month, n=10, ovals) and old (24 month age, n=5, diamonds) awake, freely moving male Fischer 344 rats following i.p. administration of nicotine. Procedures as in Fig. 7. Between under curve areas a and b : no statistically significant difference.

blockers of M_2 and M_3 receptors. Fig. 10 shows that administration of AF-DX 116, a specific blocker of M_2 receptors produced an increase in ACh release of the same magnitude in young and old rats (no anatomical and/or functional loss of this receptor in the aged brain), while 4-DAMP, a selective blocker of M_3 receptor produced, as shown in Fig. 11, an increase in ACh release considerably greater in young animals, indicating that in the aged brain there is a selective loss of this subtype of muscarinic receptor.

The age-dependent changes in the sensitiveness of the cholinergic synapse in the hippocampus to muscarinic blocker challenge due to loss of M_3 receptors cannot immediately explain the "in vivo" reduction in basal ACh release typical of the old rat. In fact, one cannot assume the loss of M_3 receptors, if located pre-synaptically, as indicative of a loss of cholinergic terminals in that, first, a similar loss of M_2 receptors, if located pre-synaptically, was not found, and, second, no substantial reduction in ACh concentration in the hippocampus was found, as a loss of terminals would imply. This assumption is still opposed, as shown in Fig. 12, by the conservation in the hippocampus of the aged rat of a normal sensitivity versus the ACh-release increasing effect of nicotine, acting on a cholinergic auto-receptor (N) credited with a pre-synaptic location and a positively regulatory role on the release of ACh. On the whole, the function of the cholinergic synapse in the hippocampus of the old rat could be impaired as a consequence of altered biochemical properties of the neuronal membrane, as evidenced especially by functional impairment of the M_3 receptor, or as a consequence of selective changes in the expression of this receptor. In any of these cases, the harmonious coordination of the various receptors in the cholinergic synapse would be altered and a reduction in functional release of ACh would follow.

CONCLUSIONS

Our experimental findings lend support to the interpretation, as argued in the Introduction, that the aging of the brain, far from consisting in a progressive impoverishment of all biochemical activities and of their functional expression, homogeneously and synchronously distributed in the various brain structures, is a highly non homogeneous and asynchronous process affecting in a discrete manner the molecular, cellular and structural organization of the brain. Because the discrete age-dependent changes taking place in the organ are not erratic but reflect some kind of program, the process moves from a dynamic qualitative state, the order typical of young stages, to another dynamic qualitative state, the disorder typical of the old stages, a disorder which can be defined as such only in its being the appearance of a novel biological state. In fact, and as examples, the changes we have illustrated in the functional organization of the cholinergic synapse in the hippocampus, due to an imbalance among the different cholinergic receptors, the possibly consequent uncoupling between concentration and release of ACh in this structure, the different degree to which some enzymatic activities are affected by aging, involve such a novel biological state which appears more as a devolutional phenomenon that an involutional one. It is as though a potential disintegrative program was intrinsic to the multicellular organization, a program which we might indicate as "gerogenesis" the issue of which must be the dismantling of the brain through successive steps, with phylogenesis being read palindromically. A program which is just the opposite of the potential integrative program intrinsic to the single generative cell, a program we know as "ontogenesis" the issue of which is the building of the brain through successive steps by reading phylogenesis palingenetically. If such a devolutional program existed, the impact of disease from environment or from "internal milieu" on the development of brain aging, such as long-life peroxidative processes or accumulation of waste products, would be accessory and inessential.

A final consideration is that if such a novel biological state was established by gerogenesis there would be little rationale in the use of the classical pharmacotherapeutical approach to treat age-dependent impairment of brain functions, in that the sensitiveness to action-specific drugs, as grounded on a "normal" molecular and cellular brain organization, would be altered and the expected benefit thus not be forthcoming.

Were this so the most promising, although at present not entirely feasible approach to the treatment of age-dependent impairments of brain functions could be, at most, prevention-retardation and, to a lesser extent, correction of gene-regulated cellular activities typical of senescence by the counterbalancing use of those biological substances which direct the cellular activities typical of pre-senescent stages.

ACKNOWLEDGEMENTS

We thank Mrs. Anna Matuella for editing the manuscript and Miss Laura Alivernini for the most valuable technical assistance.
The work was partially supported by CNR 88.01813.04, and M.P.I. 1989 Fund.

REFERENCES

1. C. E. Finch and D. G. Morgan, RNA and Protein Metabolism in the aging brain, Ann. Rev. Neurosci., 13:75 (1990).
2. O. Sugawara, M. Oshimura, M. Koi, L. A. Annab, and J. C. Barrett, Induction of cellular senescence in immortalized cells by human chromosome 1, Science, 247:707 (1990).
3. H. Yamada et al., in preparation.
4. M. C. Ingvar, P. Maeder, L. Sokoff, and C. B. Smith, Effects of ageing on local rates of cerebral protein synthesis in Sprague-Dawley rats, Brain, 108:135 (1985).

5. C. B. Smith, C. Goochee, S. I. Rapoport, and L. Sokoloff, Effects of ageing on local rates of cerebral glucose utilization in the rat, Brain, 103:351 (1980).
6. I. G. McQuarrie, S. T. Brady, and R. J. Lasek, Retardation in the slow axonal transport of cytoskeletal elements during maturation and aging, Neurobiol. Aging, 10:359 (1989).
7. F. van Leeuwan, E. van der Beek, M. Seger, P. Burbach, and R. Ivell, Age-related development of a heterozygous phenotype in solitary neurons of the homozygous Brattleboro rat, Proc. Natl. Acad. Sci. USA, 86:6417 (1989).
8. D. Harman, Aging: a theory based on free radical and radiation chemistry, J. Gerontol., 11:298 (1956).
9. G. Barja de Quiroga, R. Pérez-Campo, and M. Lòpez-Torres, Changes on cerebral antioxidant enzymes, peroxidation, and the glutathione system of frogs after aging and catalase inhibition. J. Neurosci. Res., 26:370 (1990).
10. D. Curti, F. Dagani, M. R. Galmozzi, and F. Mazzatico, Effect of aging and aceyl-l-carnitine on energetic and cholinergic metabolism in rat brain regions, Mech. Ageing Dev., 47:39 (1989).
11. E. M. Meyer, E. St. Onge, and F.T. Crews, Effects of aging on rat cortical presynaptic cholinergic processes, Neurobiol. Aging 5:315-317 (1984).
12. M. N. Gadaleta, V. Petruzzella, M. Renis, F. Fracasso, and P. Cantatore, Reduced transcription of mitochondrial DNA in the senescent rat, Eur. J. Biochem., 187:501 (1990).
13. C. F. Wu, R. Bertorelli, G. Pepeu, and S. Consolo, Endogenous acetylcholine release from brain hemispheric regions of young and senescent freely moving rats determined by microdialysis technique, Symposium "New Trends in Aging Research", Abstract n° 9 (1987).
14. P. De Boer, B. H. C. Westerink, H. Rollema, J. Zaagsma, and A. S. Horn, An M_3-like muscarinic autoreceptor regulates the "in vivo" release of acetylcholine in rat striatum, Eur. J. Pharmacol. 170:167 (1990).

PLASTICITY IN EXPRESSION OF CO-TRANSMITTERS AND AUTONOMIC NERVES IN AGING AND DISEASE

G. Burnstock

Department of Anatomy and Developmental Biology
and Centre for Neuroscience
University College London
Gower Street
London WC1E 6BT

NORMAL AUTONOMIC NEUROEFFECTOR MECHANISMS

In the past few years, there has been a major change in our understanding of autonomic control mechanisms. It is now clear that there is a multiplicity of autonomic neurotransmitters, and that both co-transmission and neuromodulation are involved (see Burnstock, 1986; Burnstock, 1990a).

Autonomic Neuromuscular Junction

The autonomic neuromuscular junction consists of extensive terminal nerve fibres which contain varicosities from which transmitter is released 'en passage' to reach smooth muscle cells that are in electrical communication with each other via gap-junctions. The varicosities do not have a fixed relationship with particular smooth muscle cells and the minimum distance between varicosities and muscle cells ranges from about 20 nm in densely innervated tissues such as vas deferens and iris to as much as 2 μm in large elastic arteries. Muscle cells do not have postjunctional specialisations. Thus, the autonomic neuromuscular junction differs from synapses (e.g. at the skeletal neuromuscular junction) where there is an established relationship with both pre- and postsynaptic specialisations.

Multiplicity of Neurotransmitters

In addition to noradrenaline (NA) and acetylcholine (ACh), other neurotransmitters are present in autonomic nerves, including adenosine 5'-triphosphate (ATP), 5-hydroxytryptamine (5-HT), dopamine, and a number of peptides, notably neuropeptide Y (NPY), vasoactive intestinal polypeptide (VIP), enkephalin, somatostatin, cholecystokinin, substance P (SP), and calcitonin gene-related peptide (CGRP).

Co-transmission of Chemical Coding

There is now compelling evidence that some, if not all, nerve cells store and release more than one transmitter (Burnstock, 1976; Burnstock, 1990a; Hökfelt et al., 1986). Systematic studies reveal specific combinations of transmitter substances ('chemical coding') for different neuron types which project to particular effecter structures and have defined central connections. This concept has been most developed for the enteric nervous system (Furness and Costa, 1987). NA and ATP act as co-transmitters in variable proportions in sympathetic nerves (Burnstock, 1990b). NPY is also stored and released by sympathetic nerves (Lundberg et al., 1983). However, in most sympathetically innervated tissues, it has little direct postjunctional action, although it is a potent neuromodulator, either enhancing the action of NA and ATP or inhibiting their release (Ellis and Burnstock, 1990). VIP appears to be stored with ACh in parasympathetic nerve fibres supplying the salivary glands (Lundberg, 1981). During low-frequency stimulation, ACh is released to increase salivary secretion from acinar cells and also to elicit weak vasodilatation of blood vessels in the gland (Bloom and Edwards, 1980). VIP is released from the same nerves, especially at high stimulation frequencies, to produce strong vasodilatation and enhancement of the actions of ACh. SP (Duckles and Buck, 1982), CGRP (Terenghi et al., 1986) and ATP (Fyffe and Perl, 1984; Jahr and Jessel, 1983) have been claimed as transmitters in primary afferent (sensory-motor) nerve fibres. SP and CGRP coexist in some sensory nerves (Lee et al., 1985) and it seems likely that ATP coexists with these peptides in some sensory-motor nerves, perhaps co-operating in axon reflex activity (Burnstock, 1977; Burnstock, 1990a). Various combinations of transmitters have been shown to coexist in subpopulations of intrinsic neurons in the heart, bladder, airways and gut (Furness and Costa, 1987; Hassall et al., 1989).

Neuromodulation

A neuromodulator is defined as a substance that modifies the process of neurotransmission. It may act as a prejunctional modulator by decreasing or increasing the amount of neurotransmitter released by a nerve varicosity, or it may act as a postjunctional modulator by altering the time course or extent of action of a transmitter. There are many reports of both pre- and postjunctional modulation occurring at the autonomic neuromuscular junction (Burnstock, 1988). Neuromodulators may be circulating hormones, local agents such as prostaglandin, histamine or bradykinin, or neurotransmitters released from other nerves nearby, or even from the same nerve varicosity (Burnstock, 1987).

Endothelial-mediated Mechanisms

The principle of endothelium-mediated control of blood flow was initiated by the discovery that ACh causes relaxation of rabbit aorta only in the presence of the endothelium (Furchgott and Zawadski, 1980). Endothelial receptors that mediate vasodilatation via the release of endothelium-derived relaxing factor (EDRF) include ATP, SP and 5-HT, as well as ACh (Vanhoutte and Rimele, 1983). Until recently, the source of these transmitter substances, which are capable of acting via the endothelium, was uncertain. The possibility that transmitters released from perivascular nerve varicosities diffuse all the way through the media of large vessels to elicit relaxation via the endothelium is unlikely. Similarly, it is doubtful that substances like ACh, ATP and 5-HT, which are rapidly degraded in the blood, could significantly affect endothelial cells from the circulation. Recent immunocytochemical studies have localised choline acetyltransferase (the enzyme involved in ACh synthesis (Parnavelas et al., 1985), 5-HT and SP within vascular endothelial cells (Burnstock et al., 1988; Loesch and Burnstock, 1988; Milner et al., 1989; Lincoln et al., 1990). In the rat coronary artery, there is an enhanced endothelial release of ACh, 5-HT and SP during hypoxia (Burnstock et al., 1988; Milner et al., 1989). Release of SP from endothelial cells has also been demonstrated in the rat hind-limb vascular preparation during increased flow (Ralevic et al., 1990). In the light of these findings, it is suggested that, while both perivascular nerves and endothelial systems interplay in the normal physiological control of vascular tone, during pathophysiological circumstances such as ischaemia, there is a more dominant role for endothelium-mediated responses as a protective mechanism against tissue hypoxia (Burnstock, 1988).

Endothelial-mediated vascular responses have also been recognised, endothelin being one of the major constrictor substances involved (Yanagisawa et al., 1988).

PLASTICITY OF AUTONOMIC NEUROEFFECTOR MECHANISMS

Development and Aging

A growing number of studies have been concerned with changes in autonomic nerves in development and aging (Baker and Santer, 1988; Black, 1978; Burnstock, 1981; Collins et al., 1980; Cowen and Burnstock, 1986; Gootman, 1986; Hervonen et al., 1986; Wills and Douglas, 1988). Many of these studies are about the innervation of blood vessels. For example, in a study of the changes in density of sympathetic adrenergic nerves in blood vessels of the rabbit, using image analysis quantitation, Cowen et al. (Cowen et al., 1982b) recognised that the pattern of change with age varied considerably between different vessels. While the early stages of development of vascular innervation were similar in all the vessels studied and reached an initial peak density at about 6 weeks after birth, the density of innervation of some vessels (e.g. femoral artery) declined thereafter; other vessels (e.g. renal artery) reached peak density at 6 months and then rapidly declined, while in the basilar artery, density of innervation continued to increase into old age (3 years). In a later study of aging rabbits, it was shown that the decrease in adrenergic innervation of middle cerebral arteries was greater on the left side of the brain than on the right (Saba et al., 1984). Quantitative fluorescence histochemistry revealed that, while the density of adrenergic innervation of the renal artery of the guinea-pig and rabbit were comparable at birth, that of the rabbit subsequently increased during postnatal development, while that of the guinea-pig declined thereafter (Gallen et al., 1982). Curiously, the adrenergic nerves supplying the guinea-pig renal artery, whether in newborn or postnatal animals, appeared to be non-functional. There is evidence of down-regulation of vascular α-adrenoceptors and receptors for 5-HT in old animals (Docherty, 1988), but other studies suggest that responses to NA are maintained in old age (Duckles, 1987).

5-HT was first demonstrated with immunohistochemistry in cerebral vessels (Griffith et al., 1982) and it was later shown that, while little change was seen in the density of 5-HT-containing cerebral perivascular nerves of the rat during postnatal development, there was a marked reduction in 5-HT-immunoreactive neuronal cell bodies in the superier cervical ganglion, the origin of most of the cerebral 5-HT-containing perivascular nerves (Cowen et al., 1987). This suggests that synthesis of 5-HT in the cell bodies of a subpopulation of sympathetic nerves in the superior cervical ganglion declines soon after birth, but that the terminal varicose perivascular projections of these neurons retain the capacity to take up, store and release 5-HT throughout life.

Changes in the development of peptide-containing perivascular nerves of guinea-pig vessels were studied between 6 weeks in utero and old age, and compared with changes in perivascular adrenergic nerves (Dhall et al., 1986). Again, variation in the pattern of development of perivascular nerves in different vessels was demonstrated. In addition, in mesenteric and carotid arteries, whereas adrenergic nerve density reached a peak 4 weeks after birth and declined thereafter, the peptide-containing nerves (VIP, CGRP and SP) reached a peak at birth and declined thereafter to about half maximum density in old age, raising the possibility that perivascular neuropeptides may play a trophic role in early development.

In a more recent developmental study from our laboratory, we have shown that whereas there is a decrease in expression of vasoconstrictor cerebrovascular neurotransmitters (NA and 5-HT) in aging rats, there is an increase in vasodilator neurotransmitters (VIP and CGRP) (Mione et al., 1988). This raises questions about the possible implications of these findings in relation to the increased incidence of cerebrovascular disorders in the elderly.

Hypertension

The distribution of NA and NPY in nerves was compared during the early development of cerebral vessels in normotensive rats and spontaneously hypertensive rats (SHR) before and after the time when hypertension becomes apparent at about 5 weeks of age (Dhital et al., 1988). Three interesting findings emerged from this study: (1) the levels of both NA and NPY were significantly higher in cerebral perivascular sympathetic nerves in SHR compared with normotensive rats; (2) in both normotensive rats and SHR, there was a discrepancy between the time course of changes in the expression of NA and NPY, i.e. the density of NA-fluorescent fibres increased rapidly between 4 and 6 weeks, while NPY-immunofluorescent nerves showed a rapid increase between 6 and 8 weeks. Since NA and NPY coexist in sympathetic perivascular nerves, this shows that the expression of co-transmitters is not necessarily identical; (3) the increase in NA and NPY in SHR does not occur in the sympathetic nerve cell bodies in the superior cervical ganglion from whch the cerebral perivascular nerves arise.

There have been reports that purinergic mechanisms are altered in SHR. Purinergic modulation of NA release from perivascular sympathetic nerves is attenuated in SHR (Kamikawa et al., 1980) and the ATP component of sympathetic co-transmission to the tail artery is claimed to be more dominant in SHR (Vidal et al., 1986). Finally, impairment of endothelium-dependent relaxation to ATP appears to develop in parallel with hypertension (Miller et al., 1987).

Diabetes

In 8-week streptozotocin-diabetic rats, the adrenergic innervation of the vasa vasorum in optic nerves is virtually absent, while in sciatic and vagal nerves it is significantly increased (Dhital et al., 1986). Perivascular nerves in penile vessels containing the potent vasodilator VIP are seriously damaged or lost in both diabetic man and in streptozotocin-diabetic rats (Crowe et al., 1983). In contrast, VIP expression is increased in the gut (Belai et al., 1987). Reduction in the expression of VIP and 5-HT, but not NPY and NA, has been demonstrated in perivascular nerves supplying the cerebral blood vessels of 8-week streptozotocin-induced diabetic rats (Lagnado et al., 1987). Attenuation of endothelium-dependent relaxation of aorta has been claimed in diabetic rat (Durante et al., 1988).

Compensatory Changes in Autonomic Nerves Following Selective Denervation

Unilateral removal of the superior cervical ganglion results in the reinnervation of denervated cerebral vessels by sprouting nerves from the contralateral superior cervical ganglion (Kåhrström et al., 1986). Other marked compensatory chages following superior cervical ganglionectomy include increased SP levels in the ipsilateral iris and ciliary body (Cole et al., 1983), increased CGRP content of pial vessels (Schon et al., 1985), and increased expression of NPY in non-adrenergic VIPergic nerves in the cerebral vasculature (Gibbins and Morris, 1988).

Chemical sympathectomy of developing rats induced by chronic guanethidine treatment leads to massive increases in CGRP levels in the rat superior cervical ganglion and heart, as well as increased brightness and density of CGRP-positive immunofluorescent nerves innervating blood vessels and the vas deferens (Aberdeen et al., 1990).

After selective denervation of primary afferent sensory neurones with capcaisin a compensatory increase in NA and NPY in sympathetic nerves supplying the uterus has been observed (M. Brauer, personal communication).

The effect of crush lesions on perivascular sympathetic nerves has shown differential rates of reinnervation in different blood vessels, suggesting the presence of characteristic levels of local neurotrophic activity (Cowen et al., 1982a).

Hormones and Autonomic Nerve Plastiscity

In late pregnancy, sympathetic innervation of guinea-pig uterine blood vessels exhibits a remarkable switch from adrenergic vasoconstrictor to cholinergic vasodilator control (Bell, 1968). Ultrastructural evidence of sympathetic nerve terminal degeneration (Sporrong et al., 1981), and impaired NA uptake (Alm et al., 1979) has been demonstrated in the guinea-pig uterus. Concomitantly with decreased NA levels, the pregnant guinea-pig uterus also shows a marked decrease of NPY (Fried et al., 1985) and VIP content (Stjernquist et al., 1985). However, unlike the uterus, ultrastructural studies of the guinea pig uterine artery did not show any degeneration of serotonergic or peptidergic (NPY, VIP, SP and CGRP) nerves in late pregnancy (Mione et al., 1990). A recent study has shown that a 4-week treatment with oestrogen, but not progesterone, leads to a marked reduction in the density and varicosity diameters of 5-HT-containing nerves supplying the rabbit basilar artery (Dhall et al., 1988). In view of the possible involvement of 5-HT in the pathogenesis of headache, this finding suggests that contraceptive pills with a high oestrogen content may be contraindicated in women prone to migraine attacks.

Atherosclerosis

Impairment of sympathetic neural function has been claimed in cholesterol-fed animals (Panek et al., 1985). It has also been suggested that surgical sympathectomy may be useful in controlling atherosclerosis in certain arterial beds (Lichtor et al., 1987). Defective cholinergic arteriolar vasodilation has been claimed in atherosclerotic rabbits (Yamamoto et al., 1988) and, in our laboratory, we have recently shown impairment of responses to perivascular nerves supplying the mesenteric, hepatic and ear arteries of Watanabe Heritable Hyperlipidaemic rabbits (Burnstock et al., 1990).

Hirschsprung's and Crohn's diseases

It has been known for some years now that in the absence of enteric ganglia in the colon of man in Hirschsprung's disease, there is a striking hyperinnervation of the musculature by both adrenergic and cholinergic nerves (Gannon et al., 1969). It is interesting that recent studies show that in contrast to these extrinsic nerves, projections of the intrinsic enteric neurons containing peptides and purines do not appear to enter the aganglionic bowel (Hamada et al., 1987). In contrast, in Crohn's disease there appears to be a compensatory increase in VIP in the disease intestine (Bishop et al., 1980).

CONCLUSIONS AND FUTURE DIRECTIONS

Changes in expression of autonomic nerves and co-transmitters that occur during development and aging, following trauma, surgery, after chronic exposure to drugs, and in a number of disease situations have been described. It is suggested that in neuropathologial analysis, compensatory increases in innervation should be considered as well as loss or damage to nerves. Studies of the mechanisms involved in the control of co-transmitter and receptor expression are now needed. Superimposed on the genetic programming of transmitter and receptor expression with development and aging, several different types of adaptive mechanisms may be involved:

Growth factors. For example, it has been suggested that following long-term sympathectomy local production of nerve growth factor could account for the increase in expression of the sensory transmitter SP in the iris (Cole et al., 1983) and CGRP in the heart and other organs (Aberdeen et al., 1990).

Levels of activity in nerves. For example, decentralisation or chronic stimulation of hypogastric nerves of adult guinea-pigs *in vivo* have profound effects on the pattern of innervation and responses of the vas deferens (Jones et al., 1983a,b; Jones and Burnstock, 1986). It seems likely that damage or surgical removal of a nerve trunk will lead to changes in levels of activity in contralateral nerves.

Removal of inhibitory control. For example, removal of the extrinsic neural connections to the heart may contribute to the appearance of 5-HT in intrinsic cardiac neurones grown in tissue culture (Hassall and Burnstock, 1987) or changes in expression of tyrosine hydroxylase in sympathetic ganglia following decentralisation (Black, 1978).

Hormones. For example, the reduction in expression of NA and increase in expression of acetylcholinesterase and associated cholinergic transmission in perivascular nerves in uterine vessels in late pregnancy (Bell, 1968; Bell, 1969; Mione et al., 1990).

Finally, an implication of the growing recognition of the high degree of plasticity of autonomic nerves in the adult concerns drug development. Since the expression of cotransmitters and receptors varies so markedly with age, sex and pathological history, it cannot be assumed that drugs tested on young, healthy male volunteers will be appropriate for all patients and this should be taken into account in designing therapeutic treatment.

REFERENCES

Aberdeen, J., Corr, L., Milner, P., Lincoln, J., and Burnstock, G., 1990, Marked increases in calcitonin gene-related peptide-containing nerves in the developing rat following long-term sympathectomy with guanethidine, Neuroscience, 35:175.

Alm, P., Owman, C., Sjöberg, N.-O., and Thorbert, G., 1979, Uptake and metabolism of [^{3}H]norepinephrine in uterine nerves of pregnant guinea pig, Am.J.Physiol., 136:C277.

Baker, D.M., and Santer, R.M., 1988, A quantitative study of the effects of age on the noradrenergic innervation of Auerbach's plexus in the rat, Mech.Ageing Dev., 42:147.

Belai, A., Lincoln, J., and Burnstock, G., 1987, Lack of release of vasoactive intestinal polypeptide and calcitonin gene-related peptide during electrical stimulation of enteric nerves in streptozotocin-diabetic rats, Gastroenterology, 93:1034.

Bell, C., 1968, Dual vasoconstrictor and vasodilator innervation of the uterine arterial supply in guinea-pig, Circ.Res., 23:279.

Bell, C., 1969, Fine structural localization of acetylcholinesterase at a cholinergic vasodilator nerve-arterial smooth muscle synapse, Circ.Res., 24:61.
Bishop, A.E., Polak, J.M., Bryant, M.G., Bloom, S.R., and Hamilton, S., 1980, Abnormalities of vasoactive polypeptide containing nerves in Crohn's disease, Gastroenterology, 79:853.
Black, I.B., 1978, Regulation of autonomic development, Ann.Rev.Neurosci., 1:183.
Bloom, S.R., and Edwards, A.V., 1980, Vasoactive intestinal polypeptide in relation to atropine resistant vasodilatation in the sub-maxillary gland of the cat, J.Physiol., 300:41.
Burnstock, G., 1976, Do some nerve cells release more than one transmitter?, Neuroscience, 1:239.
Burnstock, G., 1977, Autonomic neuroeffector junctions - reflex vasodilatation of the skin, J.Invest.Dermatol., 69:47.
Burnstock, G., 1981, Current approaches to development of the autonomic nervous system: clues to clinical problems. pp. 1-18. Chairman's concluding remarks. pp. 371-373, in: "Development of the Autonomic Nervous System. Ciba Foundation Symposium 83," K. Elliot, and G. Lawrenson, eds., Pitman Medical, London,
Burnstock, G., 1986, The changing face of autonomic neurotransmission. (The First von Euler Lecture in Physiology), Acta Physiol.Scand., 126:67.
Burnstock, G., 1987, Mechanisms of interaction of peptide and nonpeptide vascular neurotransmitter systems, J.Cardiovasc.Pharmacol., 10, Suppl.12:S74.
Burnstock, G., 1988, Regulation of local blood flow by neurohumoral substances released from perivascular nerves and endothelial cells, Acta Physiol.Scand., 133, Suppl. 571:53.
Burnstock, G., 1990a, Cotransmission. The Fifth Heymans Lecture - Gent, February 17, 1990, Arch.Int.Pharmacodyn.Ther., 304:7.
Burnstock, G., 1990b, Noradrenaline and ATP as cotransmitters in sympathetic nerves, Neurochem.Int., 17:357.
Burnstock, G., Lincoln, J., Fehér, E., Hopwood, A.M., Kirkpatrick, K., Milner, P., and Ralevic, V., 1988, Serotonin is localized in endothelial cells of coronary arteries and released during hypoxia: a possible new mechanism for hypoxia-induced vasodilatation of the rat heart, Experientia, 44:705.
Burnstock, G., Stewart-Lee, A.L., Brizzolara, A.L., Tomlinson, A., and Corr, L., 1990, Dual control of local blood flow by nerves and endothelial cells; changes in atherosclerosis, in: "Progress, Problems, and Promises for an Effective Quantitative Evaluation of Atherosclerosis in Living and Autopsied Experimental Animals and Man," M.G. Wissler, M.G. Bond, M. Mercuri, and P. Tanganelli, eds., Plenum, New York, in press.
Cole, D.F., Bloom, S.R., Burnstock, G., Butler, J.M., McGregor, G.P., Saffrey, M.J., Unger, W.G., and Zhang, S.Q., 1983, Increase in SP-like immunoreactivity in nerve fibres of rabbit iris and ciliary body one to four months following sympathetic denervation, Exp.Eye Res., 37:191.
Collins, K.J., Exton-Smith, A.H., James, N.H., and Oliver, D.J., 1980, Functional changes in autonomic nervous responses with ageing, Age Ageing, 9:17.
Cowen, T., and Burnstock, G., 1986, Development, aging and plasticity of perivascular autonomic nerves, in: "Developmental Neurobiology of the Autonomic Nervous System," P.M. Gootman, ed., Humana Press, Clifton, NJ, pp. 211-232.
Cowen, T., MacCormick, D.E.M., Toff, W.D., Burnstock, G., and Lumley, J.S.P., 1982a, The effect of surgical procedures on blood vessel innervation. A fluorescence histochemical study of degeneration and regrowth of perivascular adrenergic nerves, Blood Vessels, 19:65.
Cowen, T., Haven, A.J., Wen-Qin, C., Gallen, D.D., Franc, F., and Burnstock,

G., 1982b, Development and ageing of perivascular adrenergic nerves in the rabbit. A quantitative fluorescence histochemical study using image analysis, J.Auton.Nerv.Syst., 5:317.
Cowen, T., Alafaci, C., Crockard, H.A., and Burnstock, G., 1987, Origin and postnatal development of nerves showing 5-hydroxytryptamine-like immunoreactivity supplying major cerebral arteries of the rat, Neurosci.Lett., 78:121.
Crowe, R., Lincoln, J., Blacklay, P.F., Pryor, J.P., Lumley, J.S.P., and Burnstock, G., 1983, Vasoactive intestinal polypeptide-like immunoreactive nerves in diabetic penis. A comparison between streptozotocin-treated rats and man, Diabetes, 32:1075.
Dhall, U., Cowen, T., Haven, A.J., and Burnstock, G., 1986, Perivascular noradrenergic and peptide-containing nerves show different patterns of changes during development and ageing in the guinea-pig, J.Auton.Nerv.Syst., 16:109.
Dhall, U., Cowen, T., Haven, A.J., and Burnstock, G., 1988, Effect of oestrogen and progesterone on noradrenergic nerves and on nerves showing serotonin-like immunoreactivity in the basilar artery of the rabbit, Brain Res., 442:335.
Dhital, K., Lincoln, J., Appenzeller, O., and Burnstock, G., 1986, Adrenergic innervation of vasa and nervi nervorum of optic, sciatic, vagus and sympathetic nerve trunks in normal and streptozotocin-diabetic rats, Brain Res., 367:39.
Dhital, K.K., Gerli, R., Lincoln, J., Milner, P., Tanganelli, P., Weber, G., Fruschelli, C., and Burnstock, G., 1988, Increased density of perivascular nerves to the major cerebral vessels of the spontaneously hypertensive rat: differential changes in noradrenaline and neuropeptide Y during development, Brain Res., 444:33.
Docherty, J.R., 1988, The effects of ageing on vascular α-adrenoceptors in pithed rat and rat aorta, Eur.J.Pharmacol., 146:1.
Duckles, S.P., 1987, Influence of age on vascular adrenergic responsiveness, Blood Vessels, 24:113.
Duckles, S.P., and Buck, S.M., 1982, Substance P in the cerebral vasculature: depletion by capsaicin suggests a sensory role, Brain Res., 245:171.
Durante, W., Sen, A.K., and Sunahara, F.A., 1988, Impairment of endothelium-dependent relaxation in aortae from spontaneously diabetic rats, Br.J.Pharmacol., 94:463.
Ellis, J.L., and Burnstock, G., 1990, Neuropeptide Y neuromodulation of sympathetic co-transmission in the guinea-pig vas deferens, Br.J.Pharmacol., 100:457.
Fried, G., Hökfelt, T., Terenius, L., and Goldstein, M., 1985, Neuropeptide Y (NPY)-like immunoreactivity in guinea pig uterus is reduced during pregnancy in parallel with noradrenergic nerves, Histochemistry, 83:437.
Furchgott, R.F., and Zawadski, J.V., 1980, The obligatory role of endothelial cells in the relaxation of arterial smooth muscle by acetylcholine, Nature, 288:373.
Furness, J.B., and Costa, M., 1987, "The Enteric Nervous System", Churchill Livingstone, Edinburgh.
Fyffe, R.E.W., and Perl, E.R., 1984, Is ATP a central synaptic mediator for certain primary afferent fibres from mammalian skin?, Proc.Natl.Acad.Sci.U.S.A., 81:6890.
Gallen, D.D., Cowen, T., Griffith, S.G., Haven, A.J., and Burnstock, G., 1982, Functional and non-functional nerve-smooth muscle transmission in the renal arteries of the newborn and adult rabbit and guinea-pig, Blood Vessels, 19:237.
Gannon, B.J., Burnstock, G., Noblett, H.R., and Campbell, P.E., 1969, Histochemical diagnosis of Hirschsprung's disease, Lancet, i:894.
Gibbins, I.L., and Morris, J.L., 1988, Co-existence of immunoreactivity to neuropeptide Y and vasoactive intestinal polypeptide in non-adrenergic

axons innervating guinea-pig cerebral arteries after sympathectomy, Brain Res., 444:402.
Gootman, P.M.(Ed.), 1986, "Developmental Neurobiology of the Autonomic Nervous System", Humana Press, Clifton, NJ.
Griffith, S.G., Lincoln, J., and Burnstock, G., 1982, Serotonin as a neurotransmitter in cerebral arteries, Brain Res., 247:388.
Hamada, Y., Bishop, A.E., Federici, G., Rivosecchi, M., Talbot, I.C., and Polak, J.M., 1987, Increased neuropeptide Y-immunoreactive innervation of aganglionic bowel in Hirschsprung's disease, Virchows Arch.[A], 411:369.
Hassall, C.J.S., and Burnstock, G., 1987, Evidence for uptake and synthesis of 5-hydroxytryptamine by a subpopulation of intrinsic neurons in the guinea-pig heart, Neuroscience, 22:413.
Hassall, C.J.S., Allen, T.G.J., Pittam, B.S., and Burnstock, G., 1989, The use of cell and tissue culture techniques in the study of regulatory peptides, in: "Regulatory Peptides. Experientia Suppl," J.M. Polak, ed., Birkaeuser, Basel, pp. 113-116.
Hervonen, A., Partanen, M., Helén, P., Koistinaho, J., Alho, H., Baker, D.M., Johnson, J.E., Jr., and Santer, R.M., 1986, The sympathetic neuron as a model of neuronal aging, in: "Neurohistochemistry: Modern Methods and Applications," P. Panula, H. Paivarinta, and S. Soinila, eds., Alan Liss, New York, pp. 569-586.
Hökfelt, T., Fuxe, K., and Pernow, B.(Eds.), 1986, "Coexistence of Neuronal Messengers: A New Principle in Chemical Transmission. Progress in Brain Research, Vol. 68", Elsevier, Amsterdam.
Jahr, C.E., and Jessel, T.M., 1983, ATP excites a subpopulation of rat dorsal horn neurones, Nature, 304:730.
Jones, R., and Burnstock, G., 1986, Effects of activity on restorative processes in the autonomic nervous system, in: "Electrical Stimulation and Neuromuscular Disorders," W.A. Nix, and G. Vrbová, eds., Springer, Berlin, pp. 80-87.
Jones, R., Dennison, M.E., and Burnstock, G., 1983a, The effect of decentralisation or chronic hypogastric nerve stimulation in vivo on the innervation and responses of the guinea-pig vas deferens, Cell Tissue Res., 232:265.
Jones, R., Yokota, R., and Burnstock, G., 1983b, The long-term influence of decentralisation or preganglionic hypogastric nerve stimulation in vivo on the reinnervation of minced vas deferens in the guinea-pig, Cell Tissue Res., 232:281.
Kåhrström, J., Hardebo, J.E., Nordberg, C., and Owman, C., 1986, Experiments on cerebrovascular nerve plasticity and trophic vascular adaption in young and adult rats, in: "Neural Regulation of Brain Circulation," C. Owman, and J.E. Hardebo, eds., Elsevier, Amsterdam, pp. 589-606.
Kamikawa, Y., Cline, W.H., and Su, C., 1980, Diminished purinergic modulation of the vascular adrenergic neurotransmission in spontaneously hypertensive rats, Eur.J.Pharmacol., 66:347.
Lagnado, M.L.J., Crowe, R., Lincoln, J., and Burnstock, G., 1987, Reduction of nerves containing vasoactive intestinal polypeptide and serotonin, but not neuropeptide Y and catecholamine, in cerebral blood vessels of the 8-week streptozotocin-induced diabetic rat, Blood Vessels, 24:169.
Lee, Y., Takami, K., Kawai, Y., Girgis, S., Hillyard, C.J., Macintyre, I., Emson, P.C., and Tohyama, M., 1985, Distribution of calcitonin gene-related peptide in the rat peripheral nervous system with reference to its coexistence with substance P, Neuroscience, 15:1227.
Lichtor, T., Davis, H.R., Johns, L., Vesselinovitch, D., Wissler, R.W., and Mullan, S., 1987, The sympathetic nervous system and atherosclerosis, J.Neurosurg., 67:906.
Lincoln, J., Loesch, A., and Burnstock, G., 1990, Localization of vasopressin, serotonin and angiotensin II in endothelial cells of the renal and mesenteric arteries of the rat, Cell Tissue Res., 259:341.
Loesch, A., and Burnstock, G., 1988, Ultrastructural localisation of

serotonin and substance P in vascular endothelial cells of rat femoral and mesenteric arteries, Anat.Embryol.(Berl.), 178:137.
Lundberg, J.M., 1981, Evidence for coexistence of vasoactive intestinal polypeptide (VIP) and acetylcholine in neurons of cat exocrine glands. Morphological, biochemical and functional studies, Acta Physiol.Scand., 112, Suppl.496:1.
Lundberg, J.M., Terenius, L., Hökfelt, T., and Goldstein, M., 1983, High levels of neuropeptide Y in peripheral noradrenergic neurons in various mammals including man, Neurosci.Lett., 42:167.
Miller, M.J., Pinto, A., and Mullane, K.M., 1987, Impaired endothelium-dependent relaxations in rabbits subjected to aortic coarctation hypertension, Hypertension, 10:164.
Milner, P., Ralevic, V., Hopwood, A.M., Fehér, E., Lincoln, J., Kirkpatrick, K.A., and Burnstock, G., 1989, Ultrastructural localisation of substance P and choline acetyltransferase in endothelial cells of rat coronary artery and release of substance P and acetylcholine during hypoxia, Experientia, 45:121.
Mione, M.C., Dhital, K.K., Amenta, F., and Burnstock, G., 1988, An increase in the expression of neuropeptidergic vasodilator, but not vasoconstrictor, cerebrovascular nerves in aging rats, Brain Res., 460:103.
Mione, M.C., Cavanagh, J.F.R., Lincoln, J., Milner, P., and Burnstock, G., 1990, Pregnancy reduces noradrenaline but not neuropeptide levels in the uterine artery of the guinea-pig, Cell Tissue Res., 259:503.
Panek, R.L., Dixon, W.R., and Rutledge, C.O., 1985, Modification of sympathetic neuronal function in the rat tail artery by dietary lipid treatment, J.Pharmacol.Exp.Ther., 233:578.
Parnavelas, J.G., Kelly, W., and Burnstock, G., 1985, Ultrastructural localization of choline acetyltransferase in vascular endothelial cells in rat brain, Nature, 316:724.
Ralevic, V., Milner, P., Hudlická, O., Kristek, F., and Burnstock, G., 1990, Substance P is released from the endothelium of normal and capsaicin-treated rat hindlimb vasculature, in vivo, by increased flow, Circ.Res., 66:1178.
Saba, H., Cowen, T., Haven, A.J., and Burnstock, G., 1984, Reduction in noradrenergic perivascular nerve density in the left and right cerebral arteries of old rabbits, J.Cereb.Blood Flow Metab., 4:284.
Schon, F., Ghatei, M., Allen, J.M., Mulderry, P.K., Kelly, J.S., and Bloom, S.R., 1985, The effect of sympathectomy on calcitonin gene-related peptide levels in the rat trigemino-vascular system, Brain Res., 348:197.
Sporrong, B., Alm, P., Owman, C., Sjöberg, N.-O., and Thorbert, G., 1981, Pregnancy is associated with extensive adrenergic degeneration in the uterus. An electronmicroscopic study in the guinea-pig, Neuroscience, 6:1119.
Stjernquist, M., Alm, P., Ekman, R., Owman, C., Sjöberg, N.-O., and Sundler, F., 1985, Levels of neural vasoactive intestinal polypeptide in rat uterus are markedly changed in association with pregnancy as shown by immunocytochemistry and radioimmunoassay, Biol.Reprod., 33:157.
Terenghi, G., Polak, J.M., Rodrigo, J., Mulderry, P.K., and Bloom, S.R., 1986, Calcitonin gene-related peptide-immunoreactive nerves in the tongue, epiglottis and pharynx of the rat: Occurrence, distribution and origin, Brain Res., 365:1.
Vanhoutte, P.M., and Rimele, T.J., 1983, Role of endothelium in the control of vascular smooth muscle function, J.Physiol.(Paris), 78:681.
Vidal, M., Hicks, P.E., and Langer, S.Z., 1986, Differential effects of α,β-methylene ATP on responses to nerve stimulation in SHR and WKY tail arteries, Naunyn Schmiedebergs Arch.Pharmacol., 332:384.
Wills, M., and Douglas, J.S., 1988, Aging and cholinergic responses in bovine trachealis muscle, Br.J.Pharmacol., 93:918.

Yamamoto, H., Bossaller, C., Cartwright, J., Jr., and Henry, P.D., 1988, Videomicroscopic demonstration of defective cholinergic arteriolar vasodilation in atherosclerotic rabbit, J.Clin.Invest., 81:1752.

Yanagisawa, M., Kurihara, H., Kimura, S., Tomobe, Y., Kobayashi, M., Mitsui, Y., Yazaki, Y., Goto, K., and Masaki, T., 1988, A novel potent vasoconstrictor peptide produced by vascular endothelial cells, Nature, 332:411.

NICOTINIC CHOLINERGIC RECEPTORS IN HUMAN BRAIN: EFFECTS OF AGING AND ALZHEIMER

E. Giacobini

Department of Pharmacology
Southern Illinois University School of Medicine
Springfield, Illinois 62794

A review of studies on nicotinic cholinergic receptors in human brain is presented. The paper focusses upon changes in normal aging brain and in Alzheimer disease (AD). Studies from five different approaches are reported: a) molecular biology; b) receptor binding studies; c) studies with specific neurotoxins; d) immunocytochemistry; and e) PET scan. These studies document profound and characteristic differences between the normal aging and the pathological AD brain with regard to nicotinic cholinergic receptor localization, distribution and function.

Molecular Biology of Cholinergic Receptors

Recent molecular genetic and pharmacological studies have revealed the existence of gene families that encode several nicotinic acetylcholine receptor (nAChR) subunits in the mammalian and avian brain. At least three discrete nicotinic receptor α-subunits have been identified in brain (α_2, α_3 and α_4) which are closely related to the single α-subunit (α_1) found in skeletal muscle (Goldman et al., 1987; Neff et al., 1988; Schoepfer et al., 1988). In situ hybridization experiments show that each of these α-subunits has a discrete localization of expression in the brain, suggesting that they are components of different receptor subtypes (Duvoisin et al., 1989; Goldman et al., 1986, 1987; Neff et al., 1988; Schoepfer et al., 1988; Wada et al, 1989).

Cholinergic Receptors in Human Brain During Normal Aging

Nordberg and Winblad (1986a) reported a significant decrease in number of muscarinic binding sites for quinuclinidyl benzylate (^{3}H-QNB) of human hippocampus from 44 controls (0-100 years). In the hippocampus, a similar decrease with age was found by the same authors for nicotinic (^{3}H-tubocurarine) receptors. No change was found for α-bungarotoxin (α-BTX) binding and a significant increase was seen in ^{3}H-nicotine (^{3}H-NIC) binding. Age-related changes in thalamus were quite different than in the hippocampus, with a decrease in nicotinic (^{3}H-BTX) and a significant increase in muscarinic (^{3}H-QNB) receptor binding. These age effects are not as obvious as changes in AD (Nordberg and Winblad, 1986a).

Plasticity and Regeneration of the Nervous System
Edited by P.S. Timiras *et al.*, Plenum Press, New York, 1991

Changes in Nicotinic Receptors in Alzheimer Disease. Binding Studies

In Alzheimer and in other degenerative diseases (Gerstmann-Straussler, Parkinson dementia of GUAM type and Dementia pugilistica), pathological and biochemical lesions in the cholinergic neuron are localized mainly to the presynaptic region. Alterations include biochemical and morphological changes in the structures storing, synthesizing and releasing acetylcholine (ACh) and possibly also in cholinergic receptors regulating ACh release. By comparing cholinergic terminals in the CNS of normally aged with AD patients, it is apparent that nicotinic and muscarinic receptors are not equally affected (Giacobini et al., 1988a, 1989; Zubenko et al., 1989; Rinne et al., 1989) (Figs. 1 and 2).

Table I

Percent Changes in Nicotinic Receptors in Frontal Cortex of Alzheimer Patients as Compared With Controls

	Ligand	Author/Year
No difference	alpha-BTX	Davies and Feisullin, 1981
-65	ACh	Nordberg and Winblad, 1986a,b
-53	Nicotine	Nordberg and Winblad, 1986a,b
No difference	Nicotine	Shimohama et al., 1986
-62	ACh	Whitehouse et al., 1986
-56	Nicotine	Whitehouse et al., 1986, 1988
-56	ACh	Kellar et al., 1987
-44 (autopsy)	Nicotine	DeSarno et al., 1988b
-55 (biopsy)	Nicotine	DeSarno et al., 1988a,b
-53	ACh	London et al., 1989
-70	Methylcarbamyl-choline	Araujo et al., 1987

Surveys, including percent changes in nicotinic receptors in the frontal cortex of AD patients, reveal a striking difference from aging normal controls (Giacobini et al., 1988a, 1989). Nicotinic receptor binding studies are mainly from autopsies using three different ligands (NIC, α-BTX and ACh). Out of ten autopsy studies (1981-1989), including one from our laboratory (DeSarno et al., 1988b; Table I), eight show decreases from 44-70% and two show no difference. Whitehouse et al., (1988) confirmed previously reported findings of significant reductions in cortical nAChR binding sites in AD. The regions investigated were fronto-parietal, temporo-parietal and occipito-parietal cortex. Unlike muscarinic receptors, the concentration of nicotinic receptors is greater in deeper rather than in superficial lamina. These data confirm that the density of these nicotinic binding sites is reduced in both AD and Parkinson disease (PD) (Whitehouse et al., 1986). Araujo et al. (1988) measured nicotinic binding using ^{3}H-methylcarbamylcholine and found decreases in maximal binding capacity in cortical areas and hippocampus but not in subcortical regions of AD brains. London et al. (1989) found binding of both ^{3}H-ACh and ^{3}H-NIC to be lower in several brain regions of AD patients. Perry et al. (1989) found a 40-60% decrease in receptor binding (^{3}H-NIC) in all cortical areas examined of AD patients. This reduction correlated better with senile plaque density than with decreases in choline acetyltransferase (ChAT) activity. Quantitative autoradiography in large cryosections of human brain hemispheres suggest that nicotinic binding sites may be located on cholinergic neuron cell bodies or on terminals, being pre- and/or postsynaptically located (Adem et al., 1989). With regard to muscarinic receptors, using ^{3}H-QNB as ligand, we found no differences in specific binding in autopsy material but a significant 39% increase in the biopsies (Fig. 2). By contrast, in the same material, using ^{3}H-NIC as a ligand, we found a 44% decrease in the autopsies and a 55% decrease in the biopsies (Fig. 1). Preservation of the tissue might explain the differences between biopsy and autopsy material. However, in younger subjects it could reflect early changes in AD. These decreases in binding correlate to enzymatic [ChAT and acetylcholinesterase (AChE) activity] changes (Davies, 1979; Rossor et al., 1982; Bird et al., 1983; Giacobini et al., 1988a,b, 1989) supporting a presynaptic localization of the lesion.

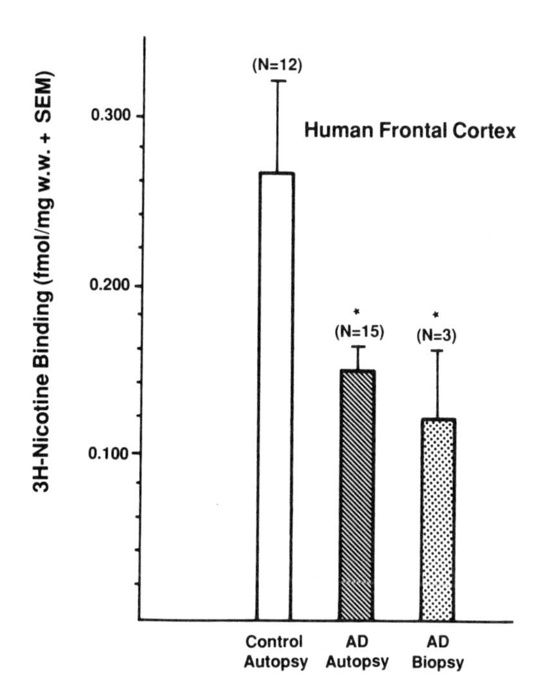

Figure 1. Comparison of ^{3}H-nicotine binding (fmol/mg w.w. ± S.E.M.) in samples of human frontal cortex from normal controls and Alzheimer patients (biopsies and autopsies). Significantly different from controls, $p < 0.05$. (Modif. from Giacobini, 1990)

An important factor of decreased cholinergic receptor function is de-regulation of ACh release. Release of ^{3}H-ACh can be measured in frontal cortex slices from human autopsy after short post-mortem delay (Nordberg et al., 1987) and from biopsies (DeSarno et al., 1988b; Giacobini et al., 1988a). A significant reduction (50-70%) in the evoked release of ACh has been reported by both laboratories. A reduction in ACh release and in number of nicotinic binding sites in the frontal

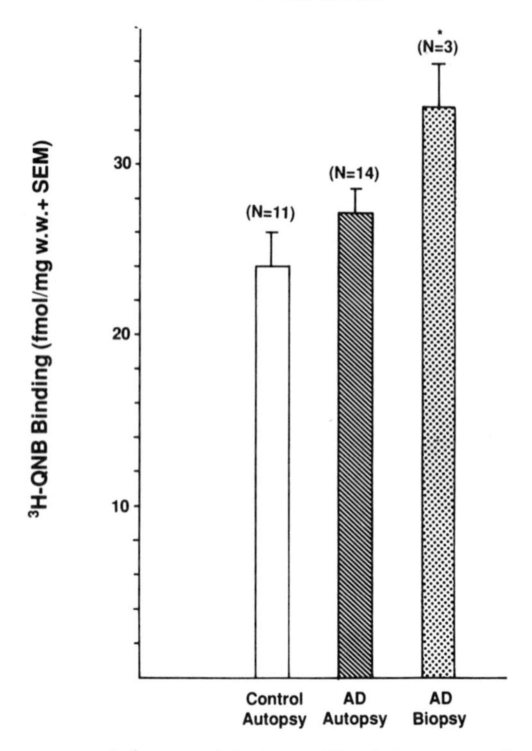

Figure 2. Comparison of ^{3}H-QNB binding (fmol/mg w.w. ± S.E.M) in samples of human frontal cortex from normal controls and Alzheimer patients (biopsies and autopsies). *Significantly different from controls, $p < 0.05$. (Modif. from Giacobini, 1990)

cortex of AD patients supports the hypothesis of a selective loss of presynaptic nicotinic receptors in AD (Giacobini et al., 1988a).

Neurotoxins as Markers of Cholinergic Receptors

Although α-BTX has proved to be a useful cholinergic ligand in the peripheral nervous system, mainly at the neuromuscular junction, it is not suitable as a marker for the CNS (Morley et al., 1983). Experiments performed with iontophoretic techniques do not support the hypothesis that α-neurotoxins such as α-BTX bind to functional nAChRs in mammalian CNS (Morley and Kemp, 1981; Chiappinelli, 1985). Consequently, characterization and quantitation of nicotinic cholinergic binding sites in the brain has met with difficulty. Recent studies have indicated differences between central and peripheral nicotinic receptors (Marks and Collins, 1982; Shimohama et al., 1985; Sugiyama and Yamashita, 1986). Chiappinelli (1985) and Wolf et al. (1987) have used a BTX that they named kappa-bungarotoxin (K-BTX) to characterize central nicotinic receptors in the chicken and in the rat. Kappa-bungarotoxin has been recently isolated from snake venom (Chiappinelli, 1985). It was shown to be a potent and selective antagonist of neuronal nicotinic receptors (Chiappinelli, 1985) including CNS sites in the chicken optic lobe (Wolf et al., 1987). We have observed that using ^{3}H-NIC, ^{125}I-α-BTX and ^{125}I-K-BTX, at least three putative categories of nicotinic receptor are present in the human brain (Sugaya et al., 1990). Our studies suggest that each type of these receptors may have different kinetics, distribution and localization. Kappa-bungarotoxin or neuronal BTX has been shown to block nicotinic synaptic transmission in a variety of neuronal preparations where α-BTX has no effect. Vidal and Changeux (1989) have demonstrated that the effect of NIC applied by iontophoresis to the prefrontal cortex of the rat is blocked by K-BTX but not by other nicotinic antagonists. The agonistic actions of ACh in cerebellar neurons is also selectively blocked by K-BTX (de la Garza et al., 1989).

Nicotinic Receptor Subtypes Seen with Specific Neurotoxins in Human Frontal Cortex. Changes in Alzheimer Disease

Radiolabeled K-neurotoxins were used to localize and quantitate neuronal nicotinic receptors in various species (Sugaya et al., 1990).

Combining ^{3}H-NIC, ^{125}I-α-BTX and ^{125}I-K-BTX as ligands, several categories of nicotinic receptors can be postulated in the human brain (Giacobini et al., 1988b; Sugaya et al., 1990). We reported the kinetics, distribution and localization of three subtypes present in the human frontal cortex. We also described for the first time specific changes related to receptor subtypes in human cortex of AD patients (Giacobini et al., 1988b; Sugaya et al., 1990).

We used autopsy brains from healthy young controls (age 21-57), and healthy elderly controls (age 64-94) as well as from AD patients (age 67-78) from our Center for Alzheimer Disease and Related Disorders, for homogenate- or slice-*in vitro* assays of nicotinic binding. The right hemisphere was isolated and stored at -90°C for binding assays, the contralateral was fixed for histologically diagnosis.

Specific binding with increasing concentrations of ^{3}H-(-)-NIC, ^{125}I-α- or ^{125}I-K-BTX to membranes of human frontal cortex was saturable. Scatchard plots were curvilinear and Hill coefficients far from unity, indicating the presence of multiple classes of binding sites for these three ligands.

TABLE II

Percent Decrease in Number of ^{3}H-(-)-Nicotine, ^{125}I-K- or ^{125}I-alpha-Bungarotoxin Binding Sites in Frontal Cortex of Alzheimer Patients As Compared with Elderly Controls

	High Affinity (% B_{max})	Low Affinity (% B_{max})
^{3}H-(-)-NIC	47	50[a]
^{125}I-K-BTX	0	52[a]
^{125}I-α-BTX	0	32[b]

n= 6-8; [a] significant $p < 0.005$; [b] not significant (from Sugaya et al., 1990)

There were significant decreases in the number (B_{max}) of high (47%) and low (50%) affinity binding sites of ^{3}H-(-)-NIC in AD patients as compared to elderly controls (Table II). Alzheimer patients showed a significant decrease of B_{max} in the low affinity binding site of ^{125}I-K-BTX but not in the high affinity site, as compared to young (57%) and elderly (52%) controls (Table II). There was only a non-significant decrease in B_{max} in the low affinity binding site of ^{125}I-α-BTX in AD patients, as compared to young and elderly controls (34.0% and 31.6%) (Table II).

Autoradiographic analysis showed that ^{125}I-K-BTX specific binding sites are concentrated mainly in the middle and deep cortical layers. The same localization has been observed by us in the monkey (Macacus rhesus). With autoradiography, a pronounced decrease in density of ^{125}I-K-BTX binding sites was seen in AD patient vs. elderly controls.

The kinetic characteristics of these three binding sites were studied by means of competition experiments. These demonstrated the presence of two categories of nicotinic binding sites, one which was displaced by K-BTX and one which was not displaced by this toxin (Fig. 3). We named the latter $Nicotine_1$ (Fig. 3). The binding site not displaced by K-BTX could be subdivided into two subtypes, one which was displaced by α-BTX and one was not. We named these two binding sites $Nicotine_2$ and $Nicotine_3$, respectively (Fig. 3).

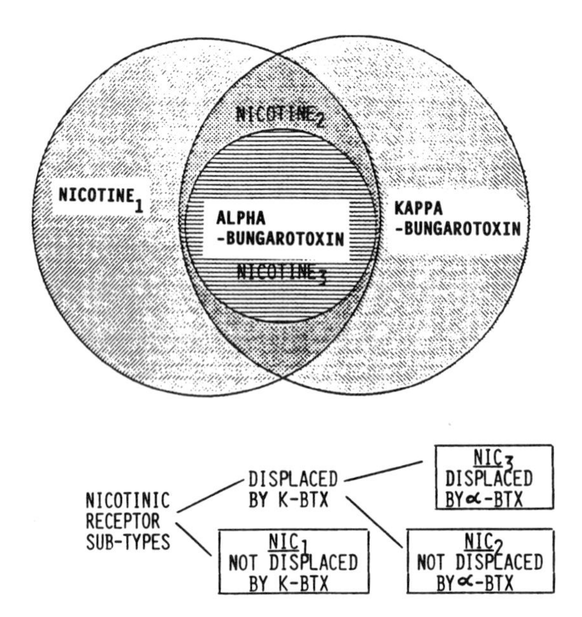

Figure 3. Diagram of subtypes of nicotinic receptors in human frontal cortex demonstrated by ^{3}H-nicotine, ^{125}I-α-bungarotoxin and ^{125}I-K-bungarotoxin as ligands. See text. (Modif. from Sugaya et al., 1990)

A functional role of cortical nicotinic receptors can be demonstrated by the effect of BTXs on the evoked fractional release of ACh from rat frontal cortex slices. In these experiments, the first electrical stimulation (S_1, 20 mA, 1 Hz, 5 min) is a pre-drug testing control and the third electrical stimulation (S_3) is a post-drug testing control. The only significant decrease was seen in the S_2/S_1 ratio of ACh release following 1 μM K-BTX but not after 1 μM α-BTX. The S_3/S_1 ratio did not change after the third stimulation. This shows that K-BTX, but not α-BTX, decreases the electrically evoked release of ACh from rat frontal cortex slices and that this effect is reversible.

Mapping of nicotinic receptors in rodent brain, using various radiolabelled agonists (Clarke et al., 1985), in situ hybridization (Goldman et al., 1987) and immunohistochemistry (Swanson et al., 1987) has demonstrated the presence of a high number of ACh receptors in the neocortex. Our data on human brain autoradiography of K-BTX binding sites also show the highest density in the cortex. We have seen variable regional distributions in the presence of (-)-NIC, α-BTX or K-BTX. Receptor subtypes in human cortex have been previously suggested based on various ligands such as ^{3}H-ACh, ^{3}H-(-)-NIC and ^{125}I-α-BTX (Adem, 1987). The regional distribution and the pharmacology of agonists and antagonist binding sites suggest that these subtypes may be present in different neuronal populations in pre- and post-synaptic locations.

Our results show that two major categories of nicotinic receptors, one K-BTX insensitive and the other one K-BTX sensitive are present in human cortex (Sugaya et al., 1990) (Fig. 3). Our experiments show that α-BTX displaces high (100%) and low (25%) affinity binding sites of K-BTX but does not decrease the electrically evoked release of ACh. The K-BTX sensitive subtype has an α-BTX binding site which includes the high affinity binding site of K-BTX. The other subtype does not have an α-BTX binding site and modulates ACh release from rat frontal cortex through a K-BTX sensitive nicotinic receptor. Thus, our data are in agreement with the molecular biological data on receptor gene families in mammalian species (Heinemann et al., 1988; Lindstrom et al., 1988) suggesting that human cortex has at least three different subtypes of nicotinic receptors, each showing specific kinetics, regional distribution and synaptic localization. These three subtypes are represented in the human frontal cortex. In addition, we have found that parietal, temporal lobe and hippocampus exhibit the same binding sites with a different receptor density and distribution. Hippocampal cortex (CA3) show the highest density in K-BTX binding sites.

Immunohistochemical Localization of Nicotinic Cholinergic Receptors

In order to achieve a detailed localization, autoradiography has been used to assess layer-specific receptor distribution (Whitehouse et al., 1988; Sugaya et al., 1990). However, cell-type specific alterations in nAChR distribution in AD cortices and the pre- or postsynaptical localization of the involved receptors remain still controversial. Recently, immunohistochemical techniques have localized AChRs in the human cerebral cortex at the light and electron microscopic level (Schroder et al., 1989a,b, 1990b). Using the monoclonal antibody (mAb) WF6 (Fels et al., 1986; Schroder et al., 1989a,b) we reported on the cellular distribution of nAChRs in the frontal cortex of AD patients as compared to age-matched and young controls (Fig. 4) (Schroder et al., 1990b).

Schroder et al. (1989a), using a mAb WF6, raised against purified torpedo nAChR, found in human neocortex that perikarya and dendrites of the projection neurons in layers III and V and fusiform cells displayed the most dense immunoreactivity. The immunoprecipitate was found on neuronal perikarya, dendrites and on the postsynaptic thickenings. In human cortex, ^{3}H-NIC autoradiography revealed the most dense labeling in "deeper layers" (Whitehouse and Kellar, 1987).

Using a double labeling technique for muscarinic (M35 antibodies) and nicotinic (WF6 antibodies) receptor proteins, Schroder et al. (1989b) found that immunofluorescence in human cortex was preponderant in pyramidal neurons located in layers II, III and V and their apical dendrites. It is interesting to note that about 30% of the cholinoceptive neurons displayed immunoreactivity for both receptor types.

Recent immunocytochemical studies of the normal human cerebral cortex have shown mainly pyramidal neurons of layers II/III and V to be nAChR-positive (Schroder et al., 1989a). The immunoprecipitate was localized in the postsynaptic densities of synaptic complexes (Schroder et al., 1989a). In this study, immunocytochemical patterns identical to those reported previously were found for both control groups. By contrast, a marked decrease of neurons was observed in the frontal cortex of AD patients, with the immunoprecipitate localized in the postsynaptic densities of synaptic complexes (Schroder et al., 1990a). The lack of a significant difference in the number of Nissl-stained neurons is in agreement with previous findings in the frontal cortex (Braak and Braak, 1986).

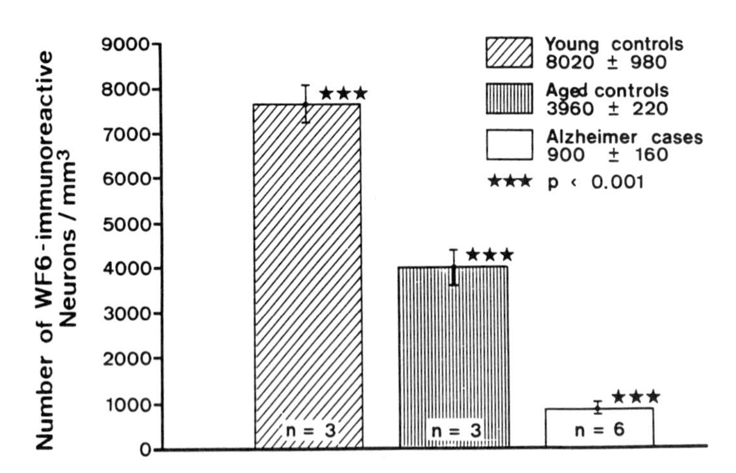

Figure 4. Mean numbers of WF6-immunoreactive cortical neurons per mm^3 for the three different groups (young controls, aged controls, Alzheimer patients) ($p < 0.001$). Error bars represent the standard error of the mean. (Modif. from Schroder et al., 1990b).

The cellular distribution of nAChR was studied by us in the frontal cortex (area 10) of AD patients and compared to age-matched and young controls using the mAb WF6 and an immunoperoxidase protocol (Schroder et al., 1990b). Statistical analysis revealed significant differences between the number of labeled neurons among all three groups tested (young contols > aged controls > Alzheimer cases) (Fig. 4). No differences were seen for cresylviolet-stained samples. These findings underline the fact that the nicotinic receptor decrease found with radioligand binding (Sugaya et al., 1990) may reflect a postsynaptic in addition to a presynaptic component.

These findings together with our previous results would indicate that nAChRs in the human cerebral cortex are located both pre- and postsynaptically as in the rat cerebral cortex (Schroder et al., 1989a). Lesions of the nucleus basalis in rats, aimed to mimic the cholinergic denervation in AD, did not produce significant changes in nAChR binding sites, with the exception of α-BTX binding sites (Downen et al., 1989), whereas ChAT activity was severely affected (Atack et al., 1989; Meyer et al., 1987; Wenk and Rokaeus, 1988). Therefore, these data do not support the suggestion that all cortical AChRs are located on the presynaptic cholinergic axons. At least two explanations need to be considered for this discrepancy. The loss of receptor protein might be independent of presynaptic changes. A second possibility is that the changes seen in the cortical neurons might reflect a secondary phenomenon of long-term cholinergic denervation, since the time of observation in animal models has been generally short (up to 13 weeks). In order to obtain further information, animal models need to be designed and studies on human brains investigating different stages of AD are necessary.

Visualization of Nicotinic Receptors by Positron Emission Tomography

Attempts to visualize nicotinic receptors in human brain by positron emission tomography (PET) have been made by Nyback et al. (1989) by studying the uptake of (-)[^{11}C]-NIC in brain. The ^{11}C-radioactivity in arterial plasma reaches a peak at 2 min and then declines to a steady state level which is maintained for 4-5 min. In the brain, radioactivity peaks at 4-6 min and declines slowly to 54 min. The regional distribution of radioactive is similar for (+) and (-) NIC with high accumulation in both cortical and subcortical regions, however, the (-) form reaches a higher activity in normal subjects than in smokers. The brain regional *in vivo* distribution of ^{11}C-NIC with PET is similar to the *in vitro* mapping of nicotinic receptor seen with conventional binding techniques. Evidence for a decreased number of nicotinic receptors in AD has been collected by Nordberg et al. (1990) using the PET technique. Both thalamus and cortex show differences in uptake of ^{11}C-NIC with lower uptake in the cortex of AD patients.

These PET studies offer the possibility of testing drug action *in vivo* and the effect of transplants in the patient.

ACKNOWLEDGEMENTS

The author wishes to thank Elizabeth Williams for technical assistance and Diana Smith for typing and editing the manuscript. Supported in part by National Institutes of Aging AG05416 and R.J. Reynolds Tobacco Company.

REFERENCES

Adem A (1987): Characterization of muscarinic and nicotinic receptors in neural and non-neural tissue - changes in Alzheimer's disease. *in*: "Comprehensive Summaries of Uppsala Dissertations from the Faculty of Pharmacy, Vol. 32", Distribution - Almqvist and Wiksell, Stockholm, Sweden, pp. 1-61.

Adem A, Nordberg A, Jossan SS, Sara V, Gillberg P-G (1989): Quantitative autoradiography of nicotinic receptors in large cryosections of human brain hemispheres. Neuroscience Lett 101:247-252.

Araujo DM, Lapchak PA, Collier B, Quirion R (1987): N-Methylcarbamyl choline, a novel nicotinic agonist, enhances acetylcholine release from rat brain slices. Soc Neurosci Abst. 16:326.2

Araujo DM, Lapchak PA, Robitaille Y, Gauthier S, Quirion R (1988): Differential alteration of various cholinergic markers in cortical and subcortical regions of human brain in Alzheimer's disease. J Neurochem 50:1914-1923.

Atack JR, Wenk GL, Wagster MV, Kellar KJ, Whitehouse PJ, Rapoport SI (1989): Bilateral changes in neocortical ^{3}H-pirenzepine and ^{3}H-oxotremorine-M binding following unilateral lesions of the rat nucleus basalis magnocellularis; an autoradiographic study. Brain Res 483:367-372.

Bird TD, Stranahan S, Sumi SM, Raskind M (1983): Alzheimer's disease: choline acetyltransferase activity in brain tissue from clinical and pathological subgroups. Ann Neurol 14:284-293.

Braak H, Braak E (1986): Ratio of pyramidal cells versus non-pyramidal cells in the human frontal isocortex and changes in ratio with aging and Alzheimer's disease. Prog Brain Res 70:185-212.

Chiappinelli VA (1985): Actions of snake venom toxins on neuronal nicotinic receptors and other neuronal receptors. Pharmac Ther 31:1-32.

Clarke PBS, Schwartz RD, Paul SM, Pert CB, Pert A (1985): Nicotinic binding in rat brain - autoradiographic comparison of ^{3}H-nicotine and ^{125}I-α-bungarotoxin. J Neurosci 5:1307-1315.

Davies P (1979): Neurotransmitter-related enzymes in senile dementia of the Alzheimer type. Brain Res 171:319-327.

Davies P, Feisullin S (1981): Postmortem stability of alpha-bungarotoxin binding sites in mouse and human brain. Brain Res 216:449-454.

de la Garza R, Freedman R, Hoffer BJ (1989): Kappa-bungarotoxin blockade of nicotine electrophysiological actions in cerebellar Purkinje neurons, Neurosci Lett 99:95-100.

DeSarno P, Giacobini E, Clark B (1988a): Changes in nicotinic receptors in human and rat CNS. Fed Proc 2:364.

DeSarno P, Giacobini E, McIlhany M, Clark B (1988b): Nicotinic receptors in human CNS: a biopsy study. in: "Proceedings 2nd Intl. Symposium on Senile Dementias", Agnoli A (ed), John Libbey Eurotext, Ltd., Montrouge, France, pp. 329-334.

Downen M, Sugaya K, Arneric SP, Giacobini E (1989): Presynaptic markers of cholinergic function in cortex following ibotenic acid lesion of the basal forebrain. in: "Advances in Behavioral Biology, Novel Approaches to the Treatment of Alzheimer Disease", Vol. 36, Meyer EM, Simpkins JW, Yamamoto J (eds), Alan R. Liss, New York, pp. 255-268.

Duvoisin RM, Deneris ES, Patrick J, Heinemann S (1989): The functional diversity of the neuronal nicotinic acetylcholine receptors is increased by a novel subunit: ß4. Neuron 3:487-496.

Fels G, Plumer-Wilk R, Schreiber M, Maelicke A (1986): A monoclonal antibody interfering with binding and response of the acetylcholine receptor. J Biol Chem 261:15746-15754.

Giacobini E (1990): The cholinergic system in Alzheimer disease. in: "Progress in Brain Research, Vol. 84", Aquilonius SS, Gillberg PG (ed), Elsevier, Amsterdam, pp. 321-332.

Giacobini E, DeSarno P, McIlhany M, Clark B (1988a): The cholinergic receptors system in the frontal lobe of Alzheimer patients. in: "Nicotinic Acetylcholine Receptors in the Nervous System - NATO ASI Series H", Vol. H25, Clementi F, Gotti C, Sher E (eds)., Springer-Verlag, Berlin, pp. 367-378.

Giacobini E, Sugaya K, DeSarno P, Chiappinelli V (1988b): Three subtypes of nicotinic receptors in human cortex. Soc Neurosci Abst p. 55.4.

Giacobini E, DeSarno P, Clark B, McIlhany M. (1989): The cholinergic receptor system of the human brain. Neurochemical and pharmacological aspects in aging and Alzheimer. in: "Progress in Brain Research, Vol. 79", Nordberg A (ed), Elsevier, Amsterdam, pp. 335-343.

Goldman D, Simmons D, Swanson LW, Patrick J, Heinemann S (1986): Mapping of brian areas expressing RNA homologous to two different acetylcholine receptor α-subunit cDNAs. Proc Natl Acad Sci 83:4076-4080.
Goldman D, Deneris E, Luyten W, Kochlar A, Patrick J, Heinemann S (1987): Members of a nicotine acetylcholine receptor gene family are expressed in different regions of the mammalian central nervous system. Cell 48:965-973.
Heinemann S, Boulter J, Deneris E, Connolly J, Gardner P, Wada E, Wada K, Ballivet M, Swanson L, Patrick J (1988): The nicotinic acetylcholine receptor gene family. in: "Nicotinic Acetylcholine Receptors in the Nervous System - NATO ASI Series H", Vol. H25, Clementi F, Gotti C, Sher E (eds), Springer-Verlag, Berlin, pp. 173-191.
Kellar KJ, Whitehouse PJ, Martino-Barrows AM, Marcus K, Price DL (1987): Muscarinic and nicotinic cholinergic binding sites in Alzheimer's disease cerebral cortex. Brain Res 436:62-68.
Lindstrom J, Whiting P, Schoepfer R, Luther M, Casey B (1988): Structure of neuronal nicotinic receptors. in: "Nicotinic Acetylcholine Receptors in the Nervous System - NATO ASI Series, Vol. H25", Clementi F, Gotti C, Sher E (eds), Springer-Verlag, Berlin, pp. 159-172.
London ED, Ball MJ, Waller SB (1989): Nicotinic binding sites in cerebral cortex and hippocampus in Alzheimer's dementia. Neurochem Res 14(8):745-750.
Marks HJ, Collins AC (1982): Characterization of nicotine binding in mouse brain and comparison with the binding of alpha-bungarotoxin and quinuclinidyl benzylate. Mol Pharmacol 22:554-564.
Meyer EM, Arendash GW, Judkins JH, Ying L, Wade C, Kem WR (1987): Effects of nucleus basalis lesions on the muscarinic and nicotinic modulation of ^{3}H-acetylcholine release in the rat cerebral cortex. J Neurochem 49:1758-1762.
Morley BJ, Kemp GE (1981): Characterization of a putative nicotinic acetylcholine receptor in mammalian brain. Brain Res Rev 3:82-104.
Morley BJ, Dwyer DS, Strang-Brown PF, Bradley RJ, Kemp GE (1983): Evidence that certain peripheral anti-acetylcholine receptor antibodies do not interact with brain bungarotoxin binding sites. Brain Res 262:109-116.
Neff P, Oneyser C, Alliod C, Couturier S, Ballivet M (1988): Genes expressed in the brain define three distinct neuronal nicotinic acctylcholine receptors. EMBO J 7:595-601.
Nordberg A, Winblad B (1986a): Brain nicotinic and muscarinic receptors in normal aging and dementia. in: "Alzheimer's and Parkinson's Disease - Advances in Behavioral Biology, Vol 29", Fisher A, Hanin I, Lachman C (eds), Plenum Press, New York, pp. 95-108.
Nordberg A, Winblad B (1986b): Reduced number of (^{3}H)-nicotine and (^{3}H)-acetylcholine binding sites in the frontal cortex of Alzheimer's brains. Neurosci Lett 72:115-119.
Nordberg A, Adem A, Nilsson L, Winblad B (1987): Cholinergic deficits in CNS and peripheral non-neuronal tissue in Alzheimer dementia. in: "Cellular and Molecular Basis of Cholinergic Function", Dowdall MJ, Hawthorne JN (eds), Ellis Horwood, New York, pp. 858-868.
Nordberg A, Adem A, Nilsson-Hakansson L, Winblad D (1990): New approaches to clinical and postmortem investigations on cholinergic mechanisms. in: "Progress in Brain Research, Vol. 84", Aquilonius SM, Gillberg P-G (eds), Elsevier, Amsterdam (In Press).
Nyback H, Nordberg A, Langstrom B, Halldin C, Hartvig P, Ahlin A, Swahn C-G, Sedvall G (1989): Attempts to visualize nicotinic receptors in the brain of monkey and man by positron emission tomography.

in: "Progress in Brain Research, Vol. 79", Nordberg A, Fuxe K, Holmstedt B, Sundwall A (eds), Elsevier, Amsterdam, pp. 313-319.

Perry EK, Smith CJ, Perry RH, Johnson M, Fairbairn AF (1989): Nicotinic (^{3}H-Nicotine) receptor binding in human brain: characterization and involvement in cholinergic neuropathology. Neurosci Res Commun 5(2):117-124.

Rinne JO, Lonnberg P, Marjamaki P, Rinne UP (1989): Brain muscarinic receptor subtypes are differently affected in Alzheimer's disease and Parkinson's diseases. Brain Res 483:402-406.

Rossor MN, Garrett NJ, Johnson AL, Mountjoy CQ, Roth M, Iversen LL (1982): A post-mortem study of the cholinergic and GABA systems in senile dementia. Brain 105:313-330.

Schoepfer R, Whiting P, Esch F, Blacher R, Shimasaki S, Lindstrom J (1988): cDNA clones coding for the structural subunit of a chicken brain nicotinic acetylcholine receptor. Neuron 1:241-248.

Schroder H, Zilles K, Maelicke A, Hajos F (1989a): Immunohisto- and cytochemical localization of cortical nicotinic cholinoceptors in rat and man. Brain Res 502:287-295.

Schroder H, Zilles K, Luiten PGM, Strosberg A, Aghchi AR (1989b): Human cortical neurons contain both nicotinic and muscarinic acetylcholine receptors: an immunocytochemical double-labeling study. Synapse 4:319-326.

Schroder H, Zilles K, Luiten PGM, Strosberg AD (1990a): Immunocytochemical visualization of muscarinic cholinoceptors in the human cerebral cortex. Brain Res. (In Press), 1990a.

Schroder H, Giacobini E, Struble RG, Zilles K, Maelicke A (1990b): Nicotinic cholinoceptive neurons of the frontal cortex are reduced in Alzheimer's disease. Neurobiol Aging (In Press).

Shimohama S, Taniguichi T, Fujiwara M, Kameyama M (1985): Biochemical characterization of the nicotinic cholinergic receptors in human brain: binding of ^{3}H-(-)-nicotine. J Neurochem 45:604-610.

Shimohama S, Taniguchi T, Fujiwara M, Kameyama M (1986): Changes in nicotinic and muscarinic cholinergic receptors in Alzheimer-type dementia. J Neurochemistry 46(1):288-293.

Sugaya K, Giacobini E, Chiappinelli VA (1990): Nicotinic acetylcholine receptor subtypes in human frontal cortex: changes in Alzheimer's disease. J Neurosci Res (In Press).

Sugiyama H, Yamashita Y (1986): Characterization of putative nicotinic acetylcholine receptors solubilized from rat brains. Brain Res 373:22-26.

Swanson LW, Simmons DH, Whiting PJ, Lindstrom J (1987): Immunohistochemical localization of neuronal nicotinic receptors in rodent central nervous system. J Neurosci 7:3334-3342.

Vidal C, Changeux J-P (1989): Pharmacological profile of nicotinic acetylcholine receptors in the rat prefrontal cortex - an electrophysiological study in a slice preparation. Neuroscience 29(2):261-270.

Wada E, Wada K, Boulter J, Deneris E, Heinemann S, Patrick J, Swanson LW (1989): Distribution of Alpha$_2$, Alpha$_3$, Alpha$_4$ and Beta$_2$ neuronal nicotinic receptor subunit mRNAs in the central nervous system: a hybridization histochemical study in the rat. J Comp Neurol 284:314-335.

Wenk GL, Rokaeus A (1988): Basal forebrain lesions differentially alter galanine levels and acetylcholinergic receptors in the hippocampus and neocortex. Brain Res 460:17-21.

Whitehouse PJ, Kellar KJ (1987): Nicotinic and muscarinic cholinergic receptors in Alzheimer's disease and related disorders. in: "Proceedings of the Fourth Meeting of the International Study Group on the Pharmacology of Memory Disorders Associated with Aging", Wurtman RJ, Corkin SH, Growdon JH (eds), Zurich Center for Brain

Sciences and Metabolism Charitable Trust, Cambridge, Mass, pp. 169-178.
Whitehouse PJ, Martino AM, Antuono PG, Lowenstein PR, Coyle JT, Price DL, Kellar KJ (1986): Nicotinic acetylcholine binding sites in Alzheimer's disease. Brain Res 371:146-151.
Whitehouse PJ, Martino AM, Wagster MV, Price DL, Mayeux R, Atack JR, Kellar KJ (1988): Reductions in ^{3}H-nicotinic acetylcholine binding in Alzheimer's disease and Parkinson's disease: an autoradiographic study. Neurology 38:720-723.
Wolf KM, Ciarleglio A, Chiappinelli VA (1987): K-Bungarotoxin - binding of a neuronal nicotinic receptor antagonist to chick optic lobe and skeletal muscle. Brain Res 89:1-10.
Zubenko GS, Moossy J, Martinez AJ, Rao GR, Kopp U, Hanin I (1989): A brain regional analysis of morphologic and cholinergic abnormalities in Alzheimer's disease. Arch Neurol 46:634-638.

MACROMOLECULAR CHANGES IN THE AGING BRAIN

AnnaMaria Giuffrida Stella

Institute of Biochemistry, Faculty of Medicine
University of Catania
Viale Andrea Doria 6, 95125, Catania, Italy

Several studies suggest that a decline in transcriptional and translational events may play a crucial role in the molecular mechanism underlying the aging process (for a review see Giuffrida and Lajtha, 1987).

Impaired transcription during aging in the brain is also suggested by decreased RNA content measured in specific brain areas or specific neuronal types (Ringborg, 1966; Johnson and Strehler, 1972; Chaconas and Finch, 1973).

Moreover it has been shown that chromatin structure in cerebral and cerebellar neurons is significantly altered in the rat brain during aging (Berkowitz et al., 1983).

Major reductions (30-50%) in the hybridization of non-repetitive DNA to nuclear RNA from aging mice have been observed (Cutler, 1975). This finding suggests that a progressive decrease in the number of different types of genes expressed may be a general characteristic of the aging process.

Aging might also be due to an impaired maturation of the primary gene transcripts caused by the interference of superoxide radicals with RNA-matrix attachment, inducing a release in the cytoplasm of immature mRNA (Schroder et al., 1987).

Moreover it has been shown that RNA synthesis by isolated neuronal nuclei from cerebral cortex is substantially reduced in senescent rats, whereas that of glial nuclei remained constant (Lindholm, 1986).

In previous studies we measured the synthesis and content of the various RNA species (Condorelli et al., 1989) and the synthesis, turnover and content of proteins in various brain regions and in subcellular fractions (Avola et al., 1988).

The results of the <u>in vivo</u> incorporation of $[^3H]$uridine into RNA extracted from different brain regions (cerebral

Plasticity and Regeneration of the Nervous System
Edited by P.S. Timiras *et al.*, Plenum Press, New York, 1991

cortex, cerebellum, hippocampus, striatum, hypothalamus, brain stem) of 4, 12 and 24 month-old rats of age demonstrated a significant decrease of RNA labeling in cerebral cortex (-15%; $p<0.05$) of 24-month-old rats while no significant differences at the various ages were observed in the other brain areas examined. RNA/DNA ratio was significantly decreased in cerebral cortex, cerebellum and striatum of 24 months-compared to 4-month-old animals (Condorelli et al., 1989).

The ratio of the specific radioactivity (dpm/mg RNA) of microsomal RNA to nuclear RNA (Mc/N) from cerebral cortex showed that at all the ages investigated the specific radioactivity of nuclear RNA (N) exceeded that of microsomal (Mc) RNA. The ratio Mc/N did not significantly change during aging.

In cerebral cortex the content of total RNA, poly$(A)^+$ mRNA and poly$(A)^-$ RNA was significantly decreased at 24 months of age, compared to 4 months. The decrease in the content of poly$(A)^+$ mRNA was more pronounced than the decrease in poly$(A)^-$ RNA in old animals.

Our results show that there is no general decrease in brain RNA synthesis but that this phenomenon is evident only in some brain regions. Indeed changes in macromolecular synthesis during senescence might be restricted to a specific group of neurons which are selectively or earlier affected by the aging process. Our results confirm the existence of significant region specific differences in the brain during aging.

CHANGES OF PROTEIN SYNTHESIS AND COMPOSITION

Protein synthesis during aging has been studied extensively but the published reports are often highly variable and, in some instances, contradictory (Ekstrom et al;, 1980; Fando et al., 1980; Dwyer et al., 1980; Makrides, 1983; Ingvar et al., 1985; Cosgrowe and Rapoport, 1987; Avola et al., 1988).

In a previous study (Avola et al., 1988) protein synthesis rate in some brain regions (cerebral cortex, cerebellum, hippocampus, hypothalamus and striatum) and also in subcellular fractions (microsomes, mitochondria, cytosol) from cerebral cortex and cerebellum was measured.

Protein synthesis was evaluated using the injection of flooding dose of labeled amino acid as precursor, according to the method of Dunlop et al.(1975).

At all the ages examined among the subcellular fractions purified from cerebral cortex, the highest incorporation rate was found in microsomes (1.4% per hour) followed by cytosolic proteins (0.85% per hour) and mitochondrial proteins (0.4%).

In the cerebellum the mean incorporation rate in the total amount proteins was quite similar to that found in cerebral cortex.

In the other regions examined the incorporation rate was 0.5% per hour in hypothalamus and hippocampus and 0.4% per hour in striatum.

However from the statistical evaluation of the data obtained in the various brain regions no significant changes in protein incorporation rate in aged animals was observed.

Our results are in agreement with those of some AA (Fando et al., 1980; Dwyer et al., 1980) who found no difference in protein synthesis in cerebellum and brain stem of old animals, but in contrast with the slight (9%) reduction observed in forebrain.

These discrepancies could be explained by the differences of protein synthesis in the different rat strains used.

Moreover it is possible that only a small fraction of specific brain proteins are more or less affected by aging in the various brain regions.

Indeed previous studies on aging in the rat brain have concluded that there are selective changes in the amounts of individual brain proteins.

Cicero et al. (1972) reported that S-100 protein is highly concentrated in the cerebellum. Originally, S-100 protein was thought to only exist in glial cells but nowadays it is known to also occur in neurons and to be capable of forming tight complexes with SPM (Mahler, 1984). Cicero et al.(1972) analysing the levels of S-100 protein in various brain regions observed a rapid accumulation of this protein associated with the developmental process. Recently it has been demonstrated by immunoassay that S-100 β protein, which is predominantly localized in glial cells, increased gradually in the cerebral cortex with age while its concentration in the cerebellum and brain stem remained relatively constant from 2 to 30 months of age (Kato et al., 1990). The amount of neuron specific enolase (14-3-2 protein) shows a slight decrease in some brain regions during aging (Cicero et al. 1972).

Cosgrowe et al. (1987) reported, using a two- dimensional electrophoretic analysis of brain proteins, that there is one protein with Mw 21 kDa, occurring in higher amount in the cerebellum than in other brain regions at 3-4 months and at 28-30 months of age.

Therefore it is difficult to find striking age-related differences by measuring the rate of brain protein synthesis or analyzing the proteins of whole brain regions. Instead it is of great importance to evaluate the possible modifications caused by aging on particular proteins from different brain regions or specific subcellular fractions. However, few studies have been performed on specific proteins with particular brain functions.

Further studies have focused on changes of mitochondrial and synaptosomal protein composition during aging.

CHANGES IN MITOCHONDRIAL MEMBRANE PROTEIN COMPOSITION DURING AGING

Changes in mitochondrial structural components with a concomitant impairment of mitochondrial function have been observed (Vorbeck et al., 1982; Hansford, 1983). A decrease in the content of cytochromes and modifications of cytochrome oxidase activity during aging (Benzi et al., 1982; Benzi and Giuffrida, 1987) suggest that the mitochondria from old animals contain less respiratory units than those of young animals, resulting in an impairment of energy transduction during aging.

It has been hypothesized that a disorganization of proteins and lipids forming the structurally and functionally active membranes, with a consequent decrease in membrane fluidity, may influence the function of membrane proteins involved in mitochondrial enzyme activities or in transport processes (Vitorica et al., 1985).

Moreover no adequate mechanism for the repair of mitochondrial DNA strand breaks has been found. This could constitute the basis for the mitochondrial genetic injury caused by oxidative damage. Indeed mitochondria are particularly vulnerable to oxidative damage because mitochondrial DNA is not protected by histones and non-histone proteins by contrast with nuclear DNA.

While there are many reports dealing with the morphology, the number of mitochondria (by electronmicroscopy or histology) in aging tissues, and with lipid composition or enzyme activities (Hansford, 1983,Benzi and Giuffrida, 1987), very little is known about the possible quantitative modifications of mitochondrial membrane proteins with advancing age.

In a previous study (Turpeenoja et al., 1988) we investigated the possible modifications in the protein composition of free non-synaptic mitochondria obtained from rat cerebellum during aging. The outer membranes (OM), inner membranes (IM) and matrix (MX) were separated. The various mitochondrial membrane proteins were separated by SDS-polyacrilamide gel electrophoresis. The percentage of each protein band in the gel was obtained by densitometry.

No significant changes in the amounts of OM and MX proteins were observed while some IM proteins were significantly decreased at 24 months of age when compared to the values found at 4 months.

The amount of the protein with molecular weight 75 kDa, corresponding to that of one subunit of NADH-dehydrogenase showed an overall decrease after 4 months, the decrease being slightly significant ($p<0.05$) at ages 16 and 20 months and becoming more significant ($p<0.01$) at the oldest age.

Two other proteins with Mw 16 and 14 kDa, corresponding to the molecular weights of subunit IV of cytochrome oxidase and cytochrome c, respectively, showed different quantitative fluctuations during aging. In particular the amount of the 14 kDa protein remained constant until 20 months of age declining sharply thereafter, being only 60% of the amount found in 4-month-old animals at the oldest age.

In further studies we measured the protein composition of free (non synaptic mitochondria) from cortex and striatum of rats at different ages (Ragusa et al., 1989).

The results obtained showed a statistically significant decrease in the amount of the two mitochondrial proteins, with molecular weights 20 kDa and 16 kDa, in both brain regions examined. For the 20 kDa protein the diminution in striatum was 69% at 12 months and 80% at 24 months compared to the values observed at 4 months of age. In the cerebral cortex the decrease was only significant (-93 %) in the oldest animals (24 months). The decrease in the quantity of 16 kDa protein was 41% in cerebral cortex and 61% in striatum at 12 months and about 50% in both areas at 24 months.

The striatal mitochondrial proteins were influenced to a greater extent by aging. The greater vulnerability of the striatum to the aging process is in agreement with previous studies (Condorelli et al., 1989).

Our results demonstrate that some specific mitochondrial proteins undergo significant quantitative modifications during aging. According to the present knowledge about mitochondrial protein composition, it is probable that the protein with molecular weight 20 kDa may represent one subunit of the complex V of the respiratory chain and the protein with molecular weight 16 kDa the subunit IV of cytochrome oxidase. However an exact characterization of these proteins cannot be done on the basis of the molecular weight alone, but needs further investigation.

We are aware of the limits of the methods used to quantify the changes in the mitochondrial proteins at different ages, however, the significant decrease obtained in the protein amount, which may have a crucial role in the mitochondrial respiratory chain, is in good agreement with the impairment of energy transduction observed in aging brain (Benzi and Giuffrida, 1985; 1987).

As some of the mitochondrial proteins are codified by mitochondrial DNA and others by nuclear DNA (Massie et al., 1975; Giuffrida Stella, 1983), which expression may be independently regulated (Bailey and Webster, 1984), it is difficult to interpret the effects of aging on mitochondrial protein changes. More specific information on the synthesis and turnover of the various subunits of mitochondrial and extramitochondrial proteins should be obtained. The impairment of either one or both nuclear and mitochondrial genomes might result in a modified protein synthesis rate or in the production of abnormal proteins. These proteins might be accumulated due to the inefficiency of the degradation system (Nohl et al., 1978). Indeed the increased amount of some proteins may be related to a decreased degradation rate leading to the accumulation of some polypeptides including abnormal and inactive forms of proteins. Therefore, the observed changes in mitochondrial inner membrane proteins may be due to alterations during aging either in the synthesis or in the degradation rates. Also the transport system and/or the permeability of membranes to allow the proteins coded by nuclear DNA to enter inside the mitochondria could be impaired in senescent animals.

CHANGES IN SYNAPTOSOMAL PROTEIN COMPOSITION DURING AGING

According to the studies of Nagy et al. (1983) the hydrophobic region of rat brain synaptosomal plasma membranes becomes more rigid and the fluidity of regions located near the surface of the membranes decreases in old animals. Studies on SPM proteins, suggest that conformational changes take place during aging (Nagy et al. 1983). These results are consistent with the membrane hypothesis of aging attributing a primary role to the membrane damage induced by free-radicals in cellular aging.

Increased cross-linking may cause a decreased ion and water permeability of the nerve cell membranes in old animals. As a consequence the cells accumulate potassium and lose water to the intracellular space. This change in cell physicochemical properties could increase the condensation of chromatin and as a consequence reduce the rate of RNA synthesis (Nagy et al. 1983).

The age-dependent modifications of synaptosomal plasma membrane (SPM) proteins in three different rat brain regions (cerebral cortex, cerebellum and striatum) at various ages (4, 12 and 24 months) were studied (Turpeenoja et al., 1990). The proteins were separated by gel-electrophoresis and the quantity of the different polypeptides was determined densitometrically from the stained gels.

The protein profile of synaptic plasma membranes (SPM) isolated from striatum, cerebral cortex and cerebellum, and the densitometric quantification of the gels showed significant age-dependent modifications in the protein composition of SPM prepared from the various rat brain regions.

The amount of the various proteins at the different ages was expressed as percentage of each band compared to all the proteins obtained in the gel and determined by densitometric scanning of the gels at λ =600 nm.

Several age-related modifications in the amount of the SPM proteins were observed. The proteins of cerebral cortex and striatum were modified by aging to a greater extent than the cerebellar ones. The amount of some polypeptides was increased while that of others was decreased probably indicating altered protein synthesis and/or degradation rates of specific proteins during aging.

In the cerebral cortex among the 16 protein group isolated, an age-related decrease in the amount of proteins with Mw 23 kDa at 12 months and of proteins with 18 kDa Mw at 24 months of age was observed.

On the contrary an increase in the amount of SPM proteins with 70 kDa Mw at 12 months and with 36 kDa Mw at 24 months was found.

In the cerebellum among the 16 SPM protein groups isolated, only three proteins were modified during aging. An age-related decrease was found at 24 months for the 83, 36 and 18 kDa proteins compared to the values obtained in 4-month-old animals.

In the striatum of the 23 protein groups obtained a statistically significant decrease at 24 months of age was found in the amount of SPM proteins with Mw 79, 62, 59 and 26 kDa.

On the contrary a statistically significant increase was obtained in the amount of two proteins with Mw 36 and 31 kDa at 24 months of age and of one protein with Mw 70 kDa at 12 months of age.

According to our study some SPM proteins are affected by aging. Among the polypeptides isolated, about seven showed statistically significant changes in the striatum during aging while in the cerebral cortex five proteins and in the cerebellum only three SPM proteins, were modified during aging. The greater vulnerability of the striatum to the aging process, compared to the other regions is also supported by previous studies (Condorelli et al. 1989).

The complex protein composition of SPM is only partially known (Mahler 1977; 1984) and several proteins are still without an exact characterization. Due to the limits of the one-dimensional electrophoretic separation for protein characterization, an exact definition of the SPM proteins which showed age-related modifications is not possible. According to their molecular weights, the protein with 83 kDa Mw may correspond to the subunit 1a of synapsin and the protein with 79 kDa Mw to the subunit Ib of synapsin (Ueda and Greengard, 1977; Nestler and Greengard, 1983). Their content decreased at 24 months in the cerebellum and at 12 and 24 months in the striatum respectively. The 70 kDa protein may contain the microtubule associated protein, MAP2c (Garner et al. 1988), its content increased in cerebral cortex and striatum at 12 months.

The SPM proteins with Mw about 60 kDa may include the nicotinic acetylcholine receptor (subunit γ 60 kDa, δ 65 kDa). The 59 kDa protein has a Mw similar to that of the β subunit of GABA-benzodiazepine receptor (Kirkness and Turner 1986).

The protein with Mw 36 kDa may correspond to the β_1 subunit of Na^+ channel (Catterall, 1988). We observed that this protein was significantly modified by aging in all the three brain regions studied. The amount of this protein increased in the cerebral cortex and striatum and decreased in the cerebellum at 24 months of age.

The 18 kDa protein has a molecular weight similar to that reported for calmodulin (Mahler 1977; 1984). The amount of this small, Ca^{2+}-binding protein is modified during aging in the rat brain (Hoskins and Scott 1984, Hoskins et al. 1985). According to our results this protein decreases drastically in the cerebral cortex and in the cerebellum.

The amount of the 23 kDa protein was significantly lower in SPM preparations from cerebral cortex at 12 months of age while it remained constant in cerebellum at all the ages investigated. This protein may be identified with D_3 subunits (Bock, 1984).

The different age-related modifications of SPM proteins isolated from the various brain regions indicate that there may be region-specific regulation of SPM protein composition. Most of the SPM proteins did not change in their relative amount as a function of age, indicating that different control mechanisms may be operating in the senescent brain (Buell and Coleman 1979, Clemens 1983).

The different forms of tubulin, present also in synaptosomal plasma membranes (Babitch 1981, Mahler 1977; 1984), seem to be unchanged in brain during aging (von Hungen et al. 1981).

Our results demonstrate that in the various brain regions the synaptosomal protein composition is affected to a different extent during aging. The identity of the SPM proteins undergoing age-related changes has yet to be determined. We can suggest that alterations in single proteins or groups of SPM proteins may contribute to the structural and functional impairment observed in the various brain regions during aging.

The age-related modifications in the protein composition of SPM may cause changes in many brain functions, such as neurotransmission, ionic transport and the bioenergetic system.

REFERENCES

Avola, R., Condorelli, D.F., Ragusa, N., Renis, M., Alberghina, M., Giuffrida Stella, A.M., and Lajtha, A., 1988, Protein synthesis rates in rat brain regions and subcellular fractions during aging, Neurochem. Res., 13:337-342.

Babitch, J.A., 1981, Synaptic plasma membrane tubulin may be an integral constituent, J. Neurochem., 37:1394-1400.

Bailey, P.J., and Webster, G.C., 1984, Lowered rate of protein synthesis by mitochondria isolated from organisms of increasing age, Mech Ageing Dev., 24:233-241.

Benzi, G., and Giuffrida, A.M., 1985, Bioenergetics of hypoxic brain during aging, Mol. Physiol., 8:535-547.

Benzi, G., and Giuffrida, A.M., 1987, Changes of synaptosomal energy metabolism induced by hypoxia during aging, Neurochem. Res., 12:149-157.

Benzi, G., Arrigoni, E., Pastoris, O., Villa, R.F., Dossena, M., Agnoli, A., and Giuffrida, A.M., 1982, Drug action on the metabolic changes induced by acute hypoxia on synaptosomes from the cerebral cortex, J. Cereb. Blood. Flow. Metabol., 2:229-239.

Berkowitz, E.M., Sanborn, A.C., and Vaughan, D.D., 1983, Chromatin structure in neuronal and neuroglial cell nuclei as function of age, J. Neurochem., 41:516-523.

Bock, E., 1984, Membrane markers of the nervous system. in: "Handbook of Neurochemistry," A. Lajtha, ed., Vol 7, 2nd ed., pp. 231-244, Plenum Press, New York.

Buell, S.J., and Coleman, P.D., 1979, Dendritic growth in the aged human brain and failure of growth in senile dementia, Science, 206:854-856

Catterall, W.A., 1988, Molecular properties of sodium and calcium channels. Abstracts of the 14th Int Congress of Biochemistry. Th, July 14. p.24.

Chaconas, G. and Finch, C.E., 1973, The effect of aging on RNA/DNA ratios in brain region of the C57BL/6J male mouse, J. Neurochem., 21:1469-1473.

Cicero, T.J., Ferrendelli, J.A., Suntzeff, V., and Moore, B.W., 1972, Regional changes in CNS levels of the S-100 and 14-3-2 proteins during development and aging of the mouse, J. Neurochem., 19:2119-2125

Clemens, J.A., 1983, Morphological changes in the hypothalamus and other brain areas influencing endocrine function during aging, in: "Neuroendocrinology of aging," J. Meites, ed., p.31 Plenum Press, New York.

Condorelli, D.F., Avola, R., Ragusa, N., Reale, S., Renis, M., Villa, R.F. and Giuffrida Stella, A.M., 1989, Age-dependent changes of nucleic acid labeling in different rat brain regions, Neurochem. Res., 14:701-706.

Cosgrowe, J.W., and Rapoport, S.I., 1987, Absence of age differences in protein synthesis by rat brain, measured with an initiating cell-free system, Neurobiol. Aging 8:27-34.

Cosgrowe, J.W., Atack, J.R., and Rapoport, S., 1987, Regional analysis of rat brain proteins during senescence, Exp. Gerontol., 22:187-198.

Cutler, R.G., 1975, Transcription of unique and reiterated DNA sequences in mouse liver and brain tissue as a function of age, Exp. Gerontol., 10:37-60.

Dunlop, D.S., van Elden, W., and Lajtha, A., 1975, A method for measuring brain protein synthesis rates in young and adult rats, J. Neurochem., 24:337-344

Dwyer, B.E., Fando, J.L., and Wasterlain, C.G. 1980, Rat brain protein synthesis declines during postdevelopmental aging, J. Neurochem., 35:746-749.

Ekstrom, R., Liu, D.S.H., and Richardson, A., 1980, Changes in brain protein synthesis during the life span of male Fischer rats, Gerontology, 26:121-128.

Fando, J.L., Salinas, M., and Wasterlain, C.G., 1980, Age-dependent changes in brain protein synthesis in the rat, Neurochem. Res., 5:373-383.

Garner, C.C., Brugg B., and Matus, A.C., 1988, A 70-kilodalton microtubule associated protein (MAP2c) related to MAP2, J. Neurochem., 50:609-615

Giuffrida Stella, A.M., 1983, Nucleic acids in developing brain, in: "Handbook of Neurochemistry, A. Lajtha, ed., vol 5, pp. 227-250, Plenum Press, New York.

Giuffrida Stella, A.M., and Lajtha, A., 1987, Macromolecular turnover in brain during aging, Gerontology, 33:136-148.

Hansford, R.G., 1983, Bioenergetics in aging, Biochim. Biophys. Acta, 726:41-80.

Hoskins, B., and Scott, J.H., 1984, Changes in activities of calmodulin-mediated enzymes in rat brain during aging, Mech. Ageing Devel., 26:231-239.

Hoskins, B., Ho, I.K., and Meydrech, E.F., 1985, Effects of aging and morphine administration on calmodulin and calmodulin-regulated enzymes in striata of mice, J. Neurochem., 44:1069-1073.

Ingvar, M.C., Maeder, P., Sokoloff, L., and Smith, C.B., 1985, Effect of ageing on local rates of cerebral protein synthesis in Sprague-Dawley rats, Brain, 108:155-170.

Johnson, R., and Strehler, B.L., 1972, Loss of genes coding for ribosomal RNA in ageing brain cells, Nature, Lon., 240:412-414.

Kato, K., Suzuki, F., Morishita, R., Asano, T. and Sato, T. 1990, Selective increase in S-100β protein by aging in rat cerebral cortex, J. Neurochem., 54:1269-1274.

Kirkness, E.F., and Turner, A.J., 1986, The γ-aminobutyrate/benzodiazepine receptor from pig brain: purification and characterization of the receptor complex from cerebral cortex and cerebellum. Biochem. J., 233:265-270

Lindholm, D.B., 1986, Decreased transcription on neuronal polyadenylated RNA during senescence in nuclei from rat brain cortex, J. Neurochem., 47:1503-1506.

Mahler, H., 1977, Proteins of the synaptic membrane, Neurochem. Res., 2:119-147

Mahler, H., 1984, Synaptic proteins, in: "Handbook of Neurochemistry," A. Lajtha, ed., Vol 7, pp 111-134, 2nd ed., Plenum Press, New York.

Makrides, S.C., 1983, Protein synthesis and degradation during aging and senescence, Biol. Rev., 58:343-422.

Massie, H.R., Baird, M.B., McMahon, M.M., 1975, Loss of mitochondrial DNA with aging, Gerontology, 21:231-237.

Nagy, K., Simon, P., and Nagy, I., 1983, Studies in synaptosomal membrane of rat brain during aging. Biochem. Biophys. Res. Commun., 117:686-694.

Nestler, E.J., and Greengard, P., 1983, Protein phosphorylation in the brain. Nature, 305:583-588.

Nohl, H., Breuninger, V., Hegner, D., 1978, Influences of mitochondrial radical formation on energy-linked respiration, Eur. J. Biochem., 90:385-390.

Ragusa, N., Turpeenoja, L., Magrì, G., Lahdesmaki, P., and Giuffrida Stella, A.M., 1989, Age-dependent modifications of mitochondrial proteins in cerebral cortex and striatum of rat brain, Neurochem. Res., 14:515-518.

Ringborg, U., 1966, Composition and content of RNA in neurons of rat hippocampus at different ages, Brain Res., 2:296-298.

Schroder, H.C., Messer, R., Bachmann, M., Bernd, A., and Muller, W.E.G., 1987, Superoxide radical-induced loss of nuclear restriction of immature mRNA: a possible cause for ageing, Mech. Ageing Dev., 41:251-266.

Turpeenoja, L., Villa, R.F., Magrì, G., and Giuffrida Stella, A.M., 1988, Changes in mitochondrial membrane proteins of rat cerebellum during aging, Neurochem. Res., 13:859-865.

Turpeenoja, L., Villa, R.F., Magrì, G., Ingrao, F., Gorini, A., and Giuffrida Stella, A.M., 1990, Modifications of synaptosomal plasma membrane protein composition in various brain regions during aging, Neurochem. Res., Submitted.

Ueda, T., and Greengard, P., 1977, Adenosine 3'-5'-monophosphate-regulated phosphoprotein system of neuronal membranes. J. Biol. Chem., 252:5155-5163.

Vitorica, J., Clark, A., Machado, A., and Satrustequi, J., 1985, Impairment of glutamate uptake and absence of alterations in the energy transducing ability of old rat brain mitochondria, Mech. Ageing Dev., 229:255-266.

von Hungen, K., Chin, R.C., Baxter, C.F., 1981, Brain tubulin microheterogeneity in the mouse during development and aging. J. Neurochem., 37:511-514.

Vorbeck, M., Martin, A.P., Lang, J.W.K., Smith, J.N., and Orr, R., 1982, Aging dependent modification of lipid composition and lipid structural order parameter of hepatic mitochondria, Arch.Biochem.Biophys.,277:351-361.

ADP-RIBOSYLATION : APPROACH TO MOLECULAR BASIS OF AGING

Paul Mandel

Centre de Neurochimie du CNRS, 5 rue Blaise Pascal
67084 Strasbourg Cedex, France

ADP-ribosylation reaction

About 25 years ago we observed[1] that a liver nuclear extract is able to produce from NAD a polyadenosine polymer. Chemical and enzymatic degradation followed by methylation experiments lead us to conclude that the polymer synthesized was polyadenosine diphosphate ribose (polyADPR) (Fig. 1). We called the enzyme present in the nuclear extract polyADP-ribose polymerase (polyADP-R)P. The enzymatic activity was expressed only in the presence of DNA. The structure of the polymer was confirmed one year later by Nishizuka et al.[2] and by Sugimura et al.[3]. We have also demonstrated the activity of the polymerase in vivo[4]. A fundamental step in our understanding of the biological role of this enzyme was the demonstration of its capacity to produce a transfer of polyADPR to nuclear proteins and thus a post-translational modification similar to that already well established : phosphorylation, methylation and acetylation[5,6]. Later oligo and mono ADPR transferases (ADPRT) were discovered in cytoplasm, in mitochondria[7,8,9,10], in erythrocyte supernatant[11] and in the plasma membrane[12] as well as in ribonucleoprotein particles carrying messenger RNA[13] (mRNP) (see also for review[14,15,16,17,18]). Procedures for the purification of the nuclear (polyADP-R)P from bovine thymus were developed in our laboratory[19] and in others (see for review[14]), providing an enzyme clearly DNA-dependent. It also appeared that a specific DNA fraction is responsible for this activation[20]. Recently we purified a cytoplasmic enzyme of the mRNP particles in a 3-step procedure, using the protein blotting technique and polyclonal antibodies ; apparently common antigens exist in the nuclear and cytoplasmic enzyme[21]. Cytoskeleton proteins also undergo ADP-ribosylation by the mRNP enzyme[22].

Plasticity and Regeneration of the Nervous System
Edited by P.S. Timiras *et al.*, Plenum Press, New York, 1991

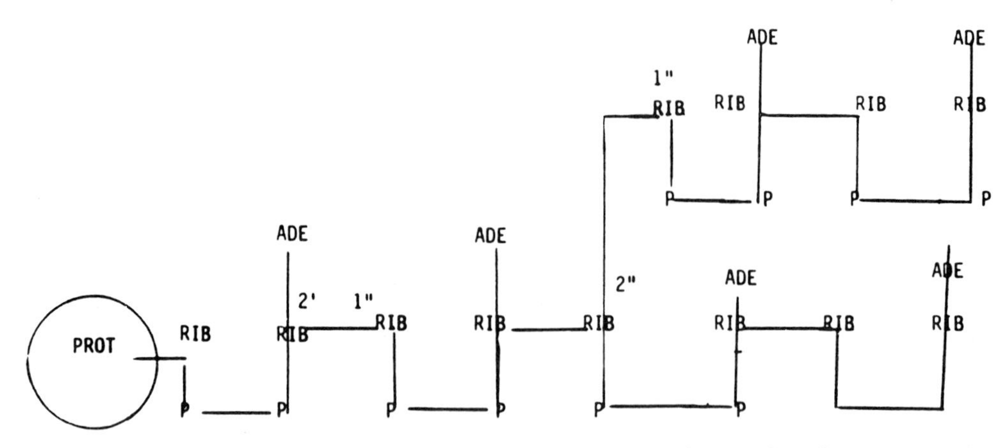

Fig. 1 Structure of protein-bound poly(ADP-Ribose).

(PolyADP-R)P appeared to be a well-preserved protein during evolution, similar to histones and other nuclear enzymes such as RNA and DNA polymerases. Nevertheless the existence of some species specific antigenic determinants was strongly suggested by immuno precipitation experiments[23]. It is noteworthy that the content of (polyADP-R)P per unit DNA evaluated by the micro complement fixation method was shown to be the highest among the chick cells examined in neurons[23].

The occurence of branched structures of polyADP-ribose was demonstrated by physicochemical methods[24] and in our laboratory by electronmicroscopy[25] (see also for review[14]). Another peculiar property of polyADPRT is its auto- polyADP-ribosylation which could be also visualized by dark-field electronmicroscopy[25].

The discovery of monoADP-ribosyltransferases in several bacterial toxins opened a large field for investigations concerning the role of ADP-ribosylation (for review see[14,15,16,17]).

Three types of enzymes ensure (polyADP-R) hydrolysis : glycohydrolases that split the ribose ribose-bounds, phosphodiesterases with phosphatase activity like those present in snake venom and in a variety of cells which hydrolyse the pyrophosphate bound of polyADPR and an ADPR lyase which removes the protein bound ADPR after the excision of the whole polyADPR chain (see for review[14,15,16,17,26]).

Nuclear (polyADP-R)P is involved in DNA replication and thus in cell proliferation (see for review[14,15,27]). This enzymatic activity is high in nuclei during G1 phase and low during the S-phase suggesting that ADP-ribosylation may be involved in the initiation of DNA synthesis. An increase of (polyADP-R)P activity measured in permeabilised cells was shown in lymphocytes following the proliferation response to phytohemoagglutinine[28,29]. Moreover when inhibitors of (polyADP-R)P : nicotinamide or 3-aminobenzamide were added at the onset of phytohemoagglutinine treatment a marked inhibition of the induced DNA synthesis and cell proliferation occured[29]. This observation and several others suggest that inhibitors of ADP-ribosylation might be useful to reduce cell proliferation, namely in tumoral cells. A decrease of proliferation of astroblasts, glioma cells and several tumoral cells could be demonstrated in cell culture when ADPRT inhibitors were present in the medium (see for review[30,26]).

Like phosphorylation, ADP-ribosylation seems to be involved as a part of regulatory cascades of a large variety of enzymes and proteins. ADP-ribosylation reactions have also been identified as control mechanisms in three well defined physiological processes in eukaryotic cells : 1. DNA repair ; 2. protein biosynthesis ; 3. adenylate cyclase action.In this respect two types of ADP-ribosylation reactions may be considered : polyADP-ribosylation which occur mainly in the nucleus and monoADP-ribosylation localized essentially in the cytoplasm and in the plasma membranes. Thus a distinction can be expected between the monomer and the polymer action which belong also to separate enzymes. We have also to keep in mind that ADP-ribosylation reactions are quite ubiquitous ; they occur in animals, plants bacteria and viruses (see for review[14,15,26]) .

ADP-ribosylation is required for an efficient cellular response following DNA damaging events produced by alkylating agents and ionising radiations. The nuclear ADPRT appears sensitive to DNA breaks and may modulate the rate of their rejoining, possibly by regulating the activity of DNA ligase II (see for review[14,15,27]).

Involvement of ADPR in cell differentiation was first suggested by Caplan et al.[31] and in our laboratory on the basis of embryological studies[32,33]. It was also shown by Shall that during cytodifferentiation of primary chick myoblasts there is a 3-fold increase in the endogenous ADPR activity but there is no change in the total potential enzyme activity (see for review[27]). Inhibitors of ADPRT activity reversibly block both muscle cell fusion and the rise of creatine phosphokinase activity associated with this cell differentiation (see for review[34]).

Several functions have also been identified for ADP-ribosylation in the cytoplasm on the plasma membrane level (see for review[26]) : the regulation of protein synthesis by the ADP-ribosylation of elongation factor II (EF2) and the activation of adenylate cyclase (see for review[26]). A great variety of proteins undergo ADP-ribosylation by bacterial enzymes (see for review[14,15,26]). Recent experiments in our laboratory demonstrated that similar ADP-ribosylation can be performed by an eucaryotic cytoplasmic ADPRT we have recently purified[21,22].

It is obvious that in view of the importance of ADP-ribosylation reaction in a number of cellular events the question of whether ADP-ribosylation undergoes changes during ageing must be raised.

Eye lens, a structural model for experimental and physiological ageing investigations

In order to explore the alterations or involvements of ADP-ribosylation in ageing it was of interest to choose an organ or tissue which offers some facilities for such investigations. The eye lens appears to be a most suitable model. The particular avascular structure of the lens and its physiological nutrition by direct exchange with the environmental liquide medium[35,36] facilitate studies in vitro of ageing processes in the field of energy production mechanisms, protein and nucleic acid metabolism.

Generated from ectodermal cells the eye lens is separated from a direct blood supply for life at an early embryonic stage. Moreover the lens is only composed of three cell types (Fig. 2) : 1) a monolayer of epithelial cells situated at the anterior part ; 2) long fiber cells which arise from the equatorial epithelial cells and progress to the central area forming concentric layers of fibers of increasing age around a central part ; 3) the nucleus, which has been produced during the embryonic and foetal stages. A very specific characteristic of the lens is that it never looses its cells[37].

Transparancy of the lens, an essential condition for a normal vision, results from a perfect homogeneity of various elements among which epithelial cells play a predominant role. Indeed, these cells constitute the laboratory for synthesis of the lens components and have to ensure the perfect regularity of the structure at the molecular and cellular level. Ageing affects epithelial cells, disorganising the synthetic system and interfering with molecular architecture and several basic functions.

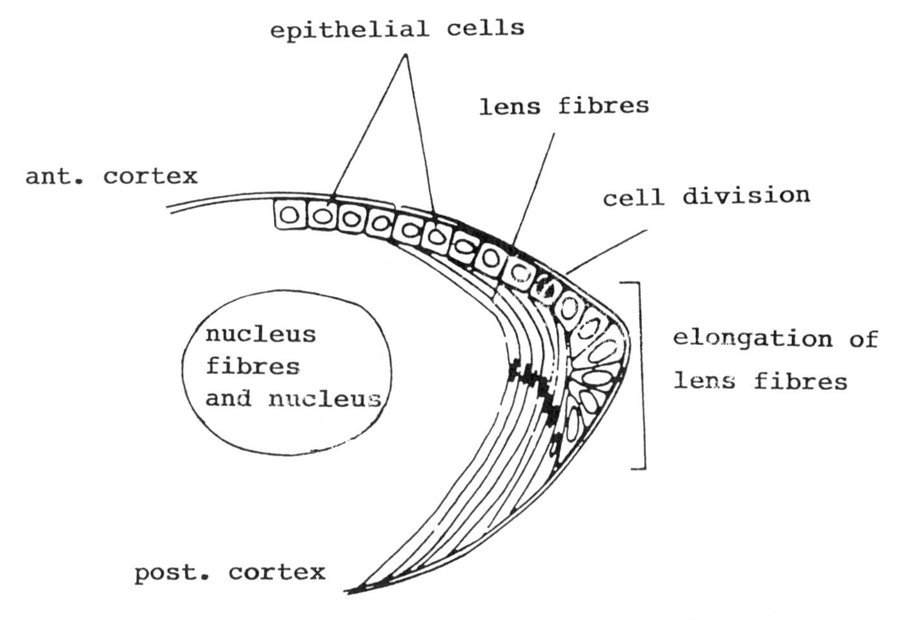

Fig. 2 Schematic presentation of lens morphology.

Water and electrolytes contents of bovine lenses show characteristic alterations during ageing. Available evidence indicates that an osmotic imbalence occurs in a significant percentage of human senile lenses[37,38,39,40].

In view of the importance of energy production in protein synthesis and post-transcriptional modifications including ADP-ribosylation we investigated glucose consumption of the lens in adults and during ageing[41,42,43,44]. It appeared clearly that the lens possesses the lowest oxygen and glucose consumption of all rat tissues explored. The enzymatic activities involved in glucose consumption and thus energy production decrease in all lens layers : epithelial cortical and nuclear during ageing[41,42,44]. Moreover the yield of energy and ATP production following glucose consumption is also the lowest within the lens among diverse tissues except the arteries[45,46,47,48].

The pentose cycle plays a crucial role in glucose metabolism as well as in nucleotide and RNA synthesis. Here again it appeared that glucose-6-

phosphate dehydrogenase activity was much lower in the cortical and obviously in the nuclear layer than in the epithelial cells and in each layer there was a decrease during ageing[44,49].

Investigations of nucleoside triphosphate and of nucleoside diphosphate glucose and galactose content revealed a drop during ageing in all lens layers and obviously much lower values in the cortical and nuclear layer when compared to the epithelial cells. ATP, GTP, CTP and UTP content was about 50% lower in the lenses of aged animals when compared to young adults[46,47,48]. A striking decrease of ATP during ageing was also found in human lens. The turnover of γ-phosphate of ATP also decreased strongly during ageing[47]. Levels of cyclic AMP and cyclic GMP slow down radically from the epithelial to the cortical layer of the bovine lens. In addition adenosine and guanylate cyclase activities are only measurable in the epithelial cells expressing their proliferative capacity. The cyclic nucleotides phosphodiesterases activities also decrease with age in the epithelial cells[50]. Finally high molecular weight mRNA is reduced during lens ageing[51,52] ; so was protein synthesis evaluated by methionine ^{35}S incorporation in proteins[53].

A variety of chemically or physically defined processes have been observed in lens specific structural proteins named crystallines during ageing. Formation of high molecular weight, aggregates, disulphate bridges crosslinks, degradation of polypeptide chains at characteristic sites, non-enzymatic glycosylation and photooxydation of tryptophanyl residues belong to the lens ageing processes. A number of these events may partially be interrelated with post translational modifications (see for review[54]).

Lens epithelial cells ADP-ribosylation during ageing

Considering the proteins alterations mentioned above and in view of the involvement of ADPRT in post-translational modifications we approached the investigation of ADP-ribosylation activity in the lens as a model of ageing. In preliminary experiments, when ADPRT activity was tested during different time periods on lens homogenates a maximum was reached after 15 minutes of incubation followed by a decrease of ADP-ribosylation. The last event was probably due to the degradation enzyme, polyADP-ribose glycohydrolase (polyADPR-G) (Fig. 3). Thus a short incubation time was retained for measurement of ADPRT activity (3 to 10 minutes).

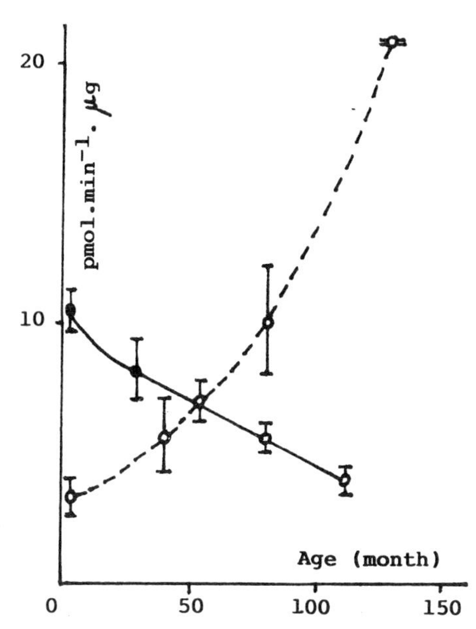

Fig. 3 Effect of aging on poly ADP-Ribosyl transferase and poly ADP-Ribose glycohydrolase activities of bovine lens epithelial cells. o------o ADP-Ribosyl transferase activity. ●———● Poly ADP-Ribose glycohydrolase activity. Brackets represent standard deviation.

It is noteworthy that no detectable ADPRT activity was found in the lens out from the epithelial cell layer. Even in the equatorial area where the nuclei of the fiber cells are still present, ADPRT activity was not appreciable although polyADPR-G was still present[55]. The apparent Km and Vmax were 156 µM and 60 pmol/min/mg protein, respectively, for lens epithelium ADPRT and 243 pmol/min/mg protein for polyADPR-G. cAMP added to incubation medium reduced polyADPR activity but was without effect on ADPR-T[55].

Since ADPRT activity was detected only in the epithelial cell layer our investigation concerning lens ageing effect on nuclear ADP-ribosylation were performed only on these cells. First of all using NAD ^{32}P it appeared that mainly high molecular weight proteins were ADP-ribsoylated (Fig. 4). In contrast to all enzymatic alterations occuring during lens and mainly lens epithelial cells ageing apparently common to other tissues, one event draw our attention : an increase of the nuclear ADPRT activity in lens epithelial cells[55] (Fig. 5). Like all other lens enzymes polyADPR-G activity increased during ageing (Fig. 6). It is noteworthy that the chain number increases while the chain length of the polyADPR attached to the proteins decreases during ageing (Table 1). Thus it appeared that the increase of ADPRT activity with age parallels shorter chains production. The frequency of branched chains in aged lenses was extremely low (not measurable).

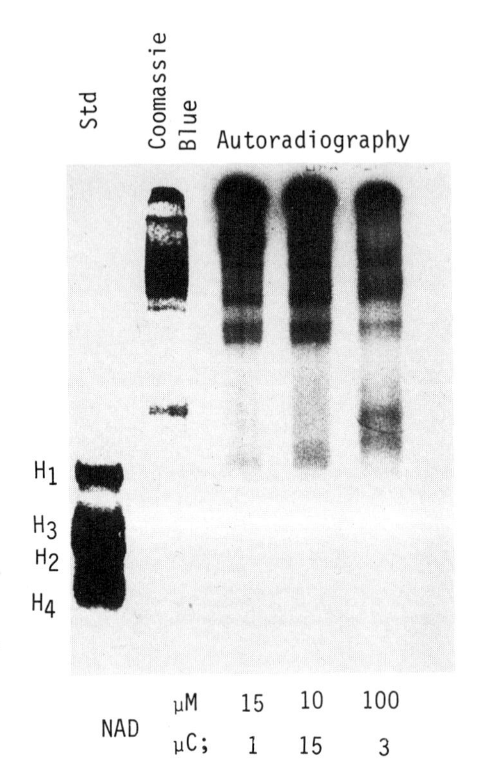

Fig. 4 Autoradiography of ADP-ribosylated epithelial cell proteins. Acid-urea polyacrylamide gel electrophoresis of ADP-ribosylated proteins of bovine lens epithelial cell nuclei.
1: Standard histones mobility H1, H2, H3, H4.
2: Coomassie blue staining of the proteins.
3, 4, 5: autoradiography of ADP-ribosylated proteins.

The question was obviously raised why ADPR-T was the only enzymatic activity which increases while all other enzymatic activities measured at the DNA level decrease during ageing as it was shown in our laboratory and in others. Pathological events were obviously excluded in the lenses we examined. We hypothesised that the increase of ADPRT during ageing follows the appearance of strand breaks. Actually damage to DNA namely occurence of single strand breaks that expose more sites for DNA synthesis or repair during ageing were suggested by Price et al.[56]. Moreover alkaline sucrose gradient sedimentation shows that the DNA of the brain of old mice sediments polydispersely as four bands but that of young mice sediments monodispersely as a single band (Chetsanga et al.[57]). Finally degradation of DNA has also been suggested on the basis of a decrease of transcription of RNA from liver chromatin using RNA polymerase (Hill and Whelan[58]). In fact we could detect strand breaks in DNA of lens epithelial cells of old bovines (84 months) when compared with young adults (23 months)[55] using DNA unvinding methods at an alkaline pH according Birnboim and Jereak[59]. There was no difference in the DNA electrophoretic profile between adult and old animals, nor was the DNA apparently lighter than 25 Kbp detected in any age group in neutral 0.5% agarose polyacrylamide gel[55]. These observations suggest an accumulation of single strand alterations rather than double stand breaks formation.

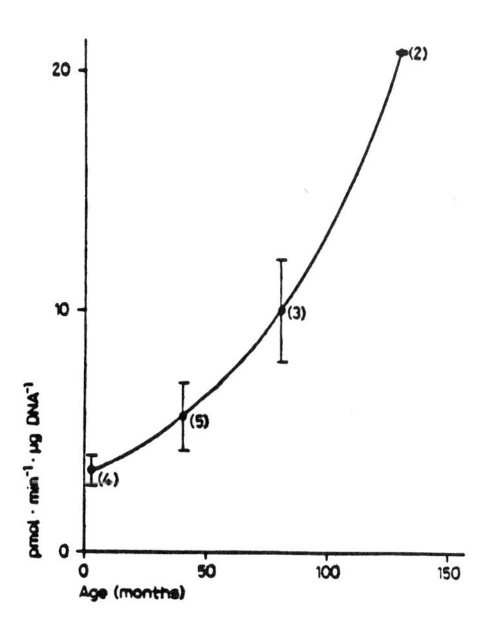

Fig. 5 ADPRT activity in lens epithelial cells during aging. Brackets represent SD with number of experiments in parentheses (more than 20 lenses per experiment).

Overall, the investigation of polyADPRT activity in lens leads first to the observation that ADPRT is only active in highly differentiated epithelial cells; even a slight loss of potentialities as occurs in equatorial lens fiber cells is already linked to a loss of ADPRT activity[55]. Moreover (polyADP-R)P is the only enzyme undergoing an increase during ageing. This event draws our attention to the existence of DNA alteration, that is, DNA breaks mainly of one strand at a rather early ageing, which can be detected by an increase of ADPRT activity. However, the most interesting information seems to be the existence of an incapacity of full DNA repair during ageing. A better knowledge of DNA repair mechanisms and of a way to overcome this deficiency may be an issue to prolong cell integrity during ageing. ADP-ribosylation of nuclear proteins may play a crucial role in these investigations.

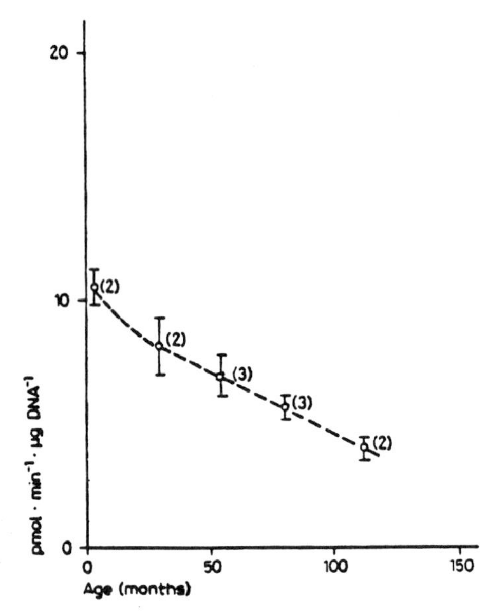

Fig. 6 Poly-ADPRT activity in lens epithelial cells during aging. Brackets represent SD with number of experiments in parentheses (more than 20 lenses per experiment).

Table 1. Poly-ADP-ribose chain length and chain number in lens epithelial cell during ageing

Age[1] months	Average chain length[2], U	Average chain number[3], pmol
3	7.5	1.23
30	3.8	1.60
54	3.1	2.10

Two experiments with more than 20 lenses.

[1] On the basis of lens weight/Age relationship.

[2] Following the ratio (phosphoribosyl-AMP) + (AMPterm)/(AMPterm).

[3] Terminal AMP of ADP-ribosylation chain.

Table 2. DNA breakage during aging

Age, months	DNA double strand p.100 (1)
23	$86.0^{a}\pm7.0$ (6)
54	$84.2^{b}\pm1.1$ (2)
94	$67.2^{c}\pm0.3$ (2)

Mean ± SE and number of experiments in parentheses. The a, c and b, c measurements are significantly different, $p < 0.05$.
(1) Percent of double-strand DNA after alkaline unwinding according to Birnboim and Jevcak[59].

References

1. P. Chambon, J.D. Weill, J. Doly, M.T. Strosser and P. Mandel, On the formation of a novel adenylic compound by enzymatic extracts of liver nuclei, Biochem. Biophys. Res. Commun. 25:638-643 (1966).
2. Nishizuka Y., K. Ueda, K. Nakazawa and O. Hayaishi, Studies on the polymer of adenosine diphosphate ribose, J. Biol. Chem. 242:3164-3171 (1967).
3. T. Sugimura, S. Fujimara, S. Hasegawa and Y. Kawamura, Polymerisation of the adenosine 5'-diphosphate ribose moiety of NAD by rat liver nuclear enzyme, Biochem. Biophys. Acata 138:438-441 (1967).
4. J. Doly and P. Mandel, Mise en évidence de la biosynthèse in vivo d'un polymère composé, le polyadénosine diphosphoribose dans les noyaux de foie de poulet, C.R. Acad. Sci. 264:2687-2690 (1967).
5. Y. Nishizuka, K. Ueda, T. Honjo, O. Hayaishi, Enzymic adenosine diphosphate ribosylation of histone and poly adenosine diphosphate ribose synthesis in rat liver, J. Biol. Chem. 243:3765-3767 (1968).
6. H. Otake, M. Miwa, S. Fujimara and T. Sugimura, Binding of ADP-ribose polymer with histone, J. Biochem. 65:145-146 (1969).
7. E. Kun, E. Kirsten, Mitochondrial ADP-ribosyltransferase system, in: "ADP-ribosylation reactions", O. Hayaishi and K. Ueda, ed., Academic Press, New York (1982).

8. L.O. Burzio, L. Saez and R. Cornejo, Poly(ADP-ribose) synthetase activity in rat testis mitochondria, Biophys. Res. Commun. 103:369-375 (1981).
9. A. Masmoudi and P. Mandel, ADP-ribosyl transferase and NAD glycohydrolase activities in rat liver mitochondria, Biochemistry 26:1965-1969 (1987).
10. A. Masmoudi, F. Islam and P. Mandel, ADP-ribosylation of highly purified rat brain mitochondria, J. Neurochem. 51:188-193 (1988).
11. J. Moss and M. Vaughan, Isolation of an avian erythrocyte protein possessing ADP-ribosyltransferase activity and capable of activating adenylate cyclase, Proc. Natl. Acad. Sci. USA 75:3621-3624 (1978).
12. P. Adamietz, K. Wielckens, R. Bredehorst, H. Lenyel and H. Hilz, Subcellular distribution of mono(ADP-ribose) protein conjugates in rat liver, Biochem. Biophys. Res. Commun. 101:96-103 (1981).
13. R. Elkaim, H. Thomassin, C. Niedergang, J.M. Egly, J. Kempf and P. Mandel, Adenosine diphosphate ribosyltransferase and protein acceptors associated with cytoplasmic free messenger ribonucleoprotein particles, Biochimie, 65:653-659 (1983).
14. O. Hayaishi and K. Ueda, "ADP-ribosylation reactions", Academic Press, New York (1982).
15. M.K. Jacobson and E.L. Jacobson, "ADP-ribose transfer reactions", Springer- Verlag, New York Berlin Heidelberg (1989).
16. H. Hilz, K. Wielckens and R. Bredehorst, Quantitation of mono(ADP-ribosyl) and poly(ADP-ribosyl) proteins, in: "ADP-ribosylation reactions", O. Hayaishi and K. Ueda, ed., Academic Press, New York (1982).
17. P. Mandel, H. Okazaki and C. Niedergang, Poly(adenosine diphosphate ribose), in: "Progress in Nucleic Acid Res. Molec. Biol." vol. 27, Academic Press, New York (1982).
18. C. Niedergang and P. Mandel, Isolation and quantitation of poly(ADP-ribose), in: "ADP-ribosylation reactions", O. Hayaishi and K. Ueda, ed., Academic Press, New York (1982).
19. H. Okazaki, C. Niedergang and P. Mandel, Purification and properties of calf thymus polyadenosine diphosphate ribose polymerase, FEBS Lett. 62:255-258 (1976).
20. M.E. Ittel, J. Jongstra-Bilen, C. Niedergang, P. Mandel and E. Delain, DNA- poly(ADP-ribose) polymerase complex : isolation of the DNA wrapping the enzyme molecule, in : "ADP-ribosylation of proteins",

F.R. Althaus, H. Hilz and S. Shall, ed., Springer-Verlag, Berlin-Heidelberg (1985).

21. C. Chypre, C. Le Calvez, F. Hog, M.O. Revel, M. Jesser and P. Mandel, Phosphorylation de la poly(ADP-ribose) polymérase cytoplasmique liée à des particules ribonucléoprotéiques libres par une protéine kinase C associée, C. R. Acad. Sci. Paris, t. 309, Série III (1989).
22. M. Jesser, C. Chypre, A. Rendon, F. Hog and D. Jung, ADP-ribosylation of cytosqueletal structures of animal cells by a cytoplasmic ADP-ribosyl transferase (in preparation).
23. H. Okazaki, J.P. Delaunoy, F. Hog, J. Bilen, C. Niedergang, E.E. Crepy, M. Ittel and P. Mandel, Studies on poly ADPR polymerase using the specific antibody, Biochem. Biophys. Res. Commun. 97:1517-1520 (1980).
24. M. Miwa and T. Sugimura, Structure and properties of poly(ADP-ribose), in: "ADP-ribosylation reactions", O. Hayaishi and K. Ueda, ed., Academic Press, New York (1982).
25. G. De Murcia, J. Jongstra-Bilen, M.E. Ittel, P. Mandel and E. Delain, Poly(ADP-ribose) polymerase auto-modification and interaction with DNA : electron microscopic visualization, EMBO J. 2:543-548 (1983).
26. F.R. Althaus and C. Richter, "ADP-ribosylation of proteins", Springer- Verlag, Berlin Heidelberg New York London Paris Tokyo (1987).
27. S. Shall, ADP-ribosylation of proteins : a ubiquitous cellular control mechanism, Biochem. Soc. Trans. 17:317-322 (1989).
28. C. Rochette-Egly, M.E. Ittel, J. Bilen and P. MANDEL, Effect of nicotinamide on RNA and DNA synthesis and on poly(ADP-ribose) polymerase activity in normal and phytohemagglutinin stimulated human lymphocytes, FEBS Lett. 120:7-11 (1980).
29. M.E. Ittel, J. Jongstra-Bilen, C. Rochette-Egly and P. Mandel, Involvement of polyADP-ribose polymerase in the initiation of phytohemagglutinin induced human lymphocyte proliferation, Biochem. Biophys. Res. Commun. 116:428-434 (1983).
30. P. Mandel, C. Niedergang, M.E. Ittel, H. Thomassin and A. Masmoudi, PolyADP-ribose polymerase and ADP-ribosylation reaction, in: "Role of RNA and DNA in brain function", A. Giuditta, B.B. Kaplan, C. Zomzely- Neurath, ed., Martinus Nijhoff (1986).
31. A.I. Caplan and M.J. Rosenberg, Interrelationship between poly(adenosine diphosphoribose) synthesis, intracellular NAD levels and muscle or cartilage differentiation from embryonic chick limb mesodermal cells, Proc. Natl. Acad. Sci. USA 72:1852-1857 (1975).

32. A. Caplan, C. Niedergang, H. Okazaki and P. Mandel, Poly ADP-ribose polymerase : self ADP-ribosylation, the stimulatioby DNA, and the effects on nuclease formation and stability, Arch. Biochem. Biophys. 198:60-69 (1979).

33. A. Caplan, C. Niedergang, H. Okazaki and P. Mandel, Poly(ADP-ribose) levels as a function of chick limb mesenchymal cell development as studied in vitro and in vivo, Developm. Biol. 72:102-109 (1979).

34. K. Ueda and O. Hayaishi, ADP-ribosylation, Ann. Rev. Biochem. 54:73-100 (1985).

35. H.K. Muller, Uber Linsenstoffwechseluntersuchungen, Ber. Dtsch. Ophtalmol. Ges. 50:167-171 (1934).

36. J. Nordmann, Contribution à l'étude de la cataracte acquise, Diss. Strasbourg (1926).

37. R.A. Weale, "The aging lens", H.K. Lewis and Co. Ltd, London (1963).

38. O. Hockwin, F. Rast, H. Rink, J. Munnighoff and H. Twenhoven, Water content of lenses of different species, Interdiscipl. Top. Gerontol. 13:239-246 (1978).

39. C.A. Paterson, Antero-posterior cation gradients in bovine lenses, Investigative. Ophtalmol. 12:861-863 (1973).

40. H. Rink, J. Munnighoff and O. Hockwin, Sodium potassium and calcium contents of bovine lenses in dependence on age, Ophtalm. Res. 9:129-135 (1977).

41. J. Nordmann and P. Mandel, Le métabolisme des glucides dans le cristallin. I. La glycolyse anaérobie, Ann. Occulist. 185:929-943 (1952).

42. P. Mandel and J. Zimmer, Etude sur la répartition des acides pyruviques et lactique dans le cristallin des bovidés, C.R. Soc. Biol. 146:762-764 (1952).

43. P. Mandel and L. Schmitt, Etude de l'activité adénosinetriphosphatasique et hexokinasique de cristallins de lapins soumis à une irradiation locale par les rayons X, Experientia 12:223-226 (1956).

44. J. Klethi and P. Mandel, Variations des activités enzymatiques du cycle oxydatif direct du cristallin au cours du vieillissement, C.R. Soc. Biol. 153:337-340 (1959).

45. J. Nordmann and P. Mandel, Etude quantitative des composés phosphorés acido-solubles du cristallin de mammifères jeunes et âgés, C.R. Acad. Sci. 325:834-835 (1952).

46. J. Klethi and P. Mandel, Les nucléotides libres du cristallin de veau, Biochem. Biophys. Acta 24:642-643 (1957).

47. J. Klethi and P Mandel, Recherches sur la biochimie du cristallin. II. Les nucléotides libres du cristallin, Bull. Soc. Chim. Biol. 50:709-723 (1968).
48. J. Klethi and P. Mandel, Recherches sur la biochimie du cristallin. III. Etudes des variations du taux des nucléotides libres en fonction de l'âge des cellules cristalliniennes, Bull. Soc. Chim. Biol. 50:1205-1214 (1968).
49. N. Virmaux and P. Mandel, Succinic dehydrogenase activity and oxidative phosphorylation of the mitochondria in the cristalline lens of bovines, Nature, 197:792 (1963).
50. J.C. Bizec, J. Klethi and P. Mandel, Cyclic guanosine 3'5' phosphate, guanylate cyclase and cyclic guanosine phosphodiesterase in the eye lens, Biochem. Biophys. Res. Commmun. 106:108-112 (1982).
51. P.F. Urban, N. Virmaux and P. Mandel, Kinetics of labelling of eye lens RNA's, Biochem. Biophys. Res. Commun. 20:10-14 (1965).
52. N. Virmaux, P.F. Urban and P. Mandel, Les acides ribonucléiques du cristallin, Doc. Ophtal. 20:13 (1966).
53. P. Mandel, U. Dardenne and A. Lessinger, Incorporation et dégradation de la méthionine par le cristallin de bovidés, C.R. Acad. Sci. 245:985-987 (1957).
54. M.S. Kanungo, "Biochemistry of Ageing", Academic Press, London (1980).
55. J.C. Bizec, J. Klethi and P. Mandel, Regulation of protein adenosine dipshophate ribosylation in bovine lens during aging, Ophtalm. Res. 21:175- 183 (1989).
56. G.B. Price, S.P. Modak and T. Makinodan, Age-associated changes in the DNA of mouse tissue, Science 171:917-920 (1971).
57. C.J. Chetsanga, V. Boyd, L. Peterson and K. Rushlow, Single-stranded regions in DNA of old mice, Nature 253:130-131 (1975).
58. B.T. Hill and R.D.H. Whelan, Studies on the degradation of aging chromatin DNA by nuclear and cytoplasmic factors and deoxyribo-nucleases, Gerontology 24:326-336 (1978).
59. H.C. Birnboim and J.J. Jevcak, Fluorimetric method for rapid detection of DNA strand breaks in human white blood cells produced by low doses of radiation, Cancer Res. 41:1889-1892 (1981).

MECHANISMS OF CELL DEATH

Regino Perez-Polo

Dept. of Human Biological Chemistry & Genetics
Univ. of Texas Medical Br. F52
Galveston, Tx. 77550

Introduction

Cell death during the ontogeny of the nervous system allows for competitive innervation of targets (Hamburger and Oppenheim, 1982; Levi-Montalcini, 1987). Neurons that successfully compete for target-derived trophic factors, such as NGF, at critical stages in development are spared cell death. Exogenous NGF, or its withdrawal by anti-NGF, has permanent effects on survival of peripheral and striatal neurons (Levi-Montalcini, 1987; Perez-Polo, 1987). The relationship between developmental neuronal cell death and its counterparts after injury or during aging in CNS is not known. Since neurons have low endogenous levels of antioxidants, NGF effects on oxidant-antioxidant balance in the CNS can affect neuronal survival after mechanical trauma, ischemia, and reperfusion injury. Since neuronal cell death is an end point of more than one physiologic process, NGF activity may be but one of several regulatory processes at work. The NGF effect on oxidant-antioxidant balance may be but one of several mechanisms by which NGF regulates neuronal survival.

Neuronal Cell Death

Free radicals, such as hydroxyl radicals, can arise due to conversion of oxygen species to H_2O_2 by superoxide dismutase. H_2O_2 and ferrous iron, in turn, yield hydroxyl radicals that initiate self sustaining lipid peroxidation reactions at the plasma membrane. Increases in iron following injury due to heme release and inflammatory events have been associated with a compromised blood-brain barrier. Specifically, H_2O_2 activates the hexose monophosphate shunt, decreases glycolytic flux, decreases $NAD^+/NADH$ and ATP, activates poly (ADP-ribose) polymerase, increases free intracellular calcium and induces single- strand breaks in DNA. One hypothesis of cell death in response to oxidative stress is that irreversible depletion of $NAD^+(H)$ and hence ATP generation occur as a consequence of ADP-ribosylation in response to single strand DNA damage. Hence any agent that could inhibit such NAD depletion could enhance recovery from oxidative injury. For injury induced "secondary cell death,"

Plasticity and Regeneration of the Nervous System
Edited by P.S. Timiras *et al.*, Plenum Press, New York, 1991

it is known that there is an increase in peroxidative activity (Demopoulos et al., 1979). The issue is complicated since sufficiently low H_2O_2 concentrations stimulate neuronal metabolism while high concentrations of H_2O_2 are toxic (Chan et al., 1984). Oxidative stress can also induce **xenobiotic** responses much like heavy metals or radiation (Spitz et al., 1987). The relative contributions of these events is <u>not</u> known.

An alternative hypothesis is based on the finding that in the absence of NGF, cultured peripheral neurons are protected from developmental neuronal death by cycloheximide, a protein synthesis inhibitor (Martin et al., 1988). Therefore neuronal cell death in development would be due to the activation of a cascade of death associated proteins and finally of **"thanatins"** or death proteins (Johnson et al., 1987). Thus, NGF would suppress the synthesis of specific gene products. Evidence for this NGF action in the periphery rests on the use of mRNA transcription and translation inhibitors. A concentration dependent effect on neuronal survival has also been proposed for NGF effects on calcium intracellular concentrations under the rubric of a **"calcium set point"** hypothesis of neuronal cell death (Johnson et al., 1987). This would propose that at low intracellular calcium concentrations, neurons are trophic factor dependant, at higher levels due to chronic synaptic activity, they are trophic factor independent and at even higher levels the calcium is toxic.

Nerve Growth Factor

NGF action has been best documented for sensory, sympathetic and **magnocellular cholinergic neurons (MCN)** of the **basal forebrain**. Notably, the vertical limb of the diagonal band (**area CH2**) and nucleus basalis of Meynert (**CH4**) are NGF-responsive in early development or after damage to the fimbria-fornix. The vertical limb projects via fimbria-fornix to the hippocampus, whereas the nucleus basalis projects mainly to the cortex. NGF mRNA levels have been determined for brain, spinal cord and the periphery and correlated with NGF protein levels as a function of development, innervation and response to injury (Ayer-LeLievre et al., 1988; Goedert et al., Lu et al., 1989). In the CNS, the highest <u>levels</u> of <u>NGF mRNA</u> are in cortex and hippocampus, the terminal regions for projections from basal forebrain cholinergic neurons, where NGF effects on ChAT induction and cell sparing following lesions have been documented (Mobley et al., 1986). There is also evidence for cholinergic non-NGF responsive neurons and vice-versa in basal forebrain and striatum (Kordower et al., 1989). During early developmental stages NGF and NGF receptor mRNA levels are highest outside the nervous system (Ernfors et al., 1988; Large et al., 1986; Thorpe et al., 1989).

Nerve Growth Factor Receptor

There are two NGF binding activities with equilibrium dissociation constants (K_d) of 10^{-11}M and 10^{-9}M respectively (Stach and Perez-Polo, 1987) corresponding to a **high affinity**, low capacity binding site **(NGFR-I)** with a **slow** dissociation rate constant and a **low affinity**, high capacity site **(NGFR-II)** with a **fast** dissociation rate constant. NGFR-I is probably the physiologically relevant receptor on neurons (Green et al., 1986; Sonnenfeld and Ishii, 1986; Stach and Perez-Polo, 1987). NGF binding activity in CNS has similar properties to that in PNS (Angelucci et al., 1988a; Cohen-Cory et al., 1989; Taglialatela et al., 1990). There is one gene for NGFR and no differential mRNA splicing (Chao et al., 1986). NGFR gene transfection into fibroblasts results in expression of type II NGF binding but into NGF-nonresponsive PC12 mutant PC12nnr, expresses type I NGF binding (Misko et al., 1987). Differences in NGFR are most likely due to the presence of an unidentified receptor-associated protein (Hosang and Shooter, 1987; Marchetti and Perez-Polo, 1987; Shan et al., 1990).

A second way to study NGFR is SDS-PAGE of ^{125}I-NGF covalently crosslinked to NGFR and immunoprecipitated (Hosan and Shooter, 1987). Most descriptions of rodent NGFR are consistent with a 70-80 KDa protein as the prevalent NGFR protein species. Human NGFR proteins of 92.5 Kda have been reported. There are often higher molecular weight species of NGFR. In hippocampus and cortex there is NGF synthesis whereas NGFR are synthesized predominantly in cell bodies of basal forebrain neurons and subsequently transported via anterograde transport to the axon terminals where they bind, internalize and retrogradely transport NGF back to the cell bodies for trophic support (Whittemore and Seiger, 1987). The expression of NGF and NGF mRNA is developmentally regulated in hippocampus and correlates with the degree of cholinergic innervation (Whittemore and Seiger, 1987). Lesioning of fimbria-fornix stimulates NGF protein and NGF mRNA in neonatal rats, while in adults only the NGF protein increases (Whittemore and Seiger, 1987). This suggests a transcription-dependent regulation of NGF synthesis in lesioned neonates while in adults the NGF protein increase is likely due to an interruption of retrograde transport and protein buildup distal to the lesion. Intraventricular injections of NGF in neonatal rats increases ChAT activity in the basal forebrain, hippocampus, cortex and the caudate nucleus (Mobley et al., 1986).

Receptors for NGF are present on cells derived from all three germ layers, consistent with the hypothesis that NGF is not exclusively a neuronotrophic factor (Levi et al., 1988; Levi-Montalcini, 1987; Thorpe et al., 1989). It has been proposed that low affinity NGF receptors (Johnson et al., 1987) may direct axonal growth along Schwann cell surfaces that are decorated with NGFR with NGF bound that is the released from these glial low affinity sites to high affinity NGFR on advancing growth cones of neurons in regeneration (Martin et al., 1988). The inability to identify NGFR in normal adult astrocytes in vivo is consistent with a possible role in early development or after injury (Taniuchi et al., 1986b). Immunohistochemical staining with MC192 anti-NGFR does not show NGFR-like antigens on glia in adult rat brain, except for Muller glia (Schatteman et al., 1988). There is NGF binding, NGFR protein and NGFR mRNA on isolated lymphocytes and rodent C6 glioma cells, (Kumar et al., 1990; Morgan et al., 1989). After injury macrophages synthesize IL-1 which stimulates Schwann cells to synthesize and secrete NGF (Heumann et al., 1987).

Model Systems

Although caution must be exercised before extrapolating conclusions gleaned from studies with transformed cells, three cell lines that have proven useful in the study of NGF are the PC12 rat pheochromocytoma, the SK-N-SH-SY5Y (SY5Y) and the LA-N-1 human neuroblastoma lines. NGF has similar effects on SY5Y, LAN1 and PC12 cells (Perez-Polo and Werrbach-Perez, 1988). NGF effects on PC12 cells have been classified temporally and on their RNA transcription dependence (Levi et al., 1988) NGF elicits cell-surface ruffling, ion fluxes across the membranes and is internalized (Levi et al., 1988). Second, NGF induces transcription-independent phosphorylation of cytoplasmic proteins and proto-oncogenes. Third, NGF induces ornithine decarboxylase and long-term transcription-dependent synthesis of proteins required for neurite growth. This spectrum of responses may not be unique to NGF but rather be specific for PC12 cells; other classes of NGF-responsive cells display different spectrums of responses (Thorpe et al., 1989). NGF receptor expression and response can be augmented by small modulator molecules like retinoic acid and acetyl-L-carnitine (Jackson et al., In Press; Taglialatela et al., submitted; Taglialatela et al., submitted).

When cell membranes from human neuroblastoma LAN 1 line, with high and low affinity NGFR, were iodinated (^{125}I), solubilized and after lentil-lectin chromatography and preparative gel electrofocusing, samples were either immunoprecipitated and subjected to SDS-PAGE and autoradiography or subjected to equilibrium binding assays using ^{131}I-NGF, it was possible to separate low from high affinity binding activities. Low affinity NGF binding was associated with a 93Kd NGF-R-like protein and high affinity NGF binding was associated with a 200 Kd NGF-R-like protein. By mixing experiments, it was shown that there was <u>receptor associated protein</u> that confers high affinity binding properties to the low affinity receptor.

Another approach took advantage of reverse phase-HPLC characterization of NGFR. First, different species of NGFR-like proteins were characterized from rodent and human NGF-responsive cells by lentil-lectin chromatography, RP-HPLC, immunoprecipitation and SDS-PAGE using monoclonal antibodies to the rodent and human NGFR protein. The higher molecular-weight species of NGFR-like molecules were eluted in the more hydrophobic fractions after RP-HPLC. Since more than one rodent NGFR-like species were precipitated by mAb 192, it may be that different molecular species share the same epitope thus providing evidence that NGFR-I and NGFR-II share a similar subunit. Among the NGFR-like species isolated by RP-HPLC on PC12 and precipitated by Mab 192, the most likely candidate for NGFR was a 76 Kda species, similar to that reported for A875 cells and the NGFR on PC12 cells, Schwann cells and rat brain (Grob et al., 1983; Taniuchi et al., 1986b; DiStefano and Johnson, 1988b). These cell types have been shown to bind NGF with low affinity NGFR-II's. A more hydrophobic 133 Kda NGFR-like protein isolated by RP-HPLC may correspond to the 158 Kda crosslinked NGF-NGFR complex reported as containing NGFR-I (Massague et al., 1982; Shan et al., 1990). The size difference between the 75-80 Kda and 133

Kda species would suggest a receptor-associated protein of 53-58 Kda protein similar to the 60 Kda NGFR-associated protein suggested by others (Hosang and Shooter, 1987; Shan et al., 1990). This is also consistent with the identification by RP-HPLC of a 93 Kda and a 148 kDa NGFR-like in two human neuroblastoma lines where the 148 kDa species here could be made up of the 93 kDa NGFR-like protein and a putative 55 kDa receptor-associated protein. The NGFR associated protein may have common features in these different cell lines. Thus, tissue or species-specific differences in the reported molecular weights for NGFR may result from differences in the degree of expression of a receptor-associated protein in agreement with other reports using other techniques (Hosang and Shooter, 1987). If one compares the properties of NGFR present on rodent and human lymphocytes they are similar. In all the studies that rely on autoradiography and SDS-PAGE analysis of immunoprecipitates of ^{125}I-NGF crosslinked to NGFR, there appear to be small differences between NGFR present in CNS and its peripheral coounterpart.

Aging and Stress Effects

Two perturbations of CNS that have been studied are effects of aging and stress on NGF binding in brain areas known to be NGF responsive. There are reported reductions in NGF and NGFR in cholinergic areas of the aged CNS (Gage et al., 1988). Since there is a reduction in trophic and principally NGF associated activity in those cholinergic regions that display aged associated pathology (Flood and Coleman, 1988), it is not surprising that there are reductions in the NGF and NGF-binding capacity of the aged rodent basal forebrain and hippocampus (Angelucci et al, 1988a). Similar deficits in NGF and NGFR protein and mRNA have also been demonstrated in the CNS although, at the present time, it is not known if these deficits are a consequence of neuronal atrophy and cell loss there or are a cause of such cell loss and atrophy (Flood and Coleman, 1988).

It is intriguing that during late aging events in the CNS there is a similar loss of corticosteroid and NGF receptors associated with cholinergic neuronal shrinking and cell death in the hippocampus (Flood and Coleman, 1988). One explanation is that age associated reductions in NGF activity in turn have as one consequence a disinhibition of the hypothalamic-pituitary-adrenocortical-axis (HPAA) and that the resultant prolonged exposure of the CNS to increased corticosterone in the rat, a consequence of loss of hippocampal plasticity, has neurotoxic effects as aging progresses. In turn, many of the aged associated deficits would then result from this disinhibition of feedback mechanisms in HPAA. Thus, stressful events might be expected to affect NGF binding capacity in hippocampus. When rats were exposed to one hour of cold stress for five consecutive days there was a significant decline in NGF binding sites in basal forebrain and hippocampus (Taglialatela et al., 1990). A similar trend is observed in frontal cortex that is not statistically significant. No changes are seen in cerebellum. Stress was here monitored by measuring corticosteroid plasma levels. In summary, following a subchronic exposure to cold stress, NGF binding declines in hippocampus and possibly frontal cortex. These results are in agreement with results in vitro that show that dexamethasone can down-regulate NGF-R in PC12 cells. This implies that NGF binding in the CNS may be responsive to short-term physiologically relevant events such as stress. While the physiological significance of these changes in NGF binding is not known, their occurrence in CNS structures intimately involved in cognitive phenomena is interesting.

REFERENCES

Angelucci, L., Ramacci, M.T., Taglialatela, G., Hulsebosch, C., Morgan, B., Werrbach-Perez, K., & Perez-Polo, J.R., 1988a, Nerve growth factor binding in aged rat central nervous system: Effect of acetyl-L-carnitine. J Neurosci Res, 20:491-496.

Ayer-LeLievre, C., Olson, L., Ebendal, T., Seiger, A., Persson, H., 1988, Expression of the β-nerve growth factor gene in hippocampal neurons. Science 240:1339-1341.

Chan, P.H., Schmidley, J.W., Fishman, R.A., & Longar, S.M., 1984, Brain injury, edema, and vascular permeability changes induced by oxygen-dervied free radicals. Neurology, 34:315-320.

Chao, M.V., Bothwell, M.A., Ross, A.H., Korprowski, H., Lanahan, A.A., Buck, C.R., & Sehgal, A., 1986, Gene transfer and molecular cloning of the human NGF receptor. Science, 232:518-521.

Cohen-Cory, S., Dreyfus, C.F., & Black, I.B., 1989, Expression of high- and low-affinity nerve growth factor receptors by purkinje cells in the developing rat cerebellum. Experimental Neurology, 105:104-109.

Demopoulos, H.B., Flamm, E.S., Seligman, M.L., Mitamura, J.A., & Ransohoff, J., 1979, Membrane perturbations in central nervous system injury theoretical basis for free radical damage and a review of the experimental data, In: Neural Trauma, AJ Popp et al., eds, Raven Press, pp.63-78.

DiStefano, P.S., & Johnson Jr, E.M., 1988b, Nerve growth factor receptors on cultured rat Schwann cells. J Neurosci, 8:231-241.

Ernfors, P., Hallbook, F., Ebendal, T., Shooter, E.M., Radeke, M.J., Misko, T.P., & Person, H., 1988, Developmental and regional expression of β-nerve growth factor receptor mRNA in the chick and rat. Neuron, 1:988-996.

Flood, D.G., & Coleman, P.D., 1988, Neuron numbers and size in aging brain: Comparisons of human, monkey and rodent data. Neurobiol. of Aging, 9:453-463.

Gage, F.H., Chen, K.S., Buzsaki, G., & Armstrong, D., 1988, Experimental approaches to age-related cognitive impairments. Neurobiol of Aging, 9:645-655.

Goedert, M., Fine, A., Dawbarn, D., Wilcock, G.K., & Chao, M.V., 1989, Nerve growth factor receptor mRNA distribution in human brain: Normal levels in basal forebrain in Alzheimer's disease. Mol Br Res, 5:1-7.

Green, S.H., Rydel, R.E., Connolly, J.L., & Greene, L.A., 1986, PC12 cell mutants that possess low-but not high-affinity nerve growth factor receptors neither respond to nor ingernalize nerve growth factor. J Cell Biol, 102:830-843.

Grob, P.M., Berlot, C.H., & Bothwell, M.A., 1983, Affinity labelling and partial purification of nerve growth factor receptors from rat pheochromocytoma and human melanoma cells. Proc Natl Acad Sci 80: 6819-6823.

Hamburger, V., & Oppenheim, R.W., 1982, Cell Death. Neurosc.Comm. 1:39-55.
Heumann, R., Lindholm, D., Bandtlow, C., Meyer, M., Radeke, M.J., Misko, T.P., Shooter, E., & Thoenen, H., 1987, Differential regulation of mRNA encoding nerve growth factor and its receptor in rat sciatic nerve during development, degeneration, and regeneration: Role of macrophages. Proc Natl Acad Sci 84:8735-8739.
Hosang, M., & Shooter, E.M., 1987, The internalization of Nerve Growth Factor by high affinity receptors on pheochromocytoma PC12 cells. EMBO J, 6:1197-1202.
Jackson, G.R., Morgan, B.C., Werrbach-Perez, K., & Perez-Polo, J.R., 1990, Antioxidant effect of retinoic acid on PC12 rat pheochromocytoma. Int.J. Develop. Neurosci. In Press.
Johnson, E.M., Taniuchi, M., Clark, H.B., Springer, J.E., Koh, S., Tayrien, M.W., & Loy, R., 1987, Demonstration of the retrograde transport of nerve growth factor receptor in the peripheral and central nervous system. J Neurosci, 7:923-929.
Kordower, J.H., Gash, D.M., Bothwell, M., Hersh, L., Mufson, E.J., 1989, Nerve growth factor receptor and choline acetyltransferase remain colocalized in the nucleus basalis (Ch4) of Alzheimer's patietns. Neurobiol. Aging 10:67-74.
Kumar, S., Huber, J., Pena, L.A., Perze-Polo, J.R., Werrbach-Perez, K., & de Vellis, J., 1990, Characterization of functional nerve growth factor-receptors in a CNS glial cell line: Monoclonal antibody 217c recognizes the nerve growth factor-receptor on C6 glioma cells. J. Neurosci. Res. 27:408-417.
Large, T.H., Bodary, S.C., Clegg, D.O., Weskamp, G., Otten, U., & Reichardt, L.F., 1986, Nerve growth factor gene expression in the developing rat brain. Science, 234:352-355.
Levi, A., Biocca, S., Cattaneo, A., & Calissano, P., 1988, The mode of action of nerve growth factor in PC12 cells. Mol. Neurobiol. 2:201-226.
Levi-Montalcini, R., 1987, The nerve growth factor: 35 years later. Science, 237:1154-1162.
Lu, B., Buck, C.R., Dreyfus, C.F., & Black, I.B., 1989, Expression of NGF and NGF receptor mRNAs in the developing brain: Evidence for local delivery and action of NGF. Exp Neurol, 104: 191-199.
Marchetti, D., & Perez-Polo, J.R., 1987, Nerve growth factor receptors in human neuroblastoma cells. J Neurochem, 49:475-486.
Martin, D.P., Schmidt, R.E., DiStefano, P.S., Lowry, O.H., Carter, J.G., & Johnson, E.M., 1988, Inhibiros of protein synthesis and RAN synthesis prevent neuronal death caused by Nerve Growth Factor deprivation. J. Cell. Biol., 106:829-844.
Massague, J., Buxser, S., Johnson, G.L., & Czech, M.P., 1982, Affinity laebling of a nerve growth factor receptor component on rat pheochromocytoma (PC12) cells. Biochim. Biophys. Acta, 693:205-212.
Misko, T.P., Radeke, M.J., & Shooter, E.M., 1987, Nerve growth factor in neuronal development and maintenance. J Exp Biol, 132:177-190.
Mobley, W.C., Rutkowski, J.L., Tennekoon, G.I., Gemski, J., Buchanan, K., & Johnston, M.V., 1986, Nerve growth factor increases choline acetyl-transferase activity in developing basal forebrain neurons. Mol Br Res, 1:53-62.
Morgan, B., Thorpe, L.W., Marchetti, D., & Perez-Polo, J.R., 1989, Expression of nerve growth factor receptors by human

peripheral blood mononuclear cells. J Neurosci Res, 23:41-45.
76.Perez-Polo JR (1987) Neuronal Factors CRC, Boca Taron, Fl. 1-202.
Perez-Polo, J.R., & Werrbach-Perez, K., 1988, Role of Nerve Growth Factor in neuronal injury and survival, In: Neural Develop. & Regen. Cellular & Molecular Aspects Gorio A, Perez-Polo JR, deVellis J, Haber B eds., Springer Heidelberg pp.399-410.
Schatteman, G.C., Gibbs, L., Lanahan, A.A., Claude, P., & Bothwell, M., 1988, Expression of NGF receptor in the developing and ault primate central nervous system. J Neurosci, 8:860-873.
Shan, D.E., Beck, C.E., Werrbach-Perez, K., & Perez-Polo, J.R., 1990, Reverse-Phase high-Performance liquid chromatogrpahy of nerve growth factor receptor-like proteins identified with monoclonal antibodies. J.Neurosci. Res. In Press.
Sonnenfeld, K.H., & Ishii, D.N., 1985, Fast and slow nerve growth factor binding sites in human neuroblastoma and rat pheochromocytoma cell lines: Relationship of sites to each other and to neurite formation. J Neurosci, 5:1717-1728.
Spitz, D.R., Dewey, W.C., & Li, G.C., 1987, Hydrogen peroxide or heat shock induces resistance to hydrogen peroxide in Chinese hamster fibroblasts. J Cell Physiol, 131: 364-373.
Stach, R.W., & Perez-Polo, R., 1987, Binding of nerve growth factor to nerve growth factor receptor. J Neurosci Res, 17:1-10.
Taglialatela, G., Angelucci, L., Ramacci, M.T., Foreman, P.J., & Perez Polo, J.R., 1990, ^{125}I-nerve growth factor binding is reduced in rat brain after stress exposure. J Neurosci Res, 25:331-335.
Taglialatela, G., Angelucci, L., Ramacci, M.T., Werrbach-Perez, K., Jackson, G.R., & Perez-Polo, J.R., 1990, Acetyl-L-carnitine stimulates the synthesis of nerve growth factor receptors in a rat pheochromocytoma cell line. Biochem. Pharmacology Submitted.
Taglialatela, G., Angelucci, L., Ramacci, M.T., Werrbach-Perez, K., Jackson, G.R., & Perez-Polo, J.R., 1990, Acetyl-L-carnitine enhances the response of PC12 cells to nerve growth factor. Developmental Brain Res., Submitted.
Taniuchi, M., Johnson, E.M., Roach, P.J., & Lawrence, J.C., 1986b, Phosphorylation of nerve growth factor receptor proteins in sympathetic neurons and PC12 cells. J Biol Chem, 261:13342-13349.
Thorpe, L.W., Stach, R.W., Morgan, B., & Perez-Polo, J.R., 1989, The biology of nerve growth factor: Its interation with cells of the immune system, In:Neural Control of Reprod. Function, Lakoski JM, Perez-Polo JR & Rassin DK, (eds) Alan Liss NY, pp:351-369.
Whittemore, S.R., & Seiger, A., 1987, The expression, localization and functional significance of β-nerve growth factor in the central nervous system. Br Res Rev, 12:439-464.

INDEX